High Performance Computing Serving Atmospheric Transport & Dispersion Modelling

High Performance Computing Serving Atmospheric Transport & Dispersion Modelling

Editor

Patrick Armand

MDPI • Basel • Beijing • Wuhan • Barcelona • Belgrade • Manchester • Tokyo • Cluj • Tianjin

Editor
Patrick Armand
French Atomic Energy and
Alternative Energies
Commission
CEA, DAM, DIF
Arpajon
France

Editorial Office
MDPI
St. Alban-Anlage 66
4052 Basel, Switzerland

This is a reprint of articles from the Special Issue published online in the open access journal *Atmosphere* (ISSN 2073-4433) (available at: www.mdpi.com/journal/atmosphere/special_issues/atmospheric_dispersion_modelling).

For citation purposes, cite each article independently as indicated on the article page online and as indicated below:

LastName, A.A.; LastName, B.B.; LastName, C.C. Article Title. *Journal Name* **Year**, *Volume Number*, Page Range.

ISBN 978-3-0365-6581-1 (Hbk)
ISBN 978-3-0365-6580-4 (PDF)

Contents

About the Editor

Patrick Armand

Patrick Armand is a fluid mechanics engineer and holds a Ph.D. from Paris Sorbonne University, where he prepared a thesis on turbulent multiphase flows. Since the beginning of his career, he has worked as a research engineer at the French Atomic Energy and Alternative Energies Commission (CEA), where he devotes himself to 3D modelling and simulations. He began by working on severe accidents in nuclear reactors; then, he became interested in the transfer of radionuclides between different compartments of the environment and, more particularly, in the atmosphere. He created and, for more than twelve years, has headed the Radiological and Chemical Impact Laboratory, where he currently works as a senior expert in atmospheric transport and dispersion.

Preface to "High Performance Computing Serving Atmospheric Transport & Dispersion Modelling"

The last decade has witnessed both the use of increasingly sophisticated physical and mathematical models of atmospheric transport and dispersion (AT&D), and considerable development in high-performance computing (HPC) based on a very large number of CPU and/or GPU processors. During this period, computational resources have not only become larger and larger, but also more and more available and accessible. They now make three-dimensional calculations possible, whose size and scale were unfeasible at the beginning of the 21st century. Simulations are very often based on efficiently parallelized versions of weather forecasting and atmospheric dispersion models. These are used in computational chains which allow the descent from the mesoscale to the microscale in the atmospheric environment, and even inside buildings or within infrastructure. Intensive simulations have several crucial aspects. On the one hand, they can achieve very high spatial and/or temporal resolutions in gigantic computational domains encompassing complex, built, natural, or mixed terrains. On the other hand, they make it possible to obtain results very quickly, in accelerated time compared with physical time, which is extremely useful for responding to emergencies. Finally, they can carry out ensembles of calculations which are essential, for example, in producing statistics for the average spatial distribution of an atmospheric release over a period of several years, to propagate uncertainties in input data, and to infer their influence on the results of possibly hazardous materials which have been dispersed, to determine the source terms at the origin of detection on a network of sensors by various methods, or to establish databases used by methods of meta-modelling or machine learning. Thus, there are countless reasons to be interested in HPC in connection with the modelling and simulation of AT&D.

This reprint brings together fifteen remarkable articles published in the Special Issue of the journal *Atmosphere*, entitled "High-Performance Computing Serving Atmospheric Transport & Dispersion Modelling". These articles address a wide variety of physical models of weather forecasting and atmospheric dispersion at all spatial and temporal scales. The applications are just as diverse, as the introductory remarks to the articles below show.

Singh et al. simulated the vertical transport of aerosols during deep convection episodes in the Himalayas using the WRF-Chem model. They highlight the importance of a fine horizontal resolution, in the order of one kilometer, to resolve convection phenomena and determine the precise distribution of rainfall.

Gunawardena et al. emulated radioactive material deposition maps obtained from WRF and FLEXPART simulations. They showed that the implemented machine learning method succeeded in reproducing the results of the physical models 10,000 times faster, even when using a limited number of realizations for learning.

Zhong et al. divided ADMS-Urban calculations over a large domain into sub-regions that were performed by multiple cores of a supercomputer. They thus managed to reduce annual air quality calculations down to the fine scale of streets in the West Midlands region (UK), from several weeks to a few hours.

Oldrini et al. describe the EMERGENCIES project for 3D high-resolution flow and dispersion modelling with PMSS over giant urban areas. They demonstrated the feasibility of using such a system in the event of the atmospheric release of hazardous materials in order to support rescue teams in their decisions. Oldrini et al. also worked on the visualization of the widespread results of the atmospheric dispersion of hazardous materials and health impacts on the enormous EMERGENCIES domain. They established that efficient parallel processing enables the production and broadcast of maps on the fly with multiple zoom levels.

Nibart et al. noted the growing use of HPC to assess urban air quality at high resolution in the context of construction projects and services such as weather forecasting. They emphasized the experience acquired on the modelling system, PMSS, which made it possible to validate the method on multiple cases and optimize its use.

Russo et al. utilized the CRESCO/ENEAGRID computing center to carry out air quality forecasts at the urban scale in the city of Rome. They specify that a super-computer was utilized to perform simulations with PMSS and collect detailed floating car data as emission inputs for the model.

Villani et al. assessed the impact of greening building facades on the concentration of atmospheric pollutants in urban areas. Based on PMSS simulations at the CRESCO/ENEAGRID computing center, they showed that these infrastructures are effective in reducing population exposure at hotspots.

Nakayama et al. implemented a method for predicting atmospheric dispersion, which combined a flow database pre-calculated with the LES model LOHDIM and local meteorological observations acquired by a Doppler LiDAR. They illustrated the relevance of the method with validation tests on a built site of the JAEA. Nakayama et al. also performed simulations of plume dispersion in downtown Oklahoma City with the LES model LODHIM using a WRF mesoscale model or observations as meteorological inputs. They showed that comparisons with measurements were better with WRF, although were still acceptable with observations.

Gowardhan et al. carried out LES simulations of turbulent flows and dispersion over complex and urban terrains using the AEOLUS model. They validated the model based on wind and tracer measurements from the Joint Urban 2003 field campaign and with the data of buoyant plumes generated by explosions.

Bieringer et al. developed a GPU-based approach to LES simulations of harmful material dispersion. They validated the model on open terrain for unstable, neutral, and stable atmospheres, and showed that their urban indoor/outdoor JOULES model provided results 150 times faster on GPU than on CPU.

Jacob et al. present an approach based on the lattice Boltzmann method (LBM) with wall model large eddy simulations (WMLES) to take into account multi-scale flows and dispersion, not only in the urban atmospheric environment, but also in indoor environments. They document a large and diverse panel of validation cases.

Elfverson and Lejon examined the scalability of the parallel OpenFOAM model in built-up environments. They solved the RANS equations with the k-ω turbulence model and used it in a domain with two million cells around the Parade square in Warsaw, where flow and dispersion results were achieved in minutes.

Schalau et al. proposed a method which preserves the mean velocity and turbulence profiles specified as boundary conditions of RANS models such as OpenFOAM. They verified that the numerical results obtained with this method agree with wind tunnel measurements for various obstacle configurations.

We hope that everyone enjoys reading this Special Issue!

Patrick Armand
Editor

atmosphere

MDPI

Article

Vertical Distribution of Aerosols during Deep-Convective Event in the Himalaya Using WRF-Chem Model at Convection Permitting Scale

Prashant Singh [1,2], Pradip Sarawade [1] and Bhupesh Adhikary [2,*]

[1] Department of Physics, University of Mumbai, Mumbai 400098, India; prashantsinghprs@gmail.com (P.S.); pradip.sarawade@physics.mu.ac.in (P.S.)
[2] International Centre for Integrated Mountain Development (ICIMOD), Khumaltar G. P.O. Box 3226, Lalitpur 44700, Nepal
* Correspondence: Bhupesh.Adhikary@icimod.org

Abstract: The Himalayan region is facing frequent cloud bursts and flood events during the summer monsoon season. The Kedarnath flooding of 2013 was one of the most devastating recent events, which claimed thousands of human lives, heavy infrastructure, and economic losses. Previous research reported that the combination of fast-moving monsoon, pre-existing westerlies, and orographic uplifting were the major reasons for the observed cloud burst over Kedarnath. Our study illustrates the vertical distribution of aerosols during this event and its possible role using the Weather Research and Forecasting model coupled with chemistry (WRF-Chem) simulations. Model performance evaluation shows that simulations can capture the spatial and temporal patterns of observed precipitation during this event. Model simulation at 25 km and 4 km horizontal grid resolution, without any changes in physical parameterization, shows a very minimal difference in precipitation. Simulation at convection-permitting scale shows detailed information related to parcel motion compared to coarser resolution. This indicates that the parameterization at different resolutions needs to be further examined for a better outcome. The modeled result shows changes of up to 20–50% in the rainfall over the area near Kedarnath due to the presence of aerosols. Simulation at both resolutions shows the significant vertical transport of natural (increases by 50%+) and anthropogenic aerosols (increases by 200%+) during the convective event, which leads to significant changes in cloud properties, rain concentration, and ice concentration in the presence of these aerosols. Simulations can detect changes in important instability indices such as convective available potential energy (CAPE), convective inhibition energy (CIN), vorticity, etc., near Kedarnath due to aerosol–radiation feedback.

Keywords: aerosols; South Asia; WRF-Chem; precipitation; CAPE; CIN

Citation: Singh, P.; Sarawade, P.; Adhikary, B. Vertical Distribution of Aerosols during Deep-Convective Event in the Himalaya Using WRF-Chem Model at Convection Permitting Scale. *Atmosphere* **2021**, *12*, 1092. https://doi.org/10.3390/atmos12091092

Academic Editor: Patrick Armand

Received: 26 July 2021
Accepted: 23 August 2021
Published: 25 August 2021

Publisher's Note: MDPI stays neutral with regard to jurisdictional claims in published maps and institutional affiliations.

1. Introduction

Dimri et al. (2017) reviewed the dynamic, thermodynamic, and physical reasons for cloud burst cases in the Himalayan region and their impact on the society in detail. Generally, the interaction of fast-moving monsoons with existing active westerlies [1] and orographic uplifting often results in havoc in the central Himalayan region [2]. Himalayan foothills make an intersection point, where northward moving monsoon, active westerlies, and orographic lifting produce high convection and thunderstorm activity [3]. A previous study has indicated that the presence of aerosols can enhance or suppress rain over Asian regions [4]. The Indo-Gangetic Plain (IGP), which is known as a hotspot of anthropogenic pollution in South Asia [5–7], as well as the deserts of Rajasthan and middle east Asia [8], supply ample aerosols to the Himalayan region.

Aerosol–cloud–precipitation interaction is considered to be a complex system and much remains to be understood [9–11]. Aerosols act as cloud condensation nuclei (CCN), which is necessary to form clouds and rain [9]. Some of the aerosols act as ice nuclei

(IN) [12,13], which may hold the water content in clouds and delay precipitation [14]. In the presence of excess aerosols, smaller cloud droplets are formed, which reduces the precipitation amount [9]. In the form of CCN and IN in clouds, aerosols affect cloud properties such as brightness, cloud cover, cloud top temperature, and cloud top pressure [15–17]. The presence of aerosol layers above or below a cloud can affect the cloud cover in either of the aforementioned ways [16,17]. Andreae and Rosenfeld (2008) reported that natural aerosol and CCN concentration is lower over land, and cloud formation over the continent mostly results from anthropogenic emissions [12]. More CCN (more aerosols) leads to smaller and narrower cloud droplet size and distribution, which results in suppressed warm rain and enhanced cold rain. Higher CCN in mixed phase clouds makes it deeper and enhances lightning activity with more flashes [18]. An increase of 1–30% in the dust and sea salt concentration affect cloud properties and precipitation significantly; the conclusion that a 15% increase in the dust concentration may delay rain by one hour was reported [13]. Up to a 70% increase in the CCN over northwestern Europe due to aged Saharan dust was reported in previous research [19]. Increased cloud fraction (~5%) and decreased cloud top pressure (~40 mb) are reported due to elevated aerosol concentration over the Atlantic [17]. Several studies have indicated that aerosols are not only limited to acting as CCN/IN but that they also affect radiation-derived parameters (such as convective available energy (CAPE) and convective inhibition energy (CIN), which are important in the prediction of severe weather [11,20].

Most of the extreme precipitation events over the western Himalayas are observed during the month of September, based on the records available from 1875–2010 [21]. From 16–17 June 2013, the rapid arrival of monsoon in northern India along with the presence of strong westerlies over the region was one of the major causes for a massive precipitation event over Kedarnath, India [22–25]. Kedarnath (30.73° N, 79.06° E; 3553 m from sea level) is a small town in the Indian state of Uttarakhand in the Himalayan region. This study focuses on the atmospheric analysis during the Kedarnath flood, which occurred from 16–17 June 2013, and was followed by significant flooding over western Nepal from 17–18 June 2013, an event which was later called the Himalayan tsunami [22]. The Kedarnath floods claimed the loss of hundreds of human lives and damaged vast infrastructure [23]. Many studies after the event tried to analyze the causes of such devastation. Some of the observations [1,23] and model-based studies [24] have analyzed meteorological conditions, orographic and climatic perspectives [22,26], and the effect of chemistry on precipitation [4].

This study attempts to understand the vertical distribution of aerosols at the synoptic and convection-permitting scales during the Kedarnath heavy precipitation event using the regional Weather Research Forecast coupled with chemistry (WRF-Chem) model. A model horizontal grid resolution of below 5×5 km is considered as a convection-permitting scale where there is no need for specific cumulus parameterization schemes in the model [27,28]. Additionally, we discuss how the presence of aerosols affected radiation and altered the severe weather indices, which are important in predicting precipitation and severe weather.

2. Data and Methodology

2.1. Observations

In situ data for various stations obtained from the Meteorological and Oceanographic Satellite Data Archival Centre (MOSDAC) of the Indian Space Research Organization (ISRO) containing precipitation, relative humidity, wind direction, wind speed, temperature, and near-surface pressure were used in this study. MOSDAC collects data from the Indian Meteorological Department (IMD) and various Automated Weather Stations (AWS) from different sources (https://www.mosdac.gov.in/, accessed on 1 April 2021). MOSDAC-AWS uses a tipping bucket rain gauge to measure accumulated rainfall [29].

Tropical Rainfall Measuring Mission (TRMM) monthly level 3 data (TRMM_3A12) available at a horizontal grid resolution of $0.5° \times 0.5°$ (https://disc.gsfc.nasa.gov/datacollection/TRMM_3A12_7.html, accessed on 1 April 2021) were used to analyze the general trends over the study region. Further analysis was performed using TRMM-TMPA (Multi-satellite Precipitation) level 3 data at a 0.25° × 0.25° spatial resolution, which was available at 3-hour temporal resolutions.

Atmospheric infrared sounder (AIRS) Aqua level 3 daily products available at a 1° × 1° resolution (AIRX3STD) downloaded from https://search.earthdata.nasa.gov/ (accessed on 1 April 2021) were used to understand the cloud properties during the heavy precipitation event. Moderate Resolution Imaging Spectroradiometer (MODIS) level 3 Terra (MOD08_D3) and Aqua (MYD08_D3) daily products available at a 1° × 1° resolution were also used along with AIRS. Both of these satellite products provided cloud fraction, cloud top pressure, and temperature, which were useful to understand the cloud properties.

2.2. Model Setup

This study uses the Weather Research and Forecasting (WRF) model [30,31] coupled with chemistry (WRF-Chem) for various simulations [32]. A total of six sets of simulations were performed to analyze the vertical transport of bulk aerosols during the Himalayan extreme precipitation event. Out of these, three sets of WRF simulations were performed at the resolution of 25 km for the domain consisting of the whole Indian subcontinent (6.5–36.0° N, 53.0–103.0° E; Figure S1) with MOZCART and MOZART and with WRF without the chemistry option. Similarly, another three sets of WRF simulations were performed at the resolution of 4 km, covering an area between 28–32.0° N, 74.25–85.75° E that covered significant portions of the IGP and the Himalayas. To stabilize the chemistry in the single domain model simulations, one week of spin-up time was used. Event analysis was conducted using the data from 3 days before to 3 days after the event. Details of the WRF simulations with MOZCART chemistry (WC25) and without chemistry at a 25 km (WRF25) horizontal grid resolution and similarly at the 4 km resolution (WC4 and WRF4, respectively) are outlined in Table 1. The Thompson Graupel Scheme, a double-moment microphysics scheme, was used in the simulations and consists of six classes of moisture species along with the ice concentration number for the prediction of cloud properties. To understand the aerosol–cloud–radiation feedback, the cloud effect on the optical depth in radiation was also activated in the simulations. Supporting Figure S1 shows the domain of the simulation at the resolution of 25 km and 4 km (red box) and at the location of Kedarnath (black dot).

The experiments were designed in such a way that the simulations for WC/WRF4 used chemical and meteorological boundary conditions from the simulations of WC/WRF25. In this kind of setup, we were able to overcome computational limitations in terms of storage and processing capacity. Both of the simulations (25 km and 4 km resolutions) used National Center for Environmental Prediction Final Analysis (NCEP-FNL) for the meteorological initial conditions [7,24], whereas boundary conditions for the simulation at WC/WRF25 used NCEP-FNL and the simulations at WC/WRF4 used meteorological data from WC/WRF25. In a similar way, the WC25 simulations used chemical boundary conditions from a global simulation model for ozone and related chemical tracers, version 4(Mozart-4) [33]. The WC4 simulations used chemical boundary conditions from WC25. For both simulations, WC4 and WC25 anthropogenic emissions were considered from those from the Emission Database for Global Atmospheric Research-Hemispheric Transport of Air Pollution (EDGAR-HTAP) [34].

Table 1. Domain and parameterization details for different simulations.

Simulations	WC25	WRF25	WC4	WRF4
Centered at	22° N, 78° E		30° N, 80° E	
Resolution	25 × 25 km		4 × 4 km	
No. of grids	130 × 203 × 40		223 × 555 × 40	
Domain	6.5–36.0° N, 53.0–103.0° E		28–32.0° N, 74.25–85.75° E	
Chemistry Scheme	MOZCART	-	MOZCART	-
Convective parameterization	Kain–Fritsch Scheme [35]			
Planetary boundary layer physics	Yonsei University Scheme (YSU) [36]			
Shortwave radiation physics	Dudhia Shortwave Scheme [37]			
Microphysics	Thompson graupel scheme [38]			
Longwave radiation physics	RRTM Longwave Scheme [39]			
Land-atmosphere interaction	Unified Noah Land Surface Model scheme [40]			
Surface layer option	MM5 Similarity Scheme [41]			
Photolysis	Madronich fast-Ultraviolet-Visible Model (F-TUV) [42]			

3. Results and Discussion

3.1. Precipitation Analysis

Normal rain was reported during 2013 monsoon season throughout India, except a few states of India such as Bihar, Arunachal Pradesh, and Jammu [25]; however, an excess of rain was reported over most of India in June 2013, except for in northeast India. Uttarakhand faced more than a 191% excess of rain during the month of June, which was the highest compared to any other state, while the overall monsoon season recorded only 12% excess rain. As per IMD records, more than 13 districts of Uttarakhand recorded excess rain during June 2013, which was an unusual event in the month of June [22].

This study used the TRMM monthly surface rain rate data product (TRMM_3A12) from January 1998 to December 2014 to determine the rain pattern in Uttarakhand, a state of India, and over Kedarnath, a Himalayan Mountain city. The year 2013 was a neutral year in terms of El Niño and La Nina. An analysis of seasonal and annual average precipitation over Uttarakhand for the year 2013 suggested that it was an average year in terms of accumulated precipitation when compared to other years from 1998–2014. On the other hand, the monthly analysis suggested that in 2013, the month of June observed the highest precipitation compared to any other year. The area-averaged rain rate taken from a few horizontal grids over Kedarnath suggests that seasonal and annual rain rate for the year 2013 was average during the analysis period, whereas the rain rate was higher compared to the state of Uttarakhand. The Kedarnath grid suggests that the rain rate during the month of June and the monsoon season of 2013 was the highest compared to any other year from 1998–2014 (Table S1 and Figure S2).

Figure 1 shows the results of 3 days of accumulated rain (from 16–18 June 2013) over the simulation domain from TRMM and different versions of the WRF model. TRMM accumulated rain (Figure 1a) shows heavy precipitation (greater than 100 mm) over the western coast, northern India, western Nepal, and the area near the southeastern Bay of Bengal. Figure 1b shows accumulated precipitation from the WRF WC25 simulation. The model reproduces major precipitation when compared to the TRMM precipitation results, with some minor differences. Figure 1c shows accumulated rain from WRF25 simulation, which again is similar to the TRMM observations with minor differences. Figure 1d shows the difference in the accumulated rain produced by the WC25 to WRF25 simulation. The results show that precipitation is reduced in the central and northeastern parts of India with the chemistry option turned on. However, northern India (majorly Uttarakhand), over the Arabian Sea and Bay of Bengal, shows increased rain in WC25.

Similar results are seen for the 4 km resolution model simulations (Figure 2). WRF with MOZCART chemistry (WC4) and WRF without chemistry (WRF4) show similar features of accumulated rain when compared to each other, while there are differences when compared to the results obtained from TRMM. It must be noted that grid resolution between the model simulation and the TRMM observations also leads to some of the observed differences. Most of the area that is focused on in the study shows a decrease in precipitation from the simulations with chemistry turned on, whereas few areas also show an increase in precipitation. Both sets of simulations (25 km and 4 km) show the effect of aerosols on precipitation amount in either direction (i.e., increase or decrease). WC25 shows that most of the area over and near Uttarakhand produces more rain in the presence of aerosols, while WC4 shows less rain in the presence of aerosols, except for in a few concentrated places.

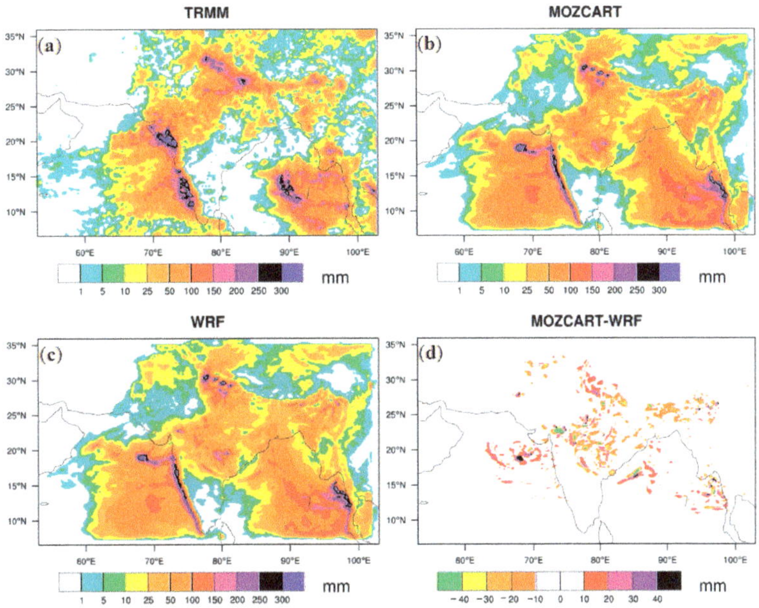

Figure 1. Three days accumulated rain from (**a**) TRMM and those from the WRF simulations (**b**) WC25 and (**c**) WRF25 and (**d**) the difference in rainfall between WC25 and WRF25.

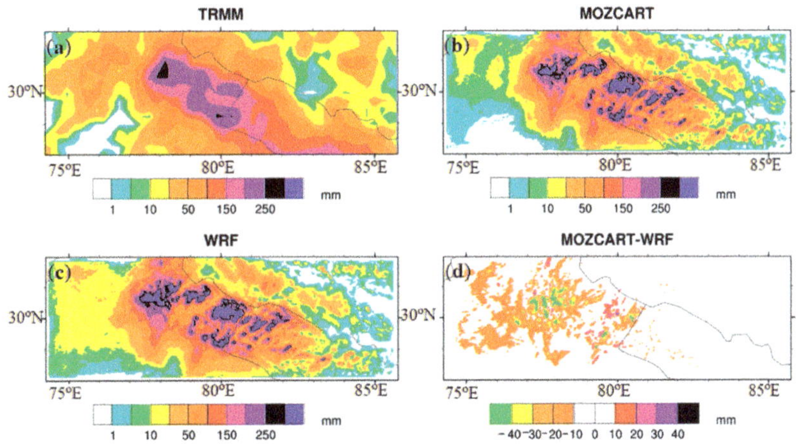

Figure 2. Three days accumulated rain from (**a**) TRMM and the simulation of WRF at a 4 km resolution for (**b**) WC4 and (**c**) WRF4 (**d**) and the difference of rain produced by WC4-WRF4.

Using a combination of different parameterization schemes in WRF simulations, the study by Chawla et al. (2018) shows that most of the combinations were able to capture spatial precipitation features for 15–18 June 2013 as they appear in TRMM with some differences [24]. A previous study using WRF-Chem simulation by Kedia et al. (2018) suggests a 20% increase in precipitation over Uttarakhand due to chemistry, which was determined using a 15-day average simulation and by observing rain over the region for analysis [4]. Average precipitation and other parameters for 15 days were shown as aggregates illustrating the impact of aerosols; however, this was illustrated without providing temporal resolution for the event. Thus, it is difficult to compare our results with previous research [4].

Figure 3 presents a time series of precipitation at four stations in the Himalayan Mountains of Uttarakhand from 15–20 June 2013. These four stations are the nearest present stations to Kedarnath. In situ observation at all four stations shows that precipitation starts from the morning of the 16 and ends on the morning of the 18. A similar trend is observed through satellites for the same time period. All four model simulations can correctly predict the start and end pattern of precipitation and peaks in precipitation with some minor temporal shifts. At all the four stations, WC25 shows early rain, whereas WRF25 shows delayed rain when the models are compared to each other. Both WC4 and WRF4 show delayed rain matching with each other and also match with the WRF25 simulation trend but predict higher rainfall compared to WRF25 for the peak rain during these days. WC25, however, better matches with the TRMM observations compared to other simulations regarding the timing of the rainfall. The study by Castorina et al. (2021) suggests an explicit resolution of the convective system and provides better simulation for extreme events. As the horizontal dimension of convective clouds varies from 0.1 km–10 km and considering that our model resolution (5 km in their study) is much higher than that, the physical parameterization of the convective system can improve the simulation [43]. However, our simulation with convective parameterization shows that the accumulated rain at 25 km and 4 km is in a similar range with some differences, which justifies the role of the parameterization of the convective system in the simulation of extreme events.

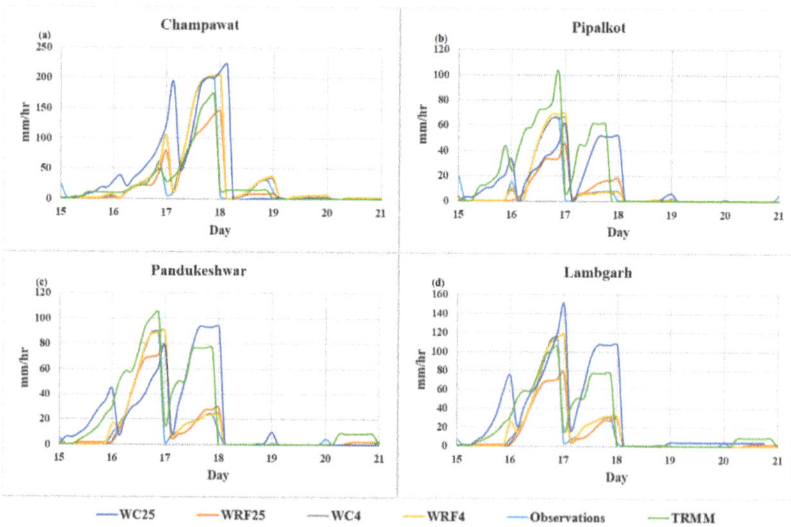

Figure 3. Precipitation from different simulations and observations from 15–20 June 2013 for (**a**) Champawat, (**b**) Pipalkoti, (**c**) Pandukeshwar, and (**d**) Lambgarh.

Supporting Table S2 shows the coefficient of determination (R2) between TRMM, during in situ observation, and all four simulations with the observations for the period from 15–20 June 2013. Over Kedarnath, the TRMM and model simulations show a low R2 value (0.36–0.41). At other places, TRMM and in situ data show R2 in the range of 0.32–0.60; R2 for TRMM and model ranges between 0.32–0.95, except for Dehradun. The R2 between in situ observation and the models ranges between 0.23–0.93, except for Jolly Grant. The R2 at Jolly Grant between TRMM and in situ observation is 0.60, and between the observation and the model, it is less than 0.31. Observed accumulated rain at Jolly Grant was as high as 200 mm/day (in situ); the model produces less than 100 mm/day during event days, and the model also does not perform well at Dehradun station. Over other stations, the model replicates the strength and period of precipitation. An observation-based report from IMD presents heavy precipitation on 16 and 17 June 2013 [25], which is evident in all model simulations and satellite observations as well (Figure 3). Other observation studies also show heavy precipitation on 16 and 17 June over most of the Uttarakhand region [22,23,26,44].

Figures 1–3 suggest that all model simulations adequately capture the spatial and temporal coverage of precipitation, with some differences in the amount of precipitation. WC25 and WRF25 show up to a 50 mm/day rain difference at some of the stations, whereas WC4 and WRF4 show a negligible precipitation difference. Additional analysis suggests more rain in the 25 km simulation without chemistry at Dhanauri (~20 mm/day), Jolly Grant (~60 mm/day), Dehradun (~70 mm/day), and Mandal (~100 mm/day). However, more rain is produced in the 25 km simulation with chemistry in places such as Kedarnath (~40 mm/day), Champawat (~60 mm/day), Nainital (~15 mm/day), Lambagrh (~40 mm/day), Pandukeshwar (~10 mm/day), and Pipalkoti (~20 mm/day). WRF and WRF-Chem were able to simulate precipitation well over most of the observation stations, which is further corroborated by the TRMM observations. Given the model's ability to replicate the rainfall over Uttarakhand, we present the model-based analysis of monsoon dynamics, cloud properties, and aerosols during this event.

3.2. Monsoon Dynamics

The Indian Meteorological Department, India, and the Department of Hydrology and Meteorology (DHM), Nepal, reported the onset of the summer monsoon on 15 and 14 June 2013, respectively. Due to low heat (high temperature leads to low-pressure zone) over northern India and the high-pressure zone over the adjacent ocean, moisture moves with the wind from the ocean to northern India during monsoons. Figure 4a,b shows the counterclockwise cyclonic motion of the wind direction, implying a strong low pressure (WC25 due to full chemistry option) zone over northern India on 15–16 June, which guided the moisture from the ocean towards the land. WC4 shows consistent winds flowing from southeast to northwest (Figure not shown). Figure 4c shows a low-pressure zone on the 17, which is moving towards the north. The same low-pressure zone further moved towards western Nepal (Figure 4d) on the 18, which created a flood situation in western Nepal (https://www.icimod.org/?q=10932, accessed on 1 April 2021); after the 19 this system disappeared (Figure 4e,f).

Singh and Chand, (2015) have shown the presence of a low-pressure zone over central India (Rajasthan and Madhya Pradesh), as simulated by their model on 16 June. Ray et al. (2014) reported the presence of a low-pressure system during the 16 and 17 over central India [23,25]. The dynamic interaction of the monsoon due to the low-pressure system over central India with the mid-latitude western disturbance results in heavy rainfall [23,25,45]. Similar wind patterns and low-pressure were simulated in all the simulations. The precipitation pattern was also simulated well in the simulations concerning observations and the previous literature.

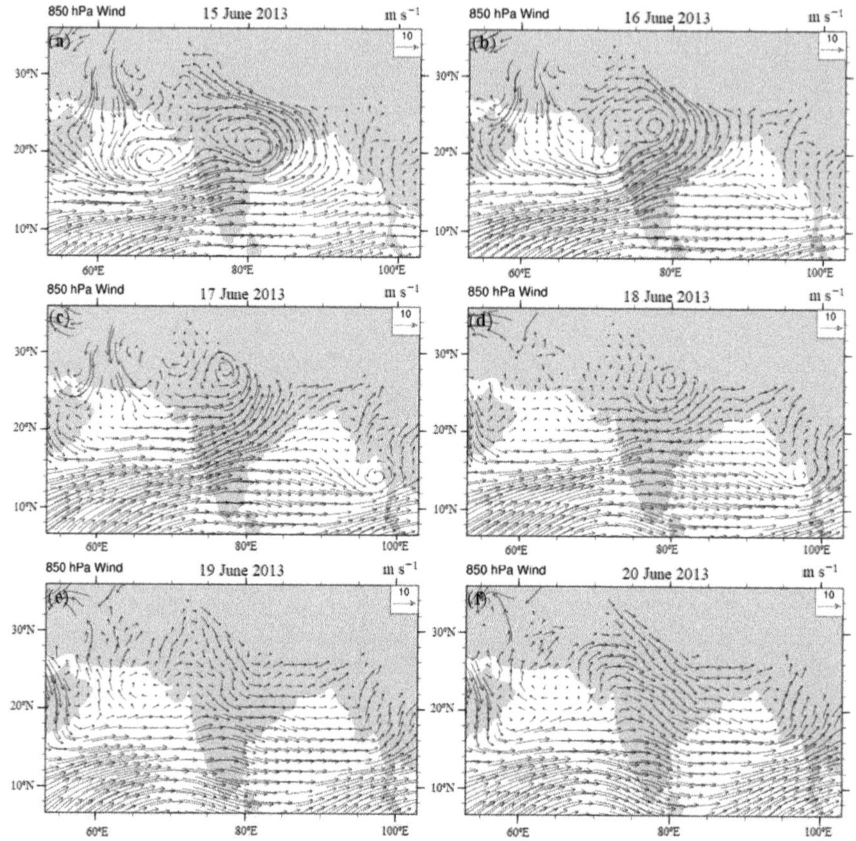

Figure 4. Daily wind speed at 850 hPa from 15–20 June (**a**–**f**). Simulation WC25.

In situ observations for all of the stations listed in the supporting table shows low pressure near the surface, about 4–6 hPa less than the average atmospheric pressure on the 17. The WC4 and WRF4 simulations represent better pressure at different stations in comparison to simulation WC25 and WRF25, mainly due to a higher model resolution that is better able to reflect topography height. Low pressure indicates an unstable atmosphere and a higher probability of precipitation. High surface relative humidity (≥90%) was observed during the 16 and 17, and the model also simulated high relative humidity for all of the stations (≥80%). Persistent higher humidity indicates a higher probability of precipitation in that area. Observations and the models both showed a rapid decrease in relative humidity from the morning of 18 June. The observed surface temperature on 17 June at all stations fell (~3 °C–5 °C) compared to the average temperature. Similarly, WC25 and WRF25 show temperature falls of ~4 °C–6 °C, while WC4 and WRF4 show a fall of ~2 °C–6 °C. All of the parameters show normal atmospheric conditions from the morning of 18 June in comparison with the event days. Literature based on observations shows 70–100% humidity, low pressure, and low wind during event days over Kedarnath [22].

3.3. Cloud Property

Figure 5 shows the average cloud fraction (CF) from the MODIS (Aqua and Terra) satellites (a,b) and the average model simulated cloud fraction (c,d) for 16 and 17 June 2013. Both the satellites and the model show similar features on 16 (Figure 5a,c): dense clouds from the Arabian Sea to northern India passing through central India along with some dense clouds in the southeastern Bay of Bengal. Both the satellites and the model-based observations show that all of Uttarakhand and western Nepal were covered by dense clouds on the 16. On the 17, the satellites do not show a dense cloud fraction over Uttarakhand and western Nepal, whereas the model does simulate dense cloud fraction over the area. However, on the 17, the satellite data do show similar cloud fraction features over the Western Ghats, Arabian Sea, and Bay of Bengal, as seen on the 16.

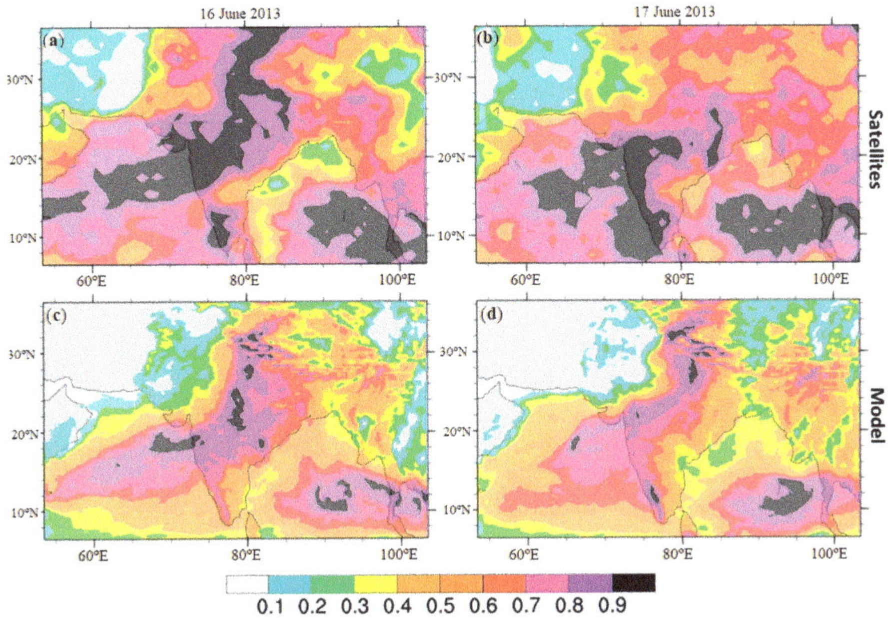

Figure 5. Average cloud fraction on 16 and 17 June (**a**) and (**b**) the MODIS satellites (Aqua and Terra combined) (**c**) and (**d**) model, respectively.

Model analysis reveals that the entirety of the Arabian Sea, Bay of Bengal, and central India is covered with low and mid-level clouds (Figure not shown). In this study, low cloud cover is considered up to 800 hPa, medium cloud cover is below 450 hPa, and high cloud cover is above 450 hPa. Along with low- and mid-level clouds, Uttarakhand, Western Nepal, a small part of the Arabian Sea, and the Bay of Bengal were covered with dense high-level clouds on 16 and 17 June. The satellites observed low cloud top pressure (CTP) over Uttarakhand compared to surrounding areas (~300 hPa) on the 16 and (~400 hPa) on 17, indicating the presence of high-level cloud cover over Uttarakhand.

Low cloud top temperature (CTT) was observed from the satellites (~−30 °C to −40 °C) on the 16 and (~ −20 °C to −30 °C) on the 17 over Uttarakhand and western Nepal. CTT from the model simulation was lower in comparison with Uttarakhand and western Nepal (~−10 °C to −20 °C) on both days. WC4 simulates further cooler CTT on the 17 (~−30 °C to −40 °C). The presence of aerosol mostly produced warmer cloud top at WC25, except in some places near Kedarnath, and WC4 simulated changes in CTT variation (±15 °C). Lower CTP and CTT indicate deep cloud with a high amount of precipitation [14].

At 25 km resolution, the presence of aerosols affected the cloud fraction by ±25% away from the major sources of emission (inferred from Figure 1d). The presence of aerosols simulated more clouds at a low level, whereas a decrease was simulated in the high-level clouds, while the mid-level clouds were affected in both ways over the domain of the simulation. Similar results were simulated at the 4 km resolution with and without aerosols for both days. The presence of aerosols played a role in creating warm clouds in the central and eastern parts of India, the same region where the aerosols negatively affected precipitation (Figure 1d). Satellite data along with the model show low CTT and CTP over Kedarnath during the 16 and 17 compared to previous days. The model indicates the presence of dense high clouds over Kedarnath, which could result in a heavy downpour.

3.4. Aerosol Concentration

Satellite and model analysis suggests the presence of dense high-level (above 6 km) clouds over Kedarnath, so we analyzed aerosols above 6 km (above ~500 hPa). Our results show that over Kedarnath, there is an elevated concentration of anthropogenic and natural aerosols at high altitudes during the event days. Figure 6a,b shows area average concentration of black carbon (BC), and Figure 6c,d shows dust over Uttarakhand (~29–31° N, ~78–81° E) from the 15–19. The night of the 17 shows the highest BC concentration at 500 hPa (>0.25 μg/m^3) and above in WC25, whereas it was (>0.05 μg/m^3) in the WC4 simulations; the columnar average BC concentration was higher by ~0.1 μg/m^3 in WC25, whereas it was lower in WC4 from the previous day. Figure 3 shows that a precipitation peak was present for a similar period over most of the stations. Similarly, natural aerosol dust shows a higher concentration at 500 hPa and above in both the WC4 and WC25 simulations (Figure 6c,d). The previous study on the vertical transport of aerosols during volcanic eruption suggests that WRF-Chem simulates the process quite realistically [46,47]. Similarly, the model-based previous study demonstrates that WRF-Chem realistically simulates vertical aerosol transport during deep convection events [33].

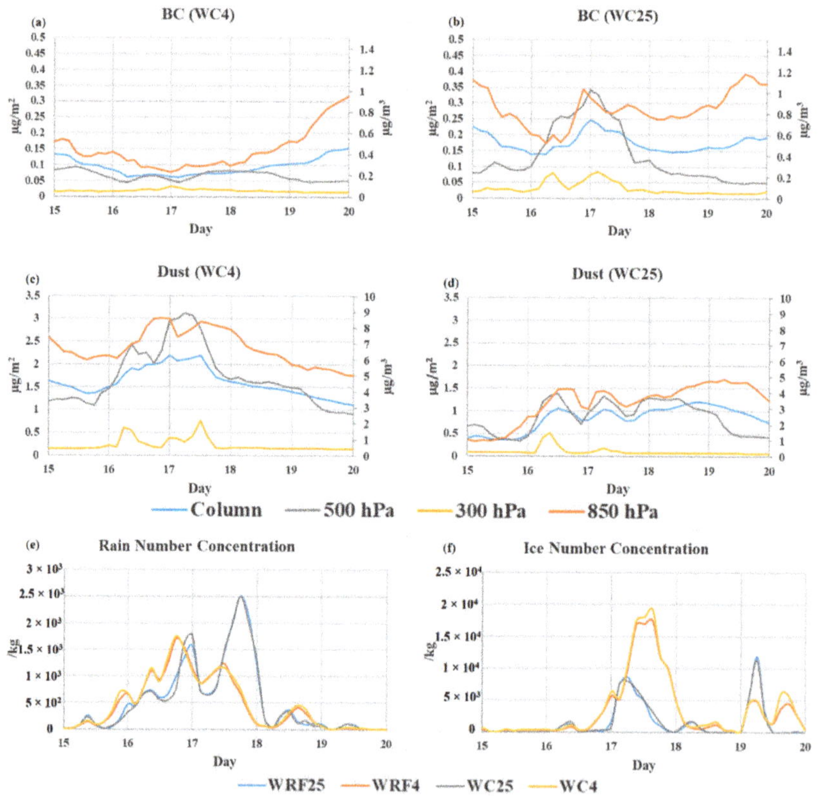

Figure 6. (**a**,**b**) BC concentration from 15–19 June 2013 at 850 hPa, 500 hPa, 300 hPa, and the columnar average at the 4 km and 25 km resolutions (**c**,**d**) for dust. (**e**) Rain number and (**f**) ice number concentration with different model simulations.

Other bulk aerosol species such as organic carbon (OC), dust (different size), sea salt (different size), and sulfates show similar elevated peaks at 500 hPa and above. Table 2 presents the differences in the concentrations of various aerosol species at 850 hPa, 500 hPa, 3 hPa, and the total column average. The methodology for calculating the differences in the concentrations is as follows:

Table 2. Δconcentration for different aerosols and Δ in absolute percentage in model column and at 850 hPa, 500 hPa, and 300 hPa for 16 and 17 June 2013 over Kedarnath at 25 km resolution.

Aerosols	Δ Column		Δ 850 hPa		Δ 500 hPa		Δ 300 hPa	
	16 June 2013							
	ug/m³	%	ug/m³	%	ug/m³	%	ug/m³	%
Black Carbon (BC)	−0.001	−0.56	−0.225	−25.10	0.156	200.26	0.029	142.23
Organic Carbon (OC)	−0.023	−2.15	−1.672	−28.86	1.038	226.53	0.177	240.49
DUST1	0.685	36.86	2.409	40.01	1.124	45.43	0.352	95.92
DUST2	1.264	29.56	4.893	34.38	2.069	38.09	0.775	104.79
DUST3	0.766	26.03	3.189	30.91	1.384	40.55	0.549	137.52
DUST4	−0.054	−5.85	−0.764	−17.66	0.357	50.64	0.149	230.30
DUST5	−0.031	−47.90	−0.165	−57.98	−0.027	−39.57	0.003	19.12
SEA SALT1	0.008	93.74	0.026	71.09	0.015	210.09	0.002	41.50
SEA SALT2	0.075	128.65	0.22	71.48	0.146	493.88	0.021	1478.02
SEA SALT3	0.058	152.67	0.194	87.73	0.083	540.52	0.011	3580.23
SEA SALT4	0.0002	873.93	0.0011	653.67	0.0001	1481.69	0.00001	43480.71
sulf	−0.028	−2.82	−1.349	−28.99	0.711	85.37	0.126	72.73
	17 June 2013							
	Δ Column		Δ 850 hPa		Δ 500 hPa		Δ 300 hPa	
	ug/m³	%	ug/m³	%	ug/m³	%	ug/m³	%
BC	0.03	20.82	−0.039	−4.36	0.154	197.75	0.031	151.07
OC	0.189	17.89	−0.473	−8.17	0.981	214.08	0.152	205.95
DUST1	0.298	16.05	1.412	23.44	0.489	19.75	0.108	29.43
DUST2	0.615	14.37	3.033	21.31	1.212	22.31	0.227	30.63
DUST3	0.423	14.38	2.063	20.00	0.999	29.29	0.148	37.05
DUST4	−0.043	−4.62	−0.773	−17.85	0.387	54.91	0.042	64.18
DUST5	−0.025	−38.44	−0.137	−48.18	−0.012	−17.96	0.001	−7.58
SEA SALT1	0.008	93.29	0.031	86.71	0.011	148.71	0.001	9.00
SEA SALT2	0.077	132.31	0.284	91.95	0.107	362.54	0.009	690.37
SEA SALT3	0.036	95.69	0.137	61.89	0.047	307.40	0.004	1261.94
SEA SALT4	0.0001	220.35	0.0002	122.56	0.00003	291.73	0.000002	7656.47
sulf	−0.056	−5.60	−1.087	−23.37	0.405	48.55	0.073	41.89

For all of the aerosol species BC, OC, dust (all sizes), sea salt (all sizes), and sulfates (sulf), we subtracted area average concentration over Uttarakhand (~29–31° N, ~78–81° E) on the event dates (16 or 17) to the nonprecipitating dates (15, 18, 19 and 20) at different pressure levels (column, 850 hPa, 500 hPa, and 300 hPa).

Change in aerosol concentration (ΔX) = area average concentration (non-precipitating days) − area average concentration (event day) (1)

Most of the bulk aerosol species BC, OC, dust, and sea salt show elevated concentration at 850 hPa and above, especially at the higher elevation of 500 hPa. DUST5 decreases while Sea Salt4 and Sea Salt1 increase insignificantly on the 16 and 17 at all elevations. The near-surface (850 hPa) concentration of DUST4, BC, and OC decreases during heavy precipitation events. At 500 hPa and above, BC, OC, DUST2, 3, 4, and SEA SALT2, 3 increase significantly during the precipitation event days. Sulfate concentration was less in a column and was near the surface, whereas at 500 hPa and above, the concentration was significantly higher during the 16–17 (Table 2). Similar results are observed with the 4 km resolution simulation (Supporting Table S3) with a lower ratio of concentration in the vertical layers compared to the 25 km resolution. The previous study on WRF-Chem sensitivity on horizontal resolution and nesting suggests that the boundary layer shows more sensitivity to 1-way nesting than 2-way nesting at a finer resolution [48], whereas the boundary layer plays an important role in atmospheric aerosols and chemistry [49]. A multimodel study suggests that the model performance to simulate aerosols does not depend on resolution, but the basic difference comes with the treatment of aerosols in a model such as that for removal and deposition parameterization [50].

Further analysis at 500 hPa and above suggest that bulk aerosols follow the fast-moving moisture content from the 15 of June from the eastern coast of India to the Himalayan region until 18 June. The concentration of aerosols was much higher than the stable atmospheric conditions in that region. The study suggests the significant uplifting of aerosols in the mid-troposphere during deep convective events as suggested in previous studies [33,49]. Figure S3 shows a lesser concentration of near-surface BC (850 hPa) during the event period, whereas Figures S4 and S5 shows a significant amount of BC over 500 hPa and 300 hPa in the active convective area during the event. Other aerosols also show similar features in the free troposphere. This indicates the strong vertical transport of aerosols during strong convective events over the effective region, which is evident in the BC flow analysis in Figures S6 and S7. That figure shows a strong updraft of BC from the plains towards to the Himalayan region during the event day, with convection anthropogenic and natural aerosol transport significantly above 500 hPa and above, which are further transported through synoptic monsoon motion in the atmosphere.

Aerosols act as IN and CCN, which are responsible for creating clouds and precipitation [12,13]. Model analysis of average columnar rain number concentration (RNC) and ice number concentration (ICC) over Kedarnath (Figure 6e,f), reveals the elevated concentration of RNC during 16 and 17 June, while ICC shows an elevated concentration on the 17 and a smaller peak on the 19. While there is a significant difference between the model simulations at 25 km and 4 km resolution, there is a minimal difference when taking chemistry (i.e., aerosols) into account. On the 16, all of the models simulated a good number of RNC, whereas ICC was not significantly high over Kedarnath. Higher RNC delays the precipitation time, which leads to higher and deeper clouds [18], as observed by the satellites and the model over Kedarnath on the 16 (Figure 5a,b).

On 17 June, above 500 hPa, a significantly high amount of ICC was present (Figure not shown), which justifies a cooler cloud top, as observed by satellites over Kedarnath. RNC, ICC (Figure 6e,f), and precipitation (Figure 3) patterns suggest two events. The first event (peak) before the 17, which was had been continuously increasing since the evening of the 15 to 16 that resulted in a downpour late at night on the 16–17. The second event with the RNC and ICC peak values that were present during the daytime of 17 led to precipitation in the late evening but with a lesser quantity. Our results suggest that a high amount of ICC held the precipitation within the cloud, which then moved with the synoptic circulation towards western Nepal on the 18, resulting in a heavy downpour there.

Analysis of the extinction coefficient profile from WC25 and WRF25 show the higher difference in extinction coefficient between the heights 2 to 8 km (calculated from Equation (1)), whereas RNC significantly increases in presence of aerosols at 4–6 km on both days over Kedarnath. During both days, the ICC decreases significantly in the presence of aerosols. Supporting Figures S8 and S9 shows decreased ICC above 6 km during the

16 and 17. The satellites were not able to capture aerosol optical depth (AOD) during event day, even near the area near the AERONET station (over Kanpur) does not have data; therefore, AOD comparison with observation was not possible. On the 17, CALIPSO shows 3% data availability, and data loss was reported to be due to a ground station anomaly (https://www-calipso.larc.nasa.gov/products/lidar/browse_images/data_event_log.php?s=production&v=V3-30&browse_date=2013-06-17, accessed on 1 April 2021), whereas the 16 and 18 passes were far from the area of interest.

Analysis suggests the presence of anthropogenic and natural aerosols (BC, OC, dust, sulfate, and sea salt) in the free troposphere above 2 km leads to an increase in RNC, which results in more precipitation over the Kedarnath region. The result shows a higher concentration of aerosols at elevated layers negatively affected the ICC numbers, which resulted in more precipitation in the region. Aerosols over the simulation domain affected the clouds and precipitation in either way.

We also considered evaluating the parameters that are important to indicate severe weather such as convective available potential energy (CAPE), convective inhibition energy (CIN), vorticity, and helicity [51,52]. CAPE was simulated high ($\geq$1000 J/kg) over the Bay of Bengal, the Arabian Sea, and many parts of the Indian subcontinent in WC25, whereas the Kedarnath region does not show high CAPE values during the event days. CIN was very less ($\leq$50 J/kg) over most of the area except over Pakistan and northern India ($\geq$100 J/kg) during event days. Strong ($\geq 12 \times 10^{-5}$/s) vorticity is simulated in models along the low-pressure zone (Figure 3). Strong helicity ($\geq$400 m^2/s^2) was also simulated over the Arabian Sea and Uttarakhand. The presence of aerosols affected the radiation over the region, which resulted in changes to the above parameters. At 25 km and 4 km, the presence of aerosols shows ($\geq\pm$300 J/kg) changes in CAPE, ($\geq\pm$50 J/kg) CIN, ($\geq\pm$100 m^2/s^2) helicity, and ($\geq \pm 6 \times 10^{-5}$/s) vorticity values over many parts in the simulation domain. Over Uttarakhand, the presence of aerosols increased ~200 J/kg in CAPE, ~20 J/kg in CIN, up to 100 m^2/s^2 in helicity, and above 4×10^{-5}/s in vorticity.

In the 4 km simulation, the aerosol effect show changes up to $\pm$200 J/kg in the CAPE value during the event day in the simulation domain. (Figure 7a), CIN values are significantly affected by the presence of aerosols, whereas similar results are simulated at the 25 km resolution simulation (Figure S10a,b). Helicity and vorticity show strong variation in the presence of aerosols over Kedarnath and nearby areas at convection-permitting scale (Figure 7c,d). Helicity and vorticity changes are not very evident at a 25 km resolution (Figure S10c,d). At the convection-permitting scale, variation in the parameters affected by orographic and thermodynamic are more evident than at low-resolution simulations. WRF-Chem based previous studies suggest that the presence of aerosols in the mid-troposphere shows a significant effect on the surrounding environment at regional and local levels, such as changes in near-surface temperature, wind speed, humidity, boundary layer, etc. [49,53].

Results indicate that the presence of aerosols does not only affect cloud and precipitation by acting as CCN and IN but also affects the severe weather indices; CAPE, CIN, helicity, and vorticity show significant changes in presence of aerosols. Over the Uttarakhand and near to Himalayan foothills, CAPE, CIN, helicity, and vorticity significantly increases in the presence of aerosols.

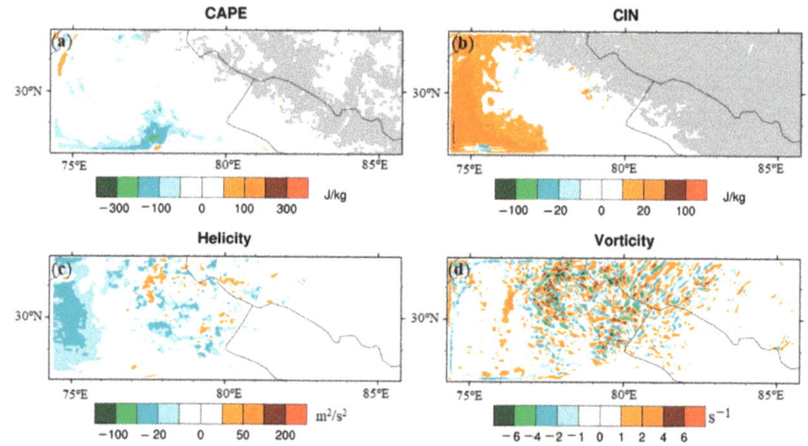

Figure 7. Effect of aerosols on (**a**) CAPE, (**b**) CIN, (**c**) helicity, and (**d**) vorticity at 4 km resolution (WC4-WRF4) during 17 June.

4. Summary and Conclusions

Long-term (1998–2014) analysis of precipitation from TRMM over Kedarnath suggested that heavy precipitation during June over Kedarnath is an unusual event. Analysis shows that June of 2013 experienced the highest precipitation compared to of the month of June in any other year from 1998–2014; otherwise, the year 2013 was recorded as a normal monsoon year. We used six sets of model simulations at 25 km and 4 km resolution (three each), and further analysis was conducted using four sets of simulations (two from each resolution). We dropped two sets of simulations (one from each resolution) since chemistry without aerosols (MOZART) and without chemistry showed no significant difference in the precipitation amount at both resolutions.

All of the model simulations captured the TRMM observed spatial coverage of precipitation with some differences. The presence of aerosols shows a significant increase in precipitation over Kedarnath and nearby areas, whereas the suppressed precipitation over the central and eastern part of India was observed at a 25 km resolution. Aerosols at the convection-permitting scale (4 km resolution) show similar results with more regional variation and differences in precipitation changes due to the presence of aerosols. In situ observations and TRMM data over many stations near Kedarnath show that the model captures the temporal trend and strength of precipitation with some differences. The model captured the consistent movement of the low-pressure system from central India on the 16 to Kedarnath on the 17 and further towards western Nepal on the 18, as reported in the literature, while it dissipated on the 19. Temperature, pressure, and humidity are also well replicated by the model. Cloud properties such as cloud fraction, cloud top temperature, and pressure are also well simulated by the model, as observed from the satellites.

Model simulation shows deep clouds (above 500 hPa) over Kedarnath and some other parts over the ocean on the 16 and 17. Model analysis suggests the presence of aerosols above 500 hPa, which may act as CCN/IC. The high amount of RNC and ICC over Kedarnath on the 16 and 17 suggests the role of aerosols in heavy precipitation events over Kedarnath. Some of the stations showed a 100 mm/day increase in the rain due to the presence of aerosols, most of the stations showed ~ $\pm$ 40 mm/day difference. Average profile analysis shows an unusual presence of aerosols over Kedarnath and significant changes in the rain and ice concentration number in the presence of aerosols. The effect of aerosols on precipitation was not only limited to CCN and IN in clouds; aerosol–radiation feedback causes significant alterations in precipitation. Analysis of CAPE, CIN, vorticity, and helicity shows positive radiation feedback in the region where precipitation was high and a negative effect in the area where precipitation was lower (Figure 1d).

Himalayan orographic lifting, a fast-moving monsoon due to low pressure generated from 15–18 June 2013, active westerlies, and aerosol effect on cloud formation due to direct CCN/IN and an indirect radiative effect made this event so devastating. Further detailed analysis of such a system is necessary to understand the vertical transport of aerosols and its effect on cloud properties and convection dynamics in the Himalayan region.

Supplementary Materials: The following are available online at https://www.mdpi.com/article/10.3390/atmos12091092/s1, Figure S1: (a) Domain of simulation at 25 km resolution, domain of simulation at 4 km resolution (red box), and Kedarnath (black dot), (b) zoom-in for the location of station 1. Kedarnath, 2. Champawat, 3. Pipalkoti, 4. Pandukeshwar, and 5. Lambgarh. Figure S2: Rain rate over Uttarakhand, nearby Kedarnath, and over Kedarnath average for the annual, monsoon, and June precipitation. Figure S3: BC1 concentration at 850 hPa from 14–19 June 2013. Figure S4: BC1 concentration at 500 hPa from 14–19 June 2013. Figure S5: BC1 concentration at 300 hPa from 14–19 June 2013. Figure S6: BC1 volume flow on 17 June 2013, 25 km resolution (gif). Figure S7: BC1 volume flow on 17 June 2013, 4 km resolution (gif). Figure S8: Difference of extinction coefficient, rain, and ice concentration profile over Kedarnath during (**a**) 16 June and (**b**) 17 June from the WC25 and WRF25 simulations. Figure S9: Difference of extinction coefficient, rain, and ice concentration profiles over Kedarnath during (**a**) 16 June and (**b**) 17 June from WC4 and WRF4 simulations (WC4-WRF4). Figure S10: Effect of aerosols on (**a**) CAPE, (**b**) CIN, (**c**) helicity, and (**d**) vorticity at 25 km resolution (WC25-WRF25) during 17 June. Table S1: Average rain rate (mm/hr) over Uttarakhand, Kedarnath, and nearby area for the period of 1998–2014 from TRMM. Table S2: R^2 of precipitation from different simulation with observed precipitation from in situ and satellite observations. Table S3: Δconcentration for different aerosols and Δ in absolute percentages in model column and at 850 hPa, 500 hPa, and 300 hPa for 16 and 17 June 2013 over Kedarnath at 4 km resolution.

Author Contributions: Data curation, P.S. (Prashant Singh) and B.A.; formal analysis, P.S. (Prashant Singh), P.S. (Pradip Sarawade) and B.A.; writing—original draft, P.S. (Prashant Singh) and B.A.; writing—review and editing, P.S. (Prashant Singh), P.S. (Pradip Sarawade), and B.A. All authors have read and agreed to the published version of the manuscript.

Funding: This research received no external funding.

Institutional Review Board Statement: Not applicable.

Informed Consent Statement: Not applicable.

Data Availability Statement: Most the website links for the publicly available data are provided in their respective places. For model data, if interested, one may contact the corresponding author through email.

Acknowledgments: This study was partially supported by the core funds of ICIMOD contributed by the governments of Afghanistan, Australia, Austria, Bangladesh, Bhutan, China, India, Myanmar, Nepal, Norway, Pakistan, Sweden, and Switzerland. The views and interpretations in this publication are those of the authors. They are not necessarily attributable to ICIMOD and do not imply the expression of any opinion by ICIMOD concerning the legal status of any country, territory, city, or area of its authority or concerning the delimitation of its frontiers or boundaries, or the endorsement of any product. Authors want to acknowledge observational data used in this study is provided by MOSDAC, SAC, ISRO.

Conflicts of Interest: The authors declare no conflict of interest.

References

1. Dimri, A.P.; Chevuturi, A.; Niyogi, D.; Thayyen, R.J.; Ray, K.; Tripathi, S.N.; Pandey, A.K.; Mohanty, U.C. Cloudbursts in Indian Himalayas: A review. *Earth-Sci. Rev.* **2017**, *168*, 1–23. [CrossRef]
2. Das, S.; Ashrit, R.; Moncrieff, M.W. Simulation of a Himalayan cloudburst event. *J. Earth Syst. Sci.* **2006**, *115*, 299–313. [CrossRef]
3. Lau, W.K.M.; Kim, K.M.; Shi, J.J.; Matsui, T.; Chin, M.; Tan, Q.; Peters-Lidard, C.; Tao, W.K. Impacts of aerosol–monsoon interaction on rainfall and circulation over Northern India and the Himalaya Foothills. *Clim. Dyn.* **2017**, *49*, 1945–1960. [CrossRef] [PubMed]
4. Kedia, S.; Vellore, R.K.; Islam, S.; Kaginalkar, A. A study of Himalayan extreme rainfall events using WRF-Chem. *Meteorol. Atmos. Phys.* **2018**, *131*, 1133–1143. [CrossRef]

5. Gautam, R.; Hsu, N.C.; Lau, K.M.; Tsay, S.C.; Kafatos, M. Enhanced pre-monsoon warming over the Himalayan-Gangetic region from 1979 to 2007. *Geophys. Res. Lett.* **2009**, *36*. [CrossRef]
6. Ojha, N.; Sharma, A.; Kumar, M.; Girach, I.; Ansari, T.U.; Sharma, S.K.; Singh, N.; Pozzer, A.; Gunthe, S.S. On the widespread enhancement in fine particulate matter across the Indo-Gangetic Plain towards winter. *Sci. Rep.* **2020**, *10*, 1–9. [CrossRef] [PubMed]
7. Saikia, A.; Pathak, B.; Singh, P.; Bhuyan, P.K.; Adhikary, B. Multi-model evaluation of meteorological drivers, air pollutants and quantification of emission sources over the upper Brahmaputra basin. *Atmosphere* **2019**, *10*, 703. [CrossRef]
8. Seinfeld, J.H.; Carmichael, G.R.; Arimoto, R.; Conant, W.C.; Brechtel, F.J.; Bates, T.S.; Cahill, T.A.; Clarke, A.D.; Doherty, S.J.; Flatau, P.J.; et al. ACE-ASIA: Regional climatic and atmospheric chemical effects of Asian dust and pollution. *Bull. Am. Meteorol. Soc.* **2004**, *85*, 367–380. [CrossRef]
9. Devara, P.C.S.; Manoj, M.G. Aerosol-cloud-precipitation interactions: A challenging problem in regional environment and climate research. *Particuology* **2013**, *11*, 25–33. [CrossRef]
10. Lebo, Z.J.; Feingold, G. On the relationship between responses in cloud water and precipitation to changes in aerosol. *Atmos. Chem. Phys.* **2014**, *14*, 11817–11831. [CrossRef]
11. Stevens, B.; Feingold, G. Untangling aerosol effects on clouds and precipitation in a buffered system. *Nature* **2009**, *461*, 607–613. [CrossRef]
12. Andreae, M.O.; Rosenfeld, D. Aerosol-cloud-precipitation interactions. Part 1. The nature and sources of cloud-active aerosols. *Earth-Sci. Rev.* **2008**, *89*, 13–41. [CrossRef]
13. Solomos, S.; Kallos, G.; Kushta, J.; Astitha, M.; Tremback, C.; Nenes, A.; Levin, Z. An integrated modeling study on the effects of mineral dust and sea salt particles on clouds and precipitation. *Atmos. Chem. Phys.* **2011**, *11*, 873–892. [CrossRef]
14. Lin, J.C.; Matsui, T.; Pielke, S.A.; Kummerow, C. Effects of biomass-burning-derived aerosols on precipitations and clouds in the Amazon Basin: A satellite-based empirical study. *J. Geophys. Res. Atmos.* **2006**, *111*. [CrossRef]
15. Bond, T.C.; Doherty, S.J.; Fahey, D.W.; Forster, P.M.; Berntsen, T.; Deangelo, B.J.; Flanner, M.G.; Ghan, S.; Kärcher, B.; Koch, D.; et al. Bounding the role of black carbon in the climate system: A scientific assessment. *J. Geophys. Res. Atmos.* **2013**, *118*, 5380–5552. [CrossRef]
16. Khain, A.; Rosenfeld, D.; Pokrovsky, A. Aerosol impact on the dynamics and microphysics of deep convective clouds. *Q. J. R. Meteorol. Soc.* **2005**, *131*, 2639–2663. [CrossRef]
17. Koren, I.; Kaufman, Y.J.; Rosenfeld, D.; Remer, L.A.; Rudich, Y. Aerosol invigoration and restructuring of Atlantic convective clouds. *Geophys. Res. Lett.* **2005**, *32*, 1–4. [CrossRef]
18. Tao, W.K.; Chen, J.-P.; Li, Z.; Wang, C.; Zhang, C. Impact of Aerosols on Convective Clouds and Precipitation. *Rev. Geophys.* **2012**, *50*, 1–62. [CrossRef]
19. Bègue, N.; Tulet, P.; Pelon, J.; Aouizerats, B.; Berger, A.; Schwarzenboeck, A. Aerosol processing and CCN formation of an intense Saharan dust plume during the EUCAARI 2008 campaign. *Atmos. Chem. Phys.* **2015**, *15*, 3497–3516. [CrossRef]
20. Rosenfeld, D.; Lohmann, U.; Raga, G.B.; O'Dowd, C.D.; Kulmala, M.; Fuzzi, S.; Reissell, A.; Andreae, M.O. Flood or Drought: How Do Aerosols Affect Precipitation? *Science* **2008**, *321*, 1309–1313. [CrossRef] [PubMed]
21. Nandargi, S.; Dhar, O.N. Extreme Rainstorm Events over the Northwest Himalayas during 1875–2010. *J. Hydrometeorol.* **2012**, *13*, 1383–1388. [CrossRef]
22. Singh, R.; Siingh, D.; Gokani, S.A.; Buchunde, P.S.; Singh, R.P.; Singh, A.K. Brief Communication: Climate, topographical and meteorological investigation of the 16–17 June 2013 Kedarnath (India) disaster causes. *Nat. Hazards Earth Syst. Sci. Discuss.* **2015**, *3*, 941–953. [CrossRef]
23. Singh, C.; Chand, R. Exceptionally heavy rainfall over Uttarakhand during 15–18 June, 2013-A case study. *Mausam* **2015**, *66*, 741–750.
24. Chawla, I.; Osuri, K.K.; Mujumdar, P.P.; Niyogi, D. Assessment of the Weather Research and Forecasting (WRF) model for simulation of extreme rainfall events in the upper Ganga Basin. *Hydrol. Earth Syst. Sci.* **2018**, *22*, 1095–1117. [CrossRef]
25. Ray, K.; Bhan, S.C.; Sunitha Devi, S. *A Meteorological Analysis of Very Heavy Rainfall Event over Uttarakhand during 14–17 June, 2013*; India Meteorological Department: Delhi, India, 2014; pp. 37–54.
26. Shukla, D.P.; Dubey, C.S.; Usham, A.L. Orographic Control of the Kedarnath disaster Orographic control of the Kedarnath disaster. *Curr. Sci.* **2013**, *105*, 1474–1476.
27. Satoh, M.; Tomita, H.; Miura, H.; Iga, S.; Nasuno, T. Development of a global cloud resolving model–A multi-scale structure of tropical convections. *J. Earth Simulator* **2005**, *3*, 11–19.
28. Grabowski, W.W. Toward Cloud Resolving Modeling of Large-Scale Tropical Circulations: A Simple Cloud Microphysics Parameterization. *Am. Meteorol. Soc.* **1998**, *55*, 3283–3298. [CrossRef]
29. Kilaru, A.; Kotamraju, S.K.; Avlonitis, N.; Sri Kavya, K.C. Rain rate intensity model for communication link design across the Indian region. *J. Atmos. Sol.-Terr. Phys.* **2016**, *145*, 136–142. [CrossRef]
30. Powers, J.G.; Klemp, J.B.; Skamarock, W.C.; Davis, C.A.; Dudhia, J.; Gill, D.O.; Coen, J.L.; Gochis, D.J.; Ahmadov, R.; Peckham, S.E.; et al. The weather research and forecasting model: Overview, system efforts, and future directions. *Bull. Am. Meteorol. Soc.* **2017**, *98*, 1717–1737. [CrossRef]
31. Skamarock, W.C.; Klemp, J.B.; Dudhi, J.; Gill, D.O.; Barker, D.M.; Duda, M.G.; Huang, X.-Y.; Wang, W.; Powers, J.G.; Dudhia, J.; et al. *A Description of the Advanced Research WRF Version 3*; NCAR Techical Note-475+ STR; NCAR: Pod, CL, USA, 2008. [CrossRef]

32. Grell, G.A.; Peckham, S.E.; Schmitz, R.; Mckeen, S.A.; Frost, G.; Skamarock, W.C.; Eder, B. Fully coupled "online" chemistry within the WRF model. *Atmos. Environ.* **2005**, *39*, 6957–6975. [CrossRef]
33. Singh, P.; Adhikary, B.; Sarawade, P. Transport of black carbon from planetary boundary layer to free troposphere on a seasonal scale over South Asia. *Atmos. Res.* **2020**, *235*, 104761. [CrossRef]
34. Janssens-Maenhout, G.; Dentener, F.; Van Aardenne, J.; Monni, S.; Pagliari, V.; Orlandini, L.; Klimont, Z.; Kurokawa, J.; Akimoto, H.; Ohara, T.; et al. *EDGAR-HTAP: A Harmonized Gridded Air Pollution Emission Dataset Based on National Inventories*; Commission Publications Office: Spra, Italy, 2012; ISBN 9789279231230.
35. Kain, J.S. The Kain–Fritsch Convective Parameterization: An Update. *J. Appl. Meteorol.* **2004**, *43*, 170–181. [CrossRef]
36. Hong, S.-Y.; Noh, Y.; Dudhia, J. A New Vertical Diffusion Package with an Explicit Treatment of Entrainment Processes. *Mon. Weather Rev.* **2006**, *134*, 2318–2341. [CrossRef]
37. Dudhia, J. Numerical Study of Convection Observed during the Winter Monsoon Experiment Using a Mesoscale Two-Dimensional Model. *J. Atmos. Sci.* **1989**, *46*, 3077–3107. [CrossRef]
38. Thompson, G.; Field, P.R.; Rasmussen, R.M.; Hall, W.D. Explicit Forecasts of Winter Precipitation Using an Improved Bulk Microphysics Scheme. Part II: Implementation of a New Snow Parameterization. *Mon. Weather Rev.* **2008**, *136*, 5095–5115. [CrossRef]
39. Mlawer, E.J.; Taubman, S.J.; Brown, P.D.; Iacono, M.J.; Clough, S.A. Radiative transfer for inhomogeneous atmospheres: RRTM, a validated correlated-k model for the longwave. *J. Geophys. Res. Atmos.* **1997**, *102*, 16663–16682. [CrossRef]
40. Tewari, M.; Chen, F.; Wang, W.; Dudhia, J.; LeMone, M.A.; Mitchell, K.; Ek, M.; Gayno, G.; Wegiel, J.; Cuenca, R.H. Implementation and verification of the unified noah land surface model in the WRF model. *Bull. Am. Meteorol. Soc.* **2004**, 2165–2170. [CrossRef]
41. Paulson, C.A. The Mathematical Representation of Wind Speed and Temperature Profiles in the Unstable Atmospheric Surface Layer. *J. Appl. Meteorol.* **1970**, *9*, 857–861. [CrossRef]
42. Madronich, S.; Weller, G. Numerical integration errors in calculated tropospheric photodissociation rate coefficients. *J. Atmos. Chem.* **1990**, *10*, 289–300. [CrossRef]
43. Castorina, G.; Caccamo, M.T.; Colombo, F.; Magazù, S. The role of physical parameterizations on the numerical weather prediction: Impact of different cumulus schemes on weather forecasting on complex orographic areas. *Atmosphere* **2021**, *12*, 616. [CrossRef]
44. Mishra, A.; Srinivasan, J. Did a cloud burst occur in Kedarnath during 16 and 17 June 2013? *Curr. Sci.* **2013**, *105*, 16–18.
45. Sikka, D.R. *Synoptic and Meso-Scale Weather Disturbances over South Asia during the Southwest Summer Monsoon Season*, 2nd ed.; Chang, C.-P., Ding, Y., Lau, N.-C., Johnson, R.H., Wang, B., Yasunari, T., Eds.; World Scientific: Singapore, 2010; ISBN 9789814343404.
46. Stuefer, M.; Freitas, S.R.; Grell, G.; Webley, P.; Peckham, S.; McKeen, S.A.; Egan, S.D. Inclusion of ash and SO2 emissions from volcanic eruptions in WRF-Chem: Development and some applications. *Geosci. Model Dev.* **2013**, *6*, 457–468. [CrossRef]
47. Rizza, U.; Brega, E.; Caccamo, M.T.; Castorina, G.; Morichetti, M.; Munaò, G.; Passerini, G.; Magazù, S. Analysis of the etna 2015 eruption using wrf- chem model and satellite observations. *Atmosphere* **2020**, *11*, 1168. [CrossRef]
48. Misenis, C.; Zhang, Y. An examination of sensitivity of WRF/Chem predictions to physical parameterizations, horizontal grid spacing, and nesting options. *Atmos. Res.* **2010**, *97*, 315–334. [CrossRef]
49. Singh, P.; Sarawade, P.; Adhikary, B. Carbonaceous aerosol from open burning and its impact on regional weather in South Asia. *Aerosol Air Qual. Res.* **2020**, *20*, 419–431. [CrossRef]
50. Eckhardt, S.; Quennehen, B.; Olivié, D.J.L.; Berntsen, T.K.; Cherian, R.; Christensen, J.H.; Collins, W.; Crepinsek, S.; Daskalakis, N.; Flanner, M.; et al. Current model capabilities for simulating black carbon and sulfate concentrations in the Arctic atmosphere: A multi-model evaluation using a comprehensive measurement data set. *Atmos. Chem. Phys.* **2015**, *15*, 9413–9433. [CrossRef]
51. Tajbakhsh, S.; Ghafarian, P.; Sahraian, F. Instability indices and forecasting thunderstorms: The case of 30 April 2009. *Nat. Hazards Earth Syst. Sci.* **2012**, *12*, 403–413. [CrossRef]
52. Singh, P. Prediction of Potential Thunderstorm Over Ocean near Sriharikota. *Int. J. Interdiscip. Res. Innov.* **2015**, *3*, 1–6.
53. Hirtl, M.; Stuefer, M.; Arnold, D.; Grell, G.; Maurer, C.; Natali, S.; Scherllin-Pirscher, B.; Webley, P. The effects of simulating volcanic aerosol radiative feedbacks with WRF-Chem during the EyjafjallajÖkull eruption, April and May 2010. *Atmos. Environ.* **2019**, *198*, 194–206. [CrossRef]

atmosphere

Article

Machine Learning Emulation of Spatial Deposition from a Multi-Physics Ensemble of Weather and Atmospheric Transport Models

Nipun Gunawardena [1,*], Giuliana Pallotta [1], Matthew Simpson [2] and Donald D. Lucas [1,*]

1 Lawrence Livermore National Laboratory, Livermore, CA 94550, USA; pallottagold1@llnl.gov
2 Scripps Institution of Oceanography, University of California San Diego, La Jolla, CA 92093, USA; mdsimpson@ucsd.edu
* Correspondence: gunawardena1@llnl.gov (N.G.); lucas26@llnl.gov (D.D.L.)

Abstract: In the event of an accidental or intentional hazardous material release in the atmosphere, researchers often run physics-based atmospheric transport and dispersion models to predict the extent and variation of the contaminant spread. These predictions are imperfect due to propagated uncertainty from atmospheric model physics (or parameterizations) and weather data initial conditions. Ensembles of simulations can be used to estimate uncertainty, but running large ensembles is often very time consuming and resource intensive, even using large supercomputers. In this paper, we present a machine-learning-based method which can be used to quickly emulate spatial deposition patterns from a multi-physics ensemble of dispersion simulations. We use a hybrid linear and logistic regression method that can predict deposition in more than 100,000 grid cells with as few as fifty training examples. Logistic regression provides probabilistic predictions of the presence or absence of hazardous materials, while linear regression predicts the quantity of hazardous materials. The coefficients of the linear regressions also open avenues of exploration regarding interpretability—the presented model can be used to find which physics schemes are most important over different spatial areas. A single regression prediction is on the order of 10,000 times faster than running a weather and dispersion simulation. However, considering the number of weather and dispersion simulations needed to train the regressions, the speed-up achieved when considering the whole ensemble is about 24 times. Ultimately, this work will allow atmospheric researchers to produce potential contamination scenarios with uncertainty estimates faster than previously possible, aiding public servants and first responders.

Keywords: deposition; machine learning; hazardous release; WRF; FLEXPART; prediction

Citation: Gunawardena, N.; Pallotta, G.; Simpson, M.; Lucas, D.D. Machine Learning Emulation of Spatial Deposition from a Multi-Physics Ensemble of Weather and Atmospheric Transport Models. *Atmosphere* **2021**, *12*, 953. https://doi.org/10.3390/atmos12080953

Academic Editor: Patrick Armand

Received: 15 June 2021
Accepted: 21 July 2021
Published: 24 July 2021

1. Introduction

From localized air-pollution caused by fireworks [1], to seasonal changes in pollution caused by cars [2], to planetary-scale dust transport from earth's deserts [3], particulate and gaseous hazardous matter can be dispersed throughout the environment from numerous natural and anthropogenic processes. One event which is important to public health and national security is the release of hazardous materials from nuclear weapons explosions, nuclear reactor breaches (such as Chernobyl or Fukushima), chemical spills, industrial accidents, and other toxic releases. These types of incidents happen suddenly and without warning, creating a plume of toxic material in the earth's atmosphere or ocean which can threaten the well-being of living organisms and environments.

In such situations, it is crucial that politicians, policy makers, and first responders have adequate knowledge about how the toxic plume will disperse and deposit throughout the environment. This can be used to determine evacuation zones and how resources are deployed to minimize the loss of public health. For example, during the 2011 Fukushima Daiichi disaster, the United States Department of Energy, the United States Environmental

Protection Agency, and other United States national agencies worked together to determine the effect of the radioactive release on international aviation routes, global food supply, and other crucial aspects of society [4].

To predict how a toxic plume disperses and deposits throughout the environment, scientists typically run computer simulations. These dispersion simulations solve physical and chemical equations to produce evolving concentration and deposition fields, but many of the processes represented in the models are uncertain or not resolved at the scales of interest. These processes are represented by empirical or semi-empirical parameterizations, and no single set of parameterizations always performs best for every scenario. Picking and choosing different sets of parameterizations provides an estimate of uncertainty and is a necessary component of the prediction process. In addition, many detailed transport and dispersion models that are currently in use are very computationally expensive, sometimes requiring several hours to complete a single simulation. Since time is of the essence during emergencies, these long run-times can be detrimental to first-responder efforts.

Therefore, in the event of a toxic environmental release, the scientists making predictions with limited computing resources often face a dilemma: using a detailed model, they can make a small number of predictions quickly but have poor knowledge of the uncertainty of those predictions, or they can make a large number of predictions slowly but have better knowledge of the uncertainty of the predictions.

A machine learning or statistical method that emulates a transport and dispersion model provides the opportunity to minimize this uncertainty versus time dilemma. To do this, the scientists would vary the inputs to the traditional weather/dispersion model to create a small number of predictions. They would then train a statistical model to produce dispersion predictions given the original input values. Finally, the statistical model could be used to create predictions for the set of inputs that were not originally run with the traditional dispersion model. That is to say, the statistical model is an emulator of the original dispersion model.

In this paper, we introduce a statistical method that rapidly predicts spatial deposition of radioactive materials over a wide area. The deposition predictions we are emulating were originally produced using material releases in the FLEXible PARTicle dispersion model (FLEXPART) [5] and meteorological fields generated from the Weather Research and Forecasting (WRF) [6] model. We created two FLEXPART-WRF ensembles for training and testing—one simulating a continuous surface release from a hypothetical industrial accident and one simulating an instantaneous elevated cloud from a hypothetical nuclear detonation. Each ensemble contained 1196 members with different weather conditions. (Each ensemble initially contained 1200 runs, but four runs did not complete due to numerical stability issues). To create the ensembles, WRF physics parameterizations were varied (i.e., a multi-physics ensemble) and used as inputs for our statistical model. Multi-physics WRF ensembles are often used to estimate weather model uncertainty, and our statistical method is able to capture this uncertainty very efficiently without having to run the full ensemble. We use a hybrid statistical model consisting of a two-dimensional grid of linear and logistic regression models for predicting spatial deposition.

The paper is organized as follows: Section 2 reviews the literature and the tools used. Section 3 describes the details of the dataset. Section 4 describes the statistical algorithm that is used as the emulator, and Section 5 presents the performance of the algorithm. Finally, Sections 6 and 7 discuss future work and summarize the current work, respectively.

2. Background

2.1. FLEXPART-WRF

There are many different methods that can be used to predict how an airborne contaminant disperses throughout the atmosphere, ranging from simple box models and Gaussian plumes to more sophisticated Lagrangian and Eulerian transport models [7]. Gaussian plume models are the simplest and the fastest to run but are often limited to very specific, idealized scenarios. Lagrangian and Eulerian models are slower but contain representations of physical and chemical processes typically needed to simulate real-world releases. One key distinction between Gaussian plume models and Lagrangian/Eulerian models is that Gaussian plume models do not incorporate spatially and temporally varying meteorological fields.

We use the FLEXPART Lagrangian dispersion model for calculating the dispersion of airborne contaminants and estimate the effects of weather uncertainty in the dispersion calculations. To transport materials through the atmosphere and deposit them on the surface, FLEXPART requires spatially and temporally varying wind and precipitation data, which can come from archived meteorological forecast/analysis/re-analysis fields or weather models. For the work presented here, we use a specific version of FLEXPART designed to work directly with WRF output [5] (FLEXPART-WRF version 3.3). Although FLEXPART also has several internal physics options, we did not vary these for this work. A detailed description of our FLEXPART setup is provided in Section 3.

Several researchers have investigated the uncertainty of atmospheric dispersion models without incorporating machine learning. Leadbetter et al. [8] classify dispersion model error into three categories: source term uncertainty, meteorological uncertainty (which we study here), and intrinsic dispersion model uncertainty. They proceed to rank the uncertainties and find that wind direction and wind speed are important. Korsakissok et al. [9] ran multiple dispersion ensembles, some hypothetical and some realistic (e.g., the Fukushima Release) and analyzed the uncertainty. Finally, Sørensen et al. [10] simulated a nuclear power plant atmospheric release (similar to our surface release scenario) and presented a methodology to quantitatively estimate the variability of the ensemble. All studies cited the need for ensemble dispersion modeling. We focus specifically on uncertainty due to meteorological modeling.

To calculate winds and estimate weather uncertainty, we use the Weather Research and Forecasting model (WRF), a tool which is used to predict weather phenomena on scales of hundreds of meters to thousands of kilometers. A detailed description of WRF is found in Skamarock et al. [6]. WRF contains several physics options known as parameterizations for simulating processes such as cumulus convection, boundary layer turbulence, and land surface interactions. In our application, we estimate weather uncertainty by using a multi-physics approach that varies these parameterizations and uses the output to drive FLEXPART. A detailed description of the WRF setup is in Section 3. We specifically use WRF code version 3.9 with the advanced research dynamical numerical core.

Several other researchers have investigated WRF multi-physics ensembles. For example, researchers produced WRF multi-physics ensembles to investigate precipitation [11], heatwaves [12,13], and climate [14,15]. In prior work, we have investigated WRF multi-physics uncertainty to investigate the release from a nuclear power plant [16]. The important thing to note is that many of these ensembles have sizes of a few dozen members to a few hundred members. Our ensemble, having 1200 members, is feasible but significantly larger than average for a WRF multi-physics ensemble.

Machine learning and statistical methods have frequently been used to emulate complicated atmospheric models. Much of the prior emulation work focused on the potential speed-up offered, with applications to uncertainty studies, though some discussed ways to improve sub-grid scale parameterizations. Jensen et al. [17] and Lucas et al. [16] used machine learning algorithms to accelerate probabilistic inverse modeling studies of atmospheric sources. Watson [18] demonstrated the use of machine learning to improve long term climate statistics. Calbó et al. [19], Mayer et al. [20], and Beddows et al. [21] used polynomial chaos expansions and Gaussian processes to emulate air quality models. Wang et al. [22] used a neural network to emulate the planetary boundary layer parameterization of WRF. Krasnopolsky et al. [23] and Pal et al. [24] demonstrated the use of machine learning to emulate radiation parameterizations for global atmospheric models. Lucas and Prinn [25], Kelp et al. [26], and Ivatt and Evans [27] used statistical and machine learning approaches to emulate atmospheric chemistry and transport models. To our best knowledge, this paper describes the first time a machine learning method is used to emulate full FLEXPART-WRF spatial deposition maps.

2.2. Linear and Logistic Regression

The two main statistical methods we used were linear regression and logistic regression. Since these simple methods are fast, easy to train, and readily interpretable, we used them over other more complex methods that we also investigated. Since linear regression and logistic regression are basic statistical tools, we will only present a brief overview here. More information about both methods can be found in many statistics and machine learning textbooks, such as Murphy [28] or Gelman and Hill [29]. Linear regression is a type of regression used to fit data that have a linear relationship. It can be written as $y = \boldsymbol{\beta}^T\mathbf{x}$, where y is the scalar output of the regression, $\boldsymbol{\beta}$ is an n-dimensional coefficient vector, T indicates the transpose operation, and $\mathbf{x}$ is the n-dimensional predictor vector. The first or last element in $\mathbf{x}$ is typically just set to 1 and is there so the fitted hyperplane has an intercept instead of being forced to pass through the origin. The "linear" in linear regression only applies to the coefficient vector—the elements of the input vector can be transformed as desired. Finally, linear regression without regularization is trained in a non-iterative fashion by minimizing the squared residuals. This contrasts with many other machine learning algorithms which are trained in an iterative fashion.

Logistic regression is a simple classification method that can be used to classify binary variables. It can be written as $p = \frac{1}{1+e^{-\boldsymbol{\beta}^T\mathbf{x}}}$. Here, p is the probability that the target class has a value of 1, $\boldsymbol{\beta}$ is an n-dimensional coefficient vector, T indicates the transpose operation, and $\mathbf{x}$ is the n-dimensional predictor vector. The function $f = \frac{1}{1+e^{-x}}$ is also known as the logistic function or sigmoid function. As with linear regression, logistic regression can have an intercept term. Unlike linear regression, logistic regression must be trained iteratively, even if regularization is not used.

3. Dataset

We trained our statistical model on two sets of FLEXPART dispersion simulations. Both sets release the radioactive particulate material cesium-137 (Cs-137), which has a half-life of 30.17 years, is highly soluble, and is subject to removal from the atmosphere by rainout and wet and dry deposition. Both sets of FLEXPART simulations use 1196 different weather conditions generated by a WRF multi-physics ensemble, as described below. The first set contains the results for a hypothetical continuous release of Cs-137 from the surface of the earth at an industrial facility representing a large-scale radiological accident. This set of simulations is referred to as the "surface release" case or the "surface" case. The second set contains simulations of a hypothetical instantaneous release of Cs-137 in the form of a mushroom cloud similar to how contaminants are created from a nuclear detonation. This set of simulations is referred to here as the "elevated release" case or "elevated" case. Any mathematical notation from this point forward can be generalized to either case unless otherwise specified.

Within a case, each ensemble member k consists of an 1×16 input vector $\mathbf{x}_k$ and an $M \times N$ target deposition map $\mathbf{Y}_k$, where M and N are the number of grid boxes in latitudinal and longitudinal directions, respectively. (The dimensionality of the input vector $\mathbf{x}_k$ will be explained later in this section). The vector $\mathbf{x}_k$ contains the physics parameterizations used by WRF and is the input to our statistical model. The deposition map $\mathbf{Y}_k$ is the output of FLEXPART-WRF given $\mathbf{x}_k$ and is used as the target data for training our statistical model. The input vectors are identical between the surface release case and the elevated release case because they are based on the same WRF ensemble, i.e., $\mathbf{x}_k{}^{\text{Surface}} = \mathbf{x}_k{}^{\text{Elevated}}$.

The FLEXPART settings remain constant for every ensemble member within a given case. Consequently, they are not included as inputs to our statistical model. Each FLEXPART simulation was set to begin at 12:00Z on 24 April 2018 and end 48 h later. An adaptive timestep was used for the sampling rate of the output, but the nominal value was 180 s. Subgrid terrain effects and turbulence were included, and land-use data were taken from

WRF. Two million Lagrangian particles were released, and the total mass for the surface and elevated cases was 1 kg and 0.28 g, respectively. We used the default precipitation scavenging coefficients for Cs-137. Table 1 shows the Cs-137 particle size distributions and masses as a function of altitude for the elevated release case, as further described in Norment [30]. Further information about the release scenarios can be found in Lucas et al. [31] and Lucas et al. [32].

While the FLEXPART settings of each ensemble member remain constant within the case, the set of physics options in WRF is different for every ensemble member. We vary the following five categories of physics parameterizations within WRF: planetary boundary layer physics (PBL), land surface model (LSM), cumulus physics (CU), microphysics (MP), and radiation (RA). Any remaining parameterizations or options remain fixed. To run WRF, one parameterization must be chosen from each physics category. While each category has several different parameterization options available, yielding well over 100,000 possible combinations of parameterizations, we selected a subset of 1200 possibilities expected to simulate the weather, as determined by expert judgment. The ensemble members were roughly chosen to maximize diversity in physics parameterizations.

In a real-world scenario, these 1200 possibilities would be forecasts, i.e., plausible scenarios for the time evolution of the weather and plumes over a two-day period given initial weather conditions that are known at the beginning of the forecast. Therefore, we assume that each ensemble member is equally likely and do not attempt to "prune" the ensemble while it is running because it is a short-term forecast. The 1200-member ensemble therefore provides an estimate of weather model uncertainty in forecasting the deposition from the hypothetical Cs-137 release events. Because we used data from 2018, we were able to verify the meteorological forecasts. In work not presented here, we ran simulations using data assimilation to produce analysis-observational fields. The ensemble simulations provide a reasonable spread around the nudged fields [32], which gives us confidence that our machine learning model can perform in realistic scenarios. Furthermore, for our short-term forecasts of two days, the WRF parameterization uncertainty is expected to dominate the variability. Very short term forecasts (e.g., 1 h) would not have a lot of variability, while longer forecasts (e.g., 7 days) have errors dominated by initial conditions, and the machine learning task would be much more difficult.

Ultimately, we selected five parameterizations for PBL, four for LSM, five for CU, four for MP, and three for RA. The specific parameterizations are shown in Table 2. This results in $5 \times 4 \times 5 \times 4 \times 3 = 1200$ different combinations of the WRF parameterizations. However, 4 of the 1200 combinations caused numerical issues in WRF, which failed to run to completion, so there are only 1196 members in the final multi-physics weather dataset. The 1196 input vectors $\mathbf{x}_k$ are vertically concatenated to create a 1196×16 input matrix $\mathbf{X}$. The 1196 output matrices $\mathbf{Y}_k$ are concatenated in the third dimension to make the $M \times N \times 1196$ output matrix $\mathbf{Y}$. The ordering of the parameterization combinations in the ensemble is shown in Figure 1.

The individual physics parameterizations are nominal categorical variables represented as numbers in WRF. In other words, the parameterizations are not ordinal—PBL parameterization 2, which represents the MYJ scheme, is not greater than PBL parameterization 1, which represents the YSU scheme. To prevent our statistical model from treating a higher numbered parameterization differently than a lower numbered parameterization, we transformed the input WRF parameterization vector using one-hot encoding [28]. This turns the five categorical variables for the parameterizations into sixteen boolean variables, which is why $\mathbf{x}_k$ has shape 1×16. For example, the LSM parameterization has four options: LSM 1, LSM 2, LSM 3, and LSM 7. When one-hot encoding, LSM 1 is represented by the vector $[0,0,0]$, LSM 2 is represented by the vector $[1,0,0]$, LSM 3 is represented by the vector $[0,1,0]$, and LSM 7 is represented by the vector $[0,0,1]$. The vectors for each parametrization are concatenated together. (For example, the ensemble member run with PBL 2, LSM 1, CU 5, MP 4, and RA 4 has a one-hot encoded input vector $[1,0,0,0,0,0,0,0,0,0,1,0,1,0,0,1]$).

The output matrix **Y** consists of 1196 simulations produced by FLEXPART-WRF . Each ensemble member $\mathbf{Y}_k$ is an $M \times N$ map of the surface deposition of Cs-137 from either the surface release or the elevated release. For the surface release case, each map contains a total of 160,000 grid cells, with 400 cells in the latitudinal direction and 400 cells in longitudinal direction using a spatial resolution of about 1.7 km per cell. For the elevated release case, each map contains 600 grid cells by 600 grid cells with a resolution of about 1.2 km. Both deposition domains range from $+32.3°$ to $+38.5°$ in the latitudinal direction and $-77.3°$ to $-84.6°$ in the longitudinal direction. The height of the surface release domain was 3000 m resolved using 11 vertical layers, and the height of the elevated release domain was 4500 m resolved using 14 layers. The latitude and longitude of the location of the surface release were $+35.4325°$ and $-80.9483°$, respectively. The latitude and longitude of the location of the elevated release were $+35.2260°$ and $-80.8486°$, respectively. This domain is centered on the southwest corner of the US state North Carolina and has many different land types, including the Appalachian Mountains and the Atlantic Ocean.

The surface deposition output of FLEXPART-WRF accounts for both wet and dry removal of Cs-137 from the atmosphere and is reported in units of Bq/m^2 using a specific activity for Cs-137 of 3.215 Bq/nanogram. We also filtered out data less than 0.01 Bq/m^2 in our analysis, as deposition values below this level are comparable to background levels from global fallout [33] and do not pose much risk to public health.

All FLEXPART-WRF runs were completed on Lawrence Livermore National Laboratory's Quartz supercomputer which has 36 compute cores, 128 GB of RAM per node, and 3018 nodes total. A single WRF run costs about 150 core-hours, and a single FLEXPART run costs about 20 core-hours. The total ensemble cost was about 180,000 core-hours. The speedup between the full ensemble and the machine learning training set cost depends on the training size, which is discussed in Section 5. For a training size of 50, the total cost would be 7500 core-hours, which is a speedup of 24 times (or a savings of 172,500 core-hours). Figures 2 and 3 show selected examples of ensemble members from the surface case and elevated case, respectively. The members were chosen to highlight the diversity of the ensemble. The examples in the figures used PBL, LSM, CU, MP, and RA parameterization combinations (1, 1, 2, 3, 4) for member 25, (2, 1, 1, 3, 4) for member 245, (2, 3, 5, 3, 4) for member 413, and (7, 7, 10, 3, 4) for member 1157.

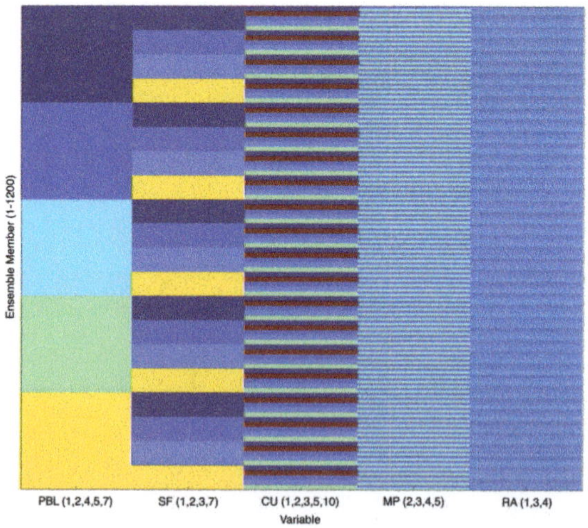

Figure 1. WRF parameterizations were varied as illustrated to create the multi-physics ensemble by iterating through the schemes in the order PBL, LSM, CU, MP, and RA.

Table 1. Profile for elevated release.

Altitude (m)	Mean Diameter (μm)	Geometric Standard Deviation	Cs-137 Mass (mg)
0	910	1.9	20.7
250	400	1.2	2.45
500	350	1.2	3.24
750	300	1.2	4.05
1000	270	1.2	5.17
1250	220	1.3	6.25
1500	170	1.4	9.67
1750	110	1.6	17.7
2000	52	2.1	51.5
2250	54	2.5	36.9
2500	48	2.4	36.6
2750	40	2.3	34.4
3000	32	2.0	26.1
3250	19	2.0	22.6

Table 2. WRF parameterizations used to create dataset, referred to here by their standard option number (between parentheses), name, and corresponding citation.

PBL	LSM	CU	MP	RA
(1) YSU [34]	(1) Thermal Diffusion [35]	(1) Kain-Fritsch [36]	(2) Lin (Purdue) [37]	(1) RRTM [38]
(2) MYJ [39]	(2) Noah [40]	(2) Betts-Miller-Janjic [39]	(3) WSM3 [41]	(3) CAM [42]
(4) QNSE [43]	(3) RUC [44]	(3) Grell-Devenyi [45]	(4) WSM5 [41]	(4) RRTMG [46]
(5) MYNN2 [47]	(7) Pleim-Xu [48]	(5) Grell-3 [49]	(5) Eta (Ferrier) [50]	
(7) ACM2 [51]		(10) CuP [52]		

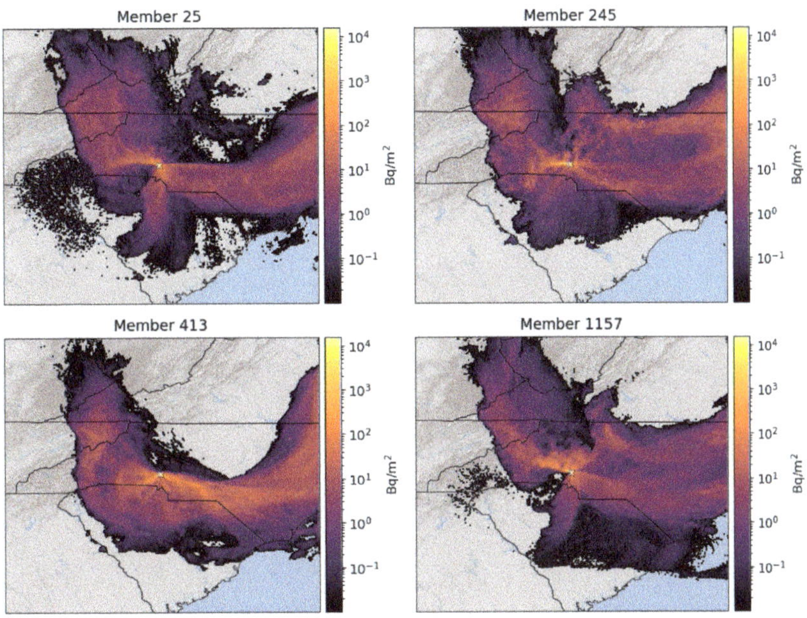

Figure 2. Examples of different deposition maps produced by FLEXPART-WRF for the surface release case. All values below 0.01 Bq/m^2 were removed. The WRF parameterizations used to create each subplot can be found in Section 3.

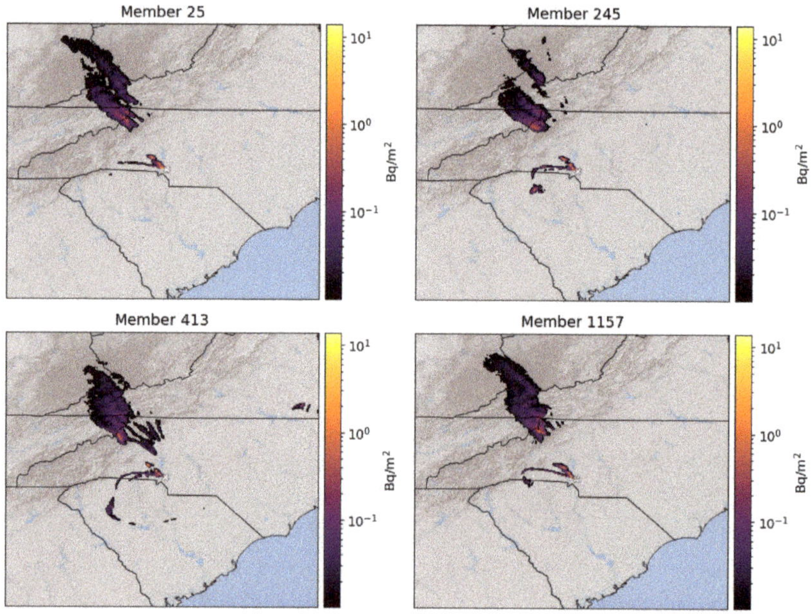

Figure 3. Examples of different deposition maps produced by FLEXPART-WRF for the elevated release case. All values below 0.01 Bq/m^2 were removed. The WRF parameterizations used to create each subplot can be found in Section 3.

4. Spatial Prediction Algorithm

The algorithm we use to emulate physics package changes in WRF is straightforward. A conceptual schematic can be seen in Figure 4. We start by creating an $M \times N$ grid $\hat{\mathbf{Y}}_k$ to represent the prediction of a given FLEXPART-WRF map $\mathbf{Y}_k$. Each grid cell $\hat{\mathbf{Y}}_{i,j,k}$ is the combined output of an independent linear regression and logistic regression model. The inputs to every linear and logistic regression model in the grid are the same: a 1×16 vector $\mathbf{x}_k$ of one-hot-encoded WRF physics categorical variables, as described in Section 3. For each grid cell, the associated logistic regression model determines the probability that the location will experience surface contamination from the hypothetical release event. If the probability at that location is greater than a pre-determined threshold value, the corresponding linear regression model determines the magnitude of the deposition. Mathematically, the value of a given grid cell $\hat{\mathbf{Y}}_{i,j,k}$ is given by Equations (1) and (2). The α and β terms represent the vector of regression coefficients for the logistic and linear regression models, respectively. The coefficients in Equation (2) are exponentiated because the linear regression is trained on the logarithm of the deposition. This linearizes the deposition values which allows the regression model to be fit; however, the logarithm is also useful for analyzing the data in general since the deposition values span many orders of magnitude.

$$\mathbf{P}_{i,j,k} = \frac{1}{1 + e^{-\boldsymbol{\alpha}_{i,j}^T \mathbf{x}_k}} \tag{1}$$

$$\hat{\mathbf{Y}}_{i,j,k} = \begin{cases} 0 & \text{if } \mathbf{P}_{i,j} \leq p_{\text{threshold}} \\ e^{\boldsymbol{\beta}_{i,j}^T \mathbf{x}_k} & \text{if } \mathbf{P}_{i,j} > p_{\text{threshold}} \end{cases} \tag{2}$$

The full $M \times N \times 1196$ dataset $\mathbf{Y}$ can be split into an $M \times N \times n$ training set $\mathbf{Y}_{\text{Train}}$ and an $M \times N \times (1196 - n)$ testing set $\mathbf{Y}_{\text{Test}}$. The linear regression models are trained on the logarithm of the deposition values, while the logistic regression models are trained on a binary indicator determining whether a grid cell has deposition or not.

We implemented our model in Python 3 using Numpy [53]. We used the linear regression and the logistic regression implementations from Scikit-Learn [54]. The logistic regression implementation in Scikit-Learn was run with the "liblinear" solver and L2 regularization with $\lambda = 1.0$. L2 regularization is necessary to obtain accurate results and ensure convergence. With 50 training examples, training regression models for every grid cell in the domain took approximately 1–1.5 min on a modern desktop computer. Making predictions for 1146 full maps took 5–6 min on the same computer, but that was achieved by re-implementing the Scikit-Learn "predict" functions using the Python just-in-time compiler Numba [55]. At approximately 315 ms per prediction on one core, the machine learning model offers an approximately two million times speedup for a single run. Some researchers have found similar speedups using ML on scientific codes [56]. Large scale experiments where the training and testing cycles had to occur thousands of times (e.g., determining training size convergence curves) were completed on Lawrence Livermore National Laboratory's Quartz Supercomputer and could take up to a few hours.

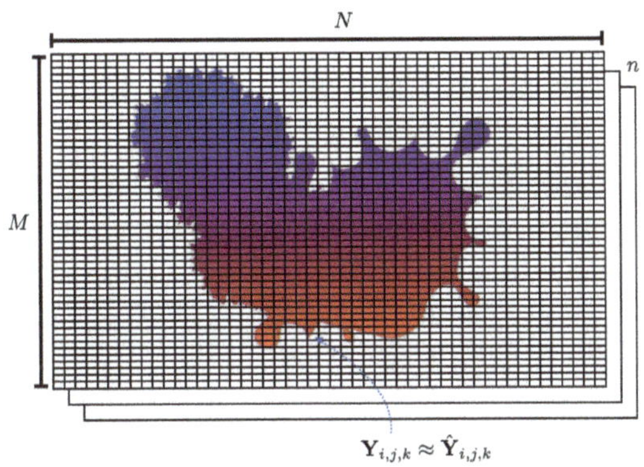

Figure 4. Conceptual diagram of the training set and model. Each deposition map produced by FLEXPART-WRF is a grid of size $M \times N$. There are n maps in the training set. A single grid cell of a single deposition map is represented by $\mathbf{Y}_{i,j,k}$ and can be approximated by our machine learning model output, $\hat{\mathbf{Y}}_{i,j,k}$. $\hat{\mathbf{Y}}_{i,j,k}$ is produced by the output of a single linear regression model and a single logistic regression model which is trained on the n data values for grid cell i, j. Further details of the model can be found in Section 4.

5. Results and Analysis

To test the effectiveness of our statistical model, we ran a suite of tests and derived performance statistics from the results. For these tests, we trained and evaluated our statistical model for eight different training sizes, with 100 runs with varying random seeds for each training size. The eight different training sizes we used were $n = 25$, $n = 50$, $n = 75$, $n = 100$, $n = 250$, $n = 500$, $n = 750$, and $n = 1000$ ensemble members. This corresponds to 2.09%, 4.18%, 6.27%, 8.36%, 20.90%, 41.81%, 62.71%, and 83.61% of our 1196-member ensemble dataset, respectively. Varying the random seed allowed each of the 100 runs for a given training size to have different members in the training set, which allowed us to see how much performance varied by training set member selection. The members of the test set for a given training size and random seed can be used in the training set for a different random seed. In other words, for a given training size and random seed, we had a training set and a testing set, but looking at all the random seeds for a given training size together was similar to k-fold cross validation. Since we used all 1196 members for this process, we did not have any truly held out test set that was not part of the 1196-member ensemble.

Figures that do not show training size variability (Figures 5–7) show the results from a 50-member training set with the same fixed random seed. The number 50 is somewhat arbitrary but shows the minimum amount of training examples that produces accurate predictions. At 50 training examples, the predictions are qualitatively good, and one starts to see significant overlap between the training and testing performance metric distributions. Figures 8–10 all show results from the cross-validation tests.

The following subsections summarize the statistical and numerical performance of the algorithm. Some subsections present summary statistics, while some subsections present individual member predictions. In subsections where individual predictions are present, the training size is also presented.

5.1. Decision Threshold

Before showing summary statistics, it is important to understand how the output of our model is a probabilistic prediction. Figures 5 and 6 both have six subplots. The top left plot shows the true output by FLEXPART-WRF for a selected ensemble member. The top middle plot shows the probability map produced by the grid of logistic regression models. The color at each pixel represents the probability that the pixel has a non-zero deposition value. The areas of this subplot that are not colored are excluded from prediction because the corresponding grid cells in the training data contain no deposition. The remaining areas use the combination of logistic and linear regressions for making predictions.

The output of the logistic regression models is used in conjunction with a user-defined decision threshold value to produce deposition predictions. As determined from the training data, grid cells with probabilities greater than the threshold are predicted to have deposition, while those less than it are not. If conservative estimates are desired, a low threshold value can be used to include low probability, but still likely, areas of contamination in the prediction. The top-right and entire bottom row of Figures 5 and 6 show the predictions at different decision thresholds. The decision threshold can also be thought of as a probability cutoff value. The term "decision threshold" is synonymous with "decision boundary", which is referred to in the literature when classifying positive and negative outcomes [28].

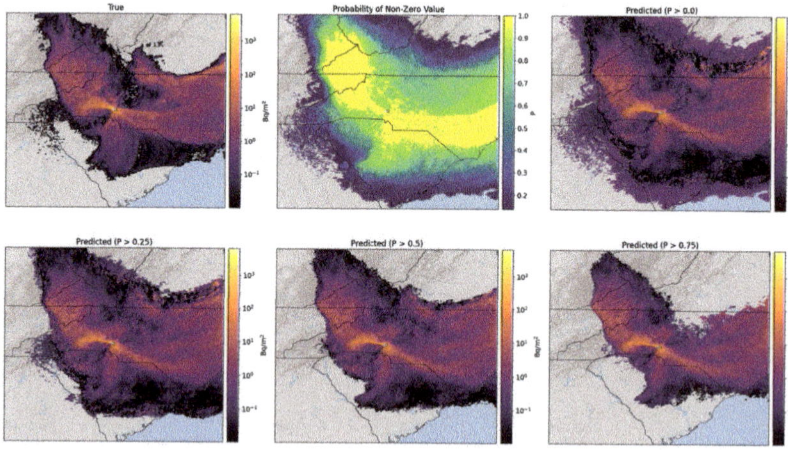

Figure 5. True FLEXPART-WRF output vs. predicted output at several decision threshold values for the surface release ensemble member 0 with $n = 50$. The WRF parameterization choices for this ensemble member were PBL 1, LSM 1, CU 1, MP 2, and RA 1. The top middle plot shows the original decision threshold map.

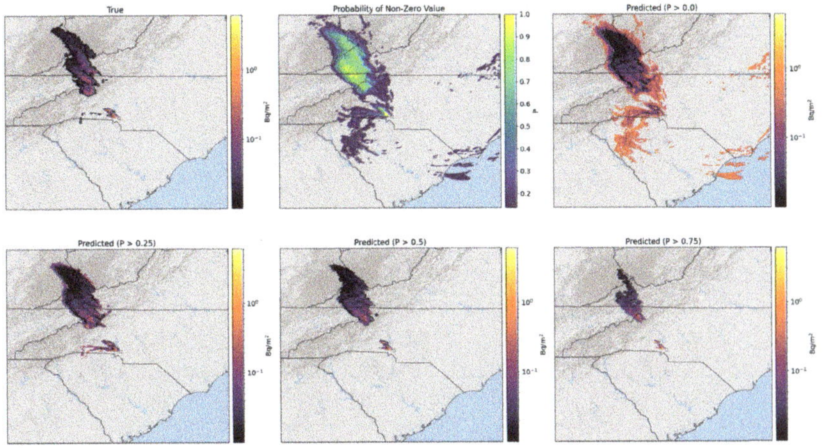

Figure 6. True FLEXPART-WRF output vs. predicted output at several decision threshold values for the elevated release ensemble member 0 with $n = 50$. The WRF parameterization choices for this ensemble member were PBL 1, LSM 1, CU 1, MP 2, and RA 1. The top middle plot shows the original decision threshold map.

Through a qualitative assessment, we determined that a decision threshold of 0.5 appears to be optimal. With values larger than 0.5, the plume shape starts becoming distorted and leaves important sections out. With values less than 0.5, noisy values at the edges of the plume are included, which are typically not accurate. These noisy values occur in grid cells where there are not many examples of deposition in the training data, and they are eliminated as more examples are included when the training size increases (see Section 5.2). These values can be seen in the bottom left subplot of Figure 5 on the northern edge of the plume. Anomalously large prediction values skew the performance statistics and are removed from the metrics if they exceed the maximum deposition value present in the training examples.

5.2. *Training Size Variability*

As with all statistical methods, the size of the training set affects the model performance. Figure 7 shows the plume prediction for ensemble member 296 as the training size increases. The members of the training set at a given size are also all included in the training set at the next largest size (i.e., the $n = 50$ training set is a proper subset of the $n = 75$ training set). The decision threshold is set to 0.5 for each training size. It is evident from the figure that as the training size increases, the deposition values and the plume boundary become less noisy. A quantitative assessment of how the predictions change with increasing training size is shown in Figures 8 and 9 for the surface case and elevated case, respectively.

These two figures show different statistical measures for predicting the members of training and testing sets as a function of training size. Because the selection of members is random and can affect the prediction performance, the experiment is repeated 100 times using different random seeds. Therefore, each "violin" in the plots displays the statistical variation stemming from member selection differences. For a given training size n, the orange training distributions are estimated from $n \times 100$ predictions, while the blue test distributions are derived from $(1196 - n) \times 100$ predictions.

The following error metrics are used to assess the predictive performance of the regression system. Two of the metrics target the logistic regressions (figure of merit in space and accuracy), three are for the linear regressions (fraction within a factor of 5, R, and fractional bias), and an aggregated metric (rank) is used to gauge the overall performance. Many other metrics are available to judge regression and classification performance (e.g., mean squared error, F1), but we wanted to use metrics that were commonly used in the atmospheric science community [57,58].

- Figure of Merit in Space (FMS): A spatial error metric which is defined as the intersection of the area of the predicted and actual plumes divided by the union of area of the predicted and actual plumes [57]. Outside of atmospheric science this is also known as the Jaccard index. This metric depends only on the absence or presence of deposition, not the magnitude, and so directly assesses the logistic regressions. This metric varies between 0 and 1, and values of 0.8 and above are generally considered good for atmospheric models.
- Fraction within a Factor 5 (FAC5): The fraction of the predicted values within a factor of 5 of the actual values is an effective metric for assessing the linear regressions. Generalized, this is defined as $FACX = \text{Fraction of data that satisfy} \frac{1}{X} \leq \frac{\text{Predicted Value}}{\text{Actual Value}} \leq X$ [57]. We present the FAC5 value for the intersection of the predicted and actual plume locations. This metric can range from 0 to 1, with values above 0.85 generally being considered good values for atmospheric models.
- Pearson's R: Pearson's correlation coefficient R measures the linear relationship between the predicted and actual magnitudes of deposition in a log space. We present R calculated for the natural log of the intersection of the predicted and actual plume locations. This metric can range from -1 to 1, with values further away from 0 being good. (A Pearson's R value of -1 implies perfect anticorrelation, which is still useful).
- Accuracy: This is the standard classification accuracy as explained in Swets [59]. It is defined as the ratio of the sum of true positives and true negatives to all classifications considered (i.e., $\frac{TP+TN}{TP+TN+FP+FN}$). In a logistic regression case such as ours, a grid cell is classified as positive or negative by whether it has deposition above or below a certain threshold value. As described in Section 3, a deposition threshold of 0.01 Bq/m^2 was used. This metric can range from 0 to 1, with values closer to 1 being considered good.
- Fractional Bias (FB): Fractional bias is defined as $FB = \frac{\overline{C_O} - \overline{C_P}}{0.5(\overline{C_O} + \overline{C_P})}$, where $\overline{C_O}$ and $\overline{C_P}$ are the mean observed values and mean predicted values, respectively. It is a normalized indicator of model bias and essentially describes the difference in means in terms of the average mean. This metric ranges from -2 to +2 and has an ideal value of 0. In Figures 8 and 9, the fractional bias plot has a different shape from all the others. One reason for this is the fractional bias range is different, and the ideal value is in the center of the range. However, even if the absolute value of the fractional bias was taken, the shape would still be different. In this case, as the training set size increases, the fractional bias statistic converges to the inherent bias that exists between our statistical model and FLEXPART-WRF , just as the others do. However, in this case, it is shown that the training size that produces the least fractional bias is $n = 50$. This does not mean that $n = 50$ is the best sample size overall, as other metrics improve with increasing sample size. Like the other metrics, the fractional bias training and test curves converge with increasing training size, though they seem to converge much faster than the others.
- Rank: Described in Maurer et al. [58], the rank score is a metric that combines several statistics into a single number used to assess the overall performance of an atmospheric model. It is defined as $\text{Rank} = R^2 + (1 - \frac{|FB|}{2}) + FAC5 + \text{Accuracy}$. R^2 is the coefficient of determination, and FB is the fractional bias. Each term in the equation ranges from 0 to 1, with 1 being best, which means the rank score ranges from 0 to 4, with 4 being best. The models studied in Maurer et al. [58] had a mean rank score of 2.06, which means our model looks very good in comparison. However, the models studied were time series models applied to individual stations, so they cannot be directly compared to our model.

In both the surface and elevated release cases, increasing the training size leads to, on average, an increase in performance on the test set and a decrease in performance on the training set. Nevertheless, as expected, the training set performance is better than the testing set performance. There is no immediately distinguishable difference in performance between the surface case and the elevated case; on some metrics the surface case performs better and on others the elevated case performs better. However, the distribution of error metrics for the elevated case is often bimodal, whereas the surface case is more unimodal. This makes intuitive sense since the elevated case often has two separate deposition patterns with different shapes, while the surface case typically only has one large pattern.

Figures 7 and 8 highlight one of the most important conclusions from this work. Very few training samples are needed to make reasonable predictions. Even a prediction using 50 training samples, or $50/1196 = 4.18\%$ of the total dataset, is capable of accurately predicting deposition values in over 100,000 grid cells. Because there is significant overlap between the training and test distributions in Figure 8, these predictions are also robust to the 50 training samples selected from the full set.

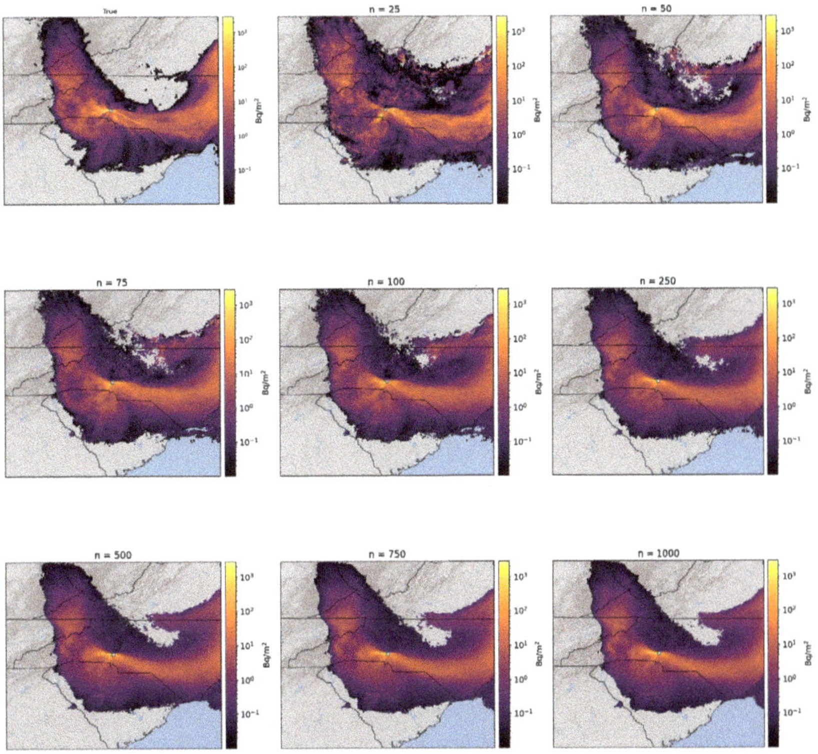

Figure 7. Spatial prediction for ensemble member 296 as the training set size increases. The true FLEXPART-WRF output is the top left subplot. The samples in the training set are randomly selected and the $p_{\text{threshold}}$ value is 0.5. Ensemble member 296 was in the test set for all training sizes. The WRF parameterization choices for this ensemble member were PBL 2, LSM 1, CU 5, MP 4, and RA 4.

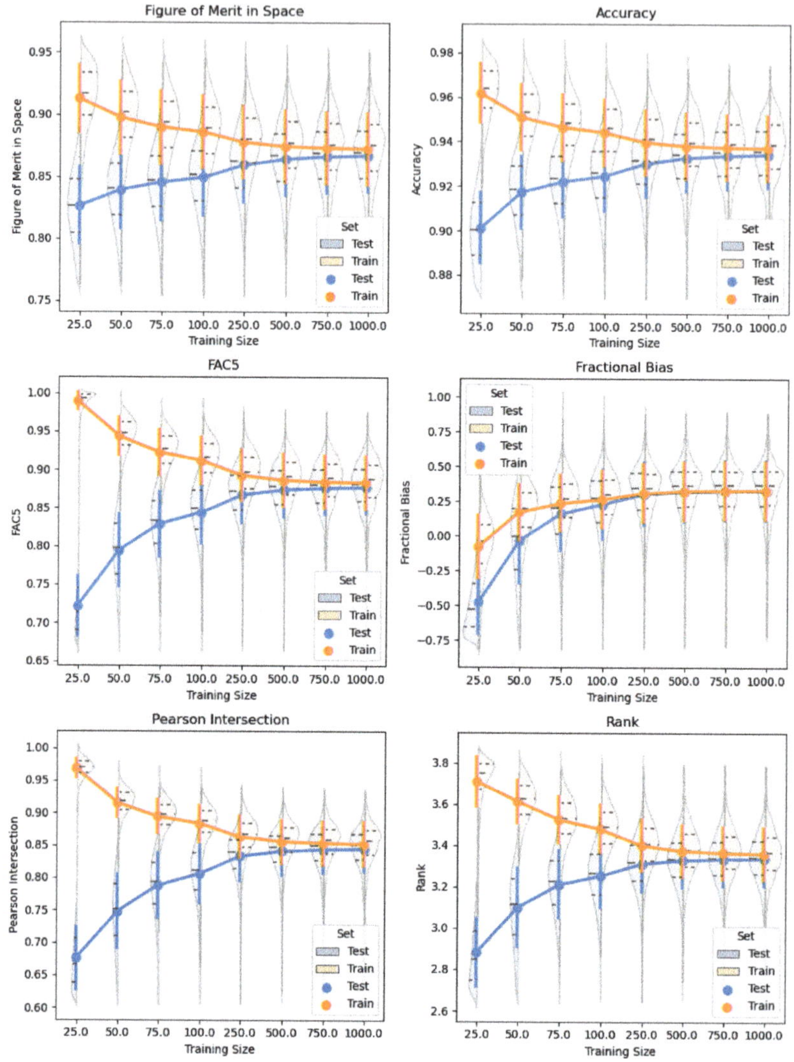

Figure 8. Spread of error metrics for all members for several different training sizes and 100 different random seeds for the surface release case. Training and test distributions are in blue and orange, respectively. A description of the metrics is provided in Section 5.2. Within the distributions, the dashed lines indicate the quartiles, the solid line is the mean, and the corresponding vertical bars are the standard deviations.

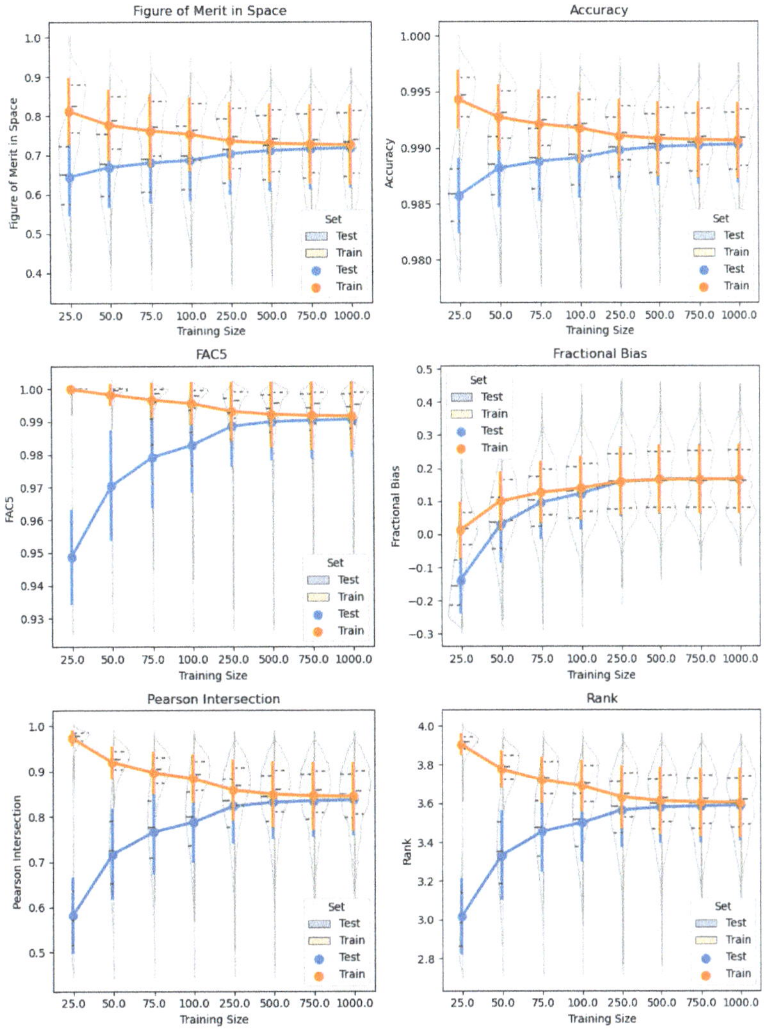

Figure 9. Same as Figure 8, except for the elevated release case.

5.3. Predictability of Individual Ensemble Members

The previous subsection described how training size affected the statistical model performance for the entire ensemble. In this section, we show how the predictions vary with training size for selected individual ensemble members. The purpose of this test is to show that some FLEXPART-WRF members are easier to predict than others, regardless of the amount of training members. Figure 10 shows the mean Pearson's R score by training size and member number for the surface release case for selected members of the test set. The members are selected by their decile average performance. We only show the members that are closest to the decile average performance because showing all 1196 members results in a visualization that is difficult to read.

For example, take the square marked by "60% (132)" on the x-axis and "250" on the y-axis. This square represents the mean Pearson's R score for member 132 calculated from every statistical model (out of 100) where member 132 was contained in the test set.

Member 132 is the member that is closest to the 60th percentile mean Pearson's R score averaged over *all* training sizes.

As already demonstrated, the general performance of the model increases as the training set size increases; however, the relative individual performance does not generally change. Part of this can be explained statistically. Our statistical model essentially fits a hyperplane in the WRF-parameter/deposition space. A hyperplane is one of the simplest possible models, and there is noise in the dataset. Some data points will be far away from the hyperplane, and increasing the training size does not move the hyperplane enough to successfully fit those points. This highlights the importance of the fact that physics-based modeling-machine learning is not able to capture all of the variation present in the dataset, even with very large training sizes. While we analyzed the WRF inputs associated with well and poorly performing members, we found no consistent pattern associated with poor predictions and WRF parameterizations. Hypothetically, if there was a relationship between WRF inputs and poorly performing members, the information could be used by WRF developers to improve accuracy for certain parameterizations. This figure also shows that low amounts of training data start producing accurate predictions. A similar analysis can be done for the elevated case but is not included here.

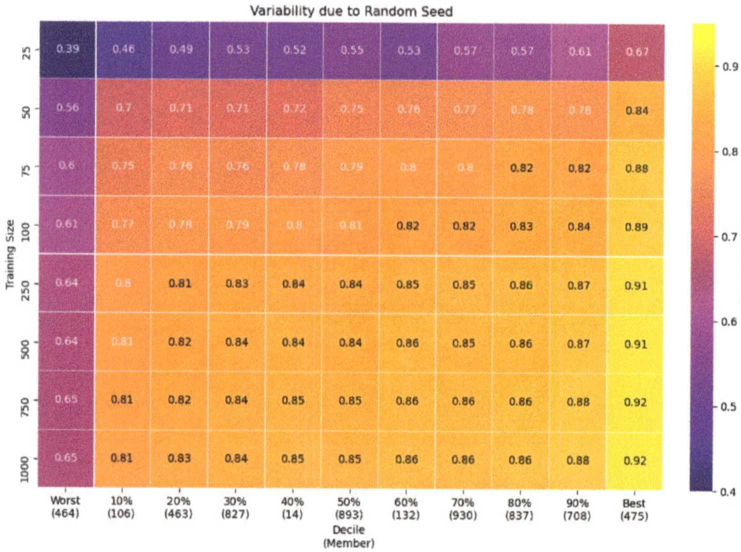

Figure 10. Mean Pearson R by training size and selected ensemble member. Some members are easier than others to predict regardless of the training size. Only instances where the ensemble member was included in the test set are used for calculations. The members were selected to be closest to the overall decile performance.

5.4. Ensemble Probability of Exceedance

One of the main goals of emulating Cs-137 spatial deposition is to account for the variability in the ensemble from weather uncertainty, so we use probability of exceedance plots to compare the variability of the predicted and true ensemble in Figure 11. The topmost and center subplots of Figure 11 show the percentage of members in the ensemble that have deposition values that exceed a threshold of 0.01 Bq/m^2 at every location. For example, if 598 ensemble members have deposition above 0.01 Bq/m^2 at grid cell (200, 200), the percentage for that cell is $598/1196 = 50\%$. Yellow colors indicate areas where many, if not all ensemble members report above-threshold deposition values. Dark purple colors indicate areas where very few ensemble members report above-threshold deposition values. Generally, the probability of exceedance drops as one moves further away from the release location. The predictions are based on 50 training samples, and both ensembles used for this plot contain all 1196 members, meaning the training and testing predictions are included for the predicted percentages.

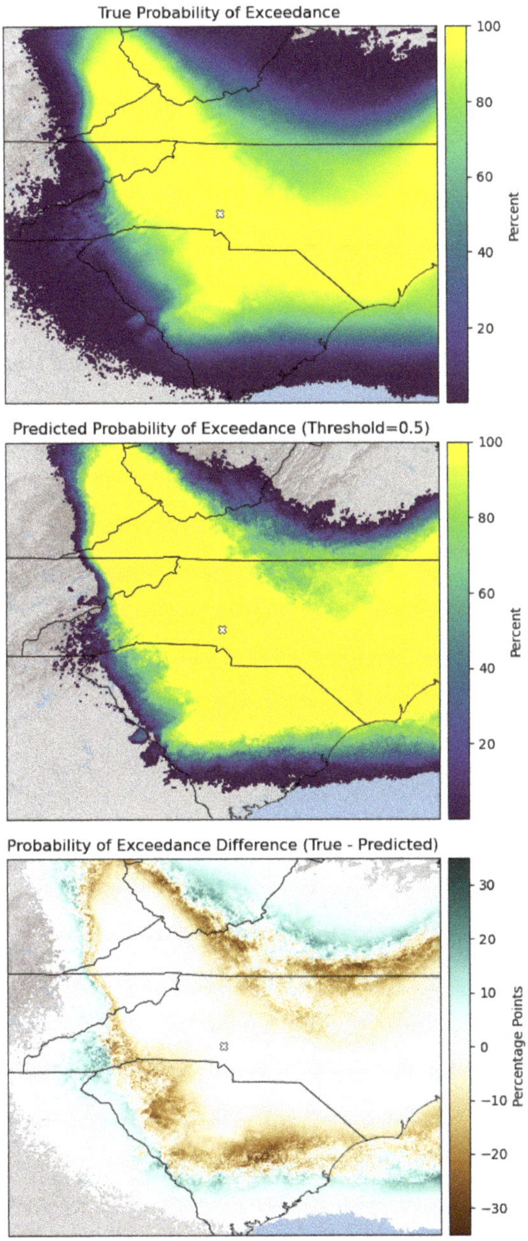

Figure 11. Percentage of members in the true (**top**) and predicted (**center**) ensembles that have deposition that exceeds 0.01 Bq/m^2 at each location. The bottom-most plot shows the difference between the two.

The topmost subplot shows the probability of exceedance of the true ensemble. As expected, the outside of the shape is made up of low percentage grid cells, as only outlier plumes make up those locations. The center subplot shows the probability of exceedance of the predicted ensemble. The predicted probability of exceedance takes up less area than the true ensemble because the outliers around the edge are challenging for the regressions to predict.

Despite the vast differences in computational resources needed to produce them, the probability of exceedance in the true and predicted ensembles appears similar. To highlight the differences, we created the bottom-most subplot of Figure 11, which shows the difference between the true ensemble percentages and the predicted ensemble percentages. Positive values, in teal, show areas where the population of members in the true ensemble is higher than the predicted ensemble. Negative values, in brown, show areas with higher predicted population than true population. Comparing the plot to Figure 12, one notices that the boundary between brown and teal happens approximately where the number of samples per pixel drops below 17, which is where the linear regression becomes underdefined. The conclusion we have drawn is that the regressions tend to overpredict values where there are sufficient samples (with some exceptions, such as in the center right of the plot) and underpredict where there are not sufficient samples.

5.5. Spatial Coefficient Analysis

One advantage our regression method holds over other machine learning models is the potential for interpretability. In this subsection we highlight one aspect of this interpretability. Our predictions are made using thousands of individual regression models, each of which has coefficients that transform the WRF parameterization input variables into a deposition value. In traditional regression approaches with non-categorical inputs, the units of all the input variables can be standardized so that the magnitude of a coefficient is related to the effect of its corresponding variable. That is, the larger the value of a coefficient, the more important the corresponding predictor is to the output. However, our WRF variables are one-hot-encoded as binary inputs, so determining their importance is not as straightforward as standard regression. Each of the regression models in our method has seventeen input terms—one for the intercept and sixteen binary encoded variables that represent five different WRF physics parameterizations. Out of these sixteen non-intercept coefficients, the first four represent the five PBL schemes, the next three represent the four LSM schemes, the next four represent the five CU schemes, the next three represent the four MP schemes, and the final two coefficients represent the three RA schemes. Taking the mean of the absolute value of a WRF physics package's coefficients gives an estimate of the importance of that variable. In other words, $\frac{1}{4}\sum_{i=1}^{4}|\beta_i|$ represents the importance of PBL, $\frac{1}{3}\sum_{i=5}^{7}|\beta_i|$ represents the importance of LSM, and so on.

Once the mean coefficient magnitudes are calculated, the argmax is used to find the WRF parameterization which is most important at a given grid cell. These results can be plotted to see which parameterizations are most important for a given area, as seen in Figure 12 for the surface release case. Figure 12 was created using models trained on 50 ensemble members and only includes grid cells that have greater than 17 samples. The intercept is not considered when determining importance. It is important to remember that with our process, the "most important variable" is not the same as "only important variable." Combinations of WRF parameterization changes can be important, resulting in the many coefficients that have a similar mean magnitude. In other words, the second most important WRF parameterization can still be very important because it has a mean coefficient magnitude slightly smaller than the most important WRF parameterization. Regardless, this analysis provides an interesting consequence of using regression models to interpret WRF physics.

Figure 12 shows that PBL variations tend to dominate other WRF parameterizations, as captured by the large areas in beige. This result is not surprising, as changing the PBL scheme in WRF is known to greatly influence atmospheric turbulence and mixing near the surface. The variable importance map also shows other interesting features, including the red areas highlighting the relatively elevated importance of cumulus convection variations over coastal and mountainous areas where precipitation occurs during the release events. Similarly, magenta areas where microphysics is important occur near areas where cumulus convection is also important, which is consistent with the correlation of these physical processes in the model. The overall spatial complexity in Figure 12 illustrates one final critical point. No single WRF parameterization is most important everywhere, so multi-physics WRF ensembles that vary a range of physical parameterizations are needed to capture weather model uncertainty.

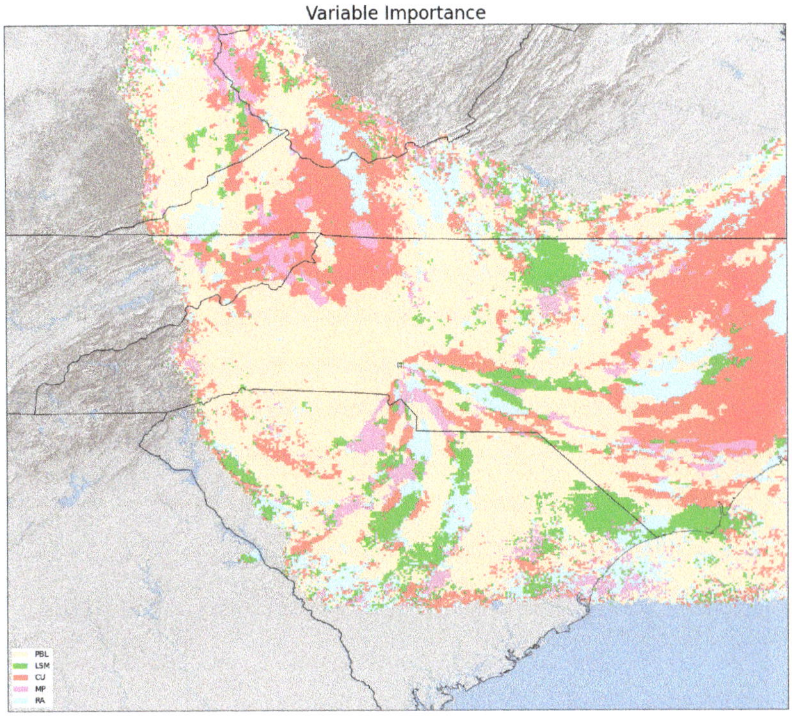

Figure 12. Primary WRF parameterization changes associated with Cs-137 deposition are color-coded and shown for every grid cell. Areas with 17 or fewer samples are excluded. The total training dataset included 50 ensemble members. In the legend, PBL stands for the planetary boundary layer physics parameterization (tan), LSM stands for land surface model (green), CU stands for cumulus physics (red), MP stands for microphysics (magenta), and RA stands for radiation (cyan).

6. Future Work

The regression prediction method we have described has some drawbacks and unknowns, which means there are several avenues for further exploration. The most significant drawback is that it does not exploit spatial correlations of nearby locations in the domain. Since each grid cell is treated as completely independent from the other grid cells, spatial correlations are not used to improve the overall prediction. This means that any predicted plume is limited to the envelope of the union of all of the training plumes, as our model cannot predict in areas that do not have any training data. However, this trait can be viewed as a positive feature of our algorithm; it will not poorly extrapolate in areas where there are no training data. To overcome this problem, spatial data can be incorporated into the overall model. Including spatial correlation information in our model may lead to a more parsimonious model or one that produces improved predictions. Including spatial correlations can also potentially be done using dimensional reduction techniques such as PCA or autoencoders. For example, the model we describe could be used to produce an initial map, and then an alternate model based off radial basis functions, multitask learning, or even linear regression can be used to refine it.

Another drawback is the subjective nature of picking a decision threshold $p_{\text{threshold}}$ in the logistic regression component. We used a value of 0.5 for all the calculations presented here, which is a reasonable value to use, but that is the result of qualitative analysis. Implementing an optimization routine to determine the best $p_{\text{threshold}}$ to use would increase the objectivity and may improve the performance of our model. The tuned threshold could also be applied at a grid-cell level, which may increase performance in the boundary regions.

As mentioned in Section 5.1, we remove outlier deposition values which are predicted to be larger than any deposition value present in the training set. This is a simple way to remove outliers and is easily implemented operationally. However, it is a naive outlier removal method. A more complex outlier removal method may be beneficial to help differentiate false extreme values from true extreme values, the latter of which can pose a large risk to public health.

When we create training sets for our method we sample randomly from the entire population of predictions. By using methods from adaptive sampling, it may be possible to dynamically produce a training set that is more representative of the population than a random sample, leading to higher performance for the trained model with fewer expensive computer simulations. In an emergency situation, this would be very useful.

The individual models that predict hazardous deposition in each grid cell do not necessarily have to be linear or logistic regression models. They can be produced by other regression and classification models such as random forests or artificial neural networks. The biggest hurdle in implementing more complex grid cell-level models is the training time. During our testing on a desktop computer, the training time for a single grid cell took between 1 and 10 ms, and training a full spatial map was on the order of minutes. Changing to a more complicated model could potentially increase training time by an order of magnitude.

Finally, this regression method should be tested with more FLEXPART-WRF simulations. It should be tested with different hazardous releases in different locations from FLEXPART-WRF , but it could also be tested on completely different physical models. More terms could also be added to the regression model to account for larger initial condition errors present in longer forecast simulations. There is nothing about our method that is inherently specific to FLEXPART-WRF , and we think this method could work for simulations that are unrelated to deposition.

7. Conclusions

In this paper, we presented a statistical method that can be used to quickly emulate complex, spatially varying radiological deposition patterns produced by the meteorological and dispersion tools WRF and FLEXPART. FLEXPART-WRF is slow to run, and a single simulation from it may have significant uncertainty due to model imperfections. To estimate uncertainty, researchers can run FLEXPART-WRF hundreds of times by varying representations of physical processes in the models, but that can take crucial hours. Instead of running FLEXPART-WRF hundreds of times, researchers can run it dozens of times, use the results to train our emulator, and then use the emulator to produce the remaining results.

Our emulator is represented by an $M \times N$ grid where the value at each grid cell is determined by the output of independent linear regression and logistic regression models. The logistic regression determines whether hazardous deposition is present at that location, and the linear regression determines the magnitude of the deposition. Since all the grid cells are independent from one another, our model can accurately predict subsets of locations.

We used two datasets for training, testing, and predicting. One was a simulated continuous surface contaminant release representing a large-scale industrial accident, and the other was a simulated instantaneous elevated contaminant release from a hypothetical nuclear explosion. For each of the two cases, there were 1196 different simulations, all representing variations in the WRF parameterizations. The WRF parameterizations were treated as categorical variables that were binary encoded and used as the inputs to the linear and logistic regression models used in our emulator.

We conducted several tests to evaluate the performance of our emulator. We found that the emulator performs well, even with only 50 samples out of the 1196-member population. While the deposition patterns have variance, they are not drastically different shapes, which is why 50 samples is sufficient to make reasonable predictions. This is promising since in an emergency situation, the amount of computationally expensive runs should be minimized. As with many machine learning models, the prediction performance on the test set increases with increasing training size. We also found that for each case there are some members that perform better than others, regardless of the training size.

In general, we think that the emulator that we have presented here is successful in predicting complex spatial patterns produced by FLEXPART-WRF with relatively few training samples. We think there are several areas that can be explored to improve our emulator, and we hope to complete some of them in the future.

Author Contributions: N.G. contributed to statistical model preparation and analysis, visualization, and draft writing and editing. G.P. contributed to data analysis and validation. M.S. contributed to conceptualization, methodology, data creation, funding acquisition, and validation. D.D.L. contributed to statistical model analysis, data creation, draft writing and editing, validation, and project administration. All authors have read and agreed to the published version of the manuscript.

Funding: This work was performed under the auspices of the U.S. Department of Energy by Lawrence Livermore National Laboratory under Contract DE-AC52-07NA27344. It was funded by LDRD 17-ERD-045. Released under LLNL-JRNL-808577.

Institutional Review Board Statement: Not applicable.

Informed Consent Statement: Not applicable.

Data Availability Statement: The surface release data presented in this study are openly available at ftp://gdo148.ucllnl.org/pub/spatial, accessed on 22 July 2021. The elevated release data are available upon request.

Conflicts of Interest: The authors declare no conflict of interest. The funders had no role in the design of the study; in the collection, analyses, or interpretation of data; in the writing of the manuscript; or in the decision to publish the results.

Notation

The following is a table of the notation used in the document. All notation is generalizable to the surface release case and the elevated release case.

Symbol	Meaning
$\mathbf{x}$	General single instance of predictor vector
$\mathbf{x}_k$	Specific single instance of predictor vector
$\mathbf{X}$	Complete set of input vectors
$\mathbf{X}_{\text{Train}}$	Training set of input vectors
$\mathbf{X}_{\text{Test}}$	Testing set of input vectors
$\mathbf{Y}$	Complete set of target matrices
$\mathbf{Y}_k$	Specific single instance of target matrix
$\mathbf{Y}_{i,j,k}$	Specific location of single instance of target matrix
$\mathbf{Y}_{\text{Train}}$	Training set of target matrices
$\mathbf{Y}_{\text{Test}}$	Testing set of target matrices
$\hat{\mathbf{Y}}$	Complete set of estimation matrices
$\hat{\mathbf{Y}}_k$	Specific single instance of estimation matrix

$\hat{\mathbf{Y}}_{i,j,k}$	Specific location of single instance of estimation matrix
M	Number of rows in $\mathbf{Y}$ or $\hat{\mathbf{Y}}$
N	Number of columns in $\mathbf{Y}$ or $\hat{\mathbf{Y}}$
n	Size of training set
$\boldsymbol{\alpha}_{i,j}$	Coefficients for the logistic regression model at a location
$\boldsymbol{\beta}_{i,j}$	Coefficients for the linear regression model at a location
$\mathbf{P}_{i,j}$	Probability of non-zero deposition at a location
$p_{\text{threshold}}$	Decision threshold to be applied to prediction

References

1. Moreno, T.; Querol, X.; Alastuey, A.; Minguillón, M.C.; Pey, J.; Rodriguez, S.; Miró, J.V.; Felis, C.; Gibbons, W. Recreational atmospheric pollution episodes: Inhalable metalliferous particles from firework displays. *Atmos. Environ.* **2007**, *41*, 913–922. [CrossRef]
2. Styer, P.; McMillan, N.; Gao, F.; Davis, J.; Sacks, J. Effect of outdoor airborne particulate matter on daily death counts. *Environ. Health Perspect.* **1995**, *103*, 490–497. [CrossRef] [PubMed]
3. Griffin, D.W.; Kellogg, C.A. Dust storms and their impact on ocean and human health: Dust in Earth's atmosphere. *EcoHealth* **2004**, *1*, 284–295. [CrossRef]
4. Bader, J.A. Dealing with Multiple Disasters in Japan. In *Obama and China's Rise: An Insider's Account of America's Asia Strategy*; Brookings Institution Press: Washington, DC, USA, 2012; pp. 130–139.
5. Brioude, J.; Arnold, D.; Stohl, A.; Cassiani, M.; Morton, D.; Seibert, P.; Angevine, W.; Evan, S.; Dingwell, A.; Fast, J.D.; et al. The Lagrangian particle dispersion model FLEXPART-WRF version 3.1. *Geosci. Model Dev.* **2013**, *6*, 1889–1904. [CrossRef]
6. Skamarock, W.C.; Klemp, J.B.; Dudhia, J.; Gill, D.O.; Barker, D.M.; Wang, W.; Powers, J.G. *A Description of the Advanced Research WRF Version 3*; NCAR Technical note-475+ STR; University Corporation for Atmospheric Research: Boulder, CO, USA, 2008.
7. Hutchinson, M.; Oh, H.; Chen, W.H. A review of source term estimation methods for atmospheric dispersion events using static or mobile sensors. *Inf. Fusion* **2017**, *36*, 130–148. [CrossRef]
8. Leadbetter, S.; Andronopoulos, S.; Bedwell, P.; Chevalier-Jabet, K.; Geertsema, G.; Gering, F.; Hamburger, T.; Jones, A.; Klein, H.; Korsakissok, I.; et al. Ranking uncertainties in atmospheric dispersion modelling following the accidental release of radioactive material. *Radioprotection* **2020**, *55*, S51–S55. [CrossRef]
9. Korsakissok, I.; Périllat, R.; Andronopoulos, S.; Bedwell, P.; Berge, E.; Charnock, T.; Geertsema, G.; Gering, F.; Hamburger, T.; Klein, H.; et al. Uncertainty propagation in atmospheric dispersion models for radiological emergencies in the pre- and early release phase: Summary of case studies. *Radioprotection* **2020**, *55*, S57–S68. [CrossRef]
10. Sørensen, J.H.; Bartnicki, J.; Buhr, A.; Feddersen, H.; Hoe, S.; Israelson, C.; Klein, H.; Lauritzen, B.; Lindgren, J.; Schönfeldt, F.; et al. Uncertainties in atmospheric dispersion modelling during nuclear accidents. *J. Environ. Radioact.* **2020**, *222*, 106356. [CrossRef]
11. Kirthiga, S.; Narasimhan, B.; Balaji, C. A multi-physics ensemble approach for short-term precipitation forecasts at convective permitting scales based on sensitivity experiments over southern parts of peninsular India. *J. Earth Syst. Sci.* **2021**, *130*, 1–29. [CrossRef]
12. Imran, H.M.; Kala, J.; Ng, A.; Muthukumaran, S. An evaluation of the performance of a WRF multi-physics ensemble for heatwave events over the city of Melbourne in southeast Australia. *Clim. Dyn.* **2018**, *50*, 2553–2586. [CrossRef]
13. Stegehuis, A.I.; Vautard, R.; Ciais, P.; Teuling, A.J.; Miralles, D.G.; Wild, M. An observation-constrained multi-physics WRF ensemble for simulating European mega heat waves. *Geosci. Model Dev.* **2015**, *8*, 2285–2298. [CrossRef]
14. Katragkou, E.; García-Díez, M.; Vautard, R.; Sobolowski, S.; Zanis, P.; Alexandri, G.; Cardoso, R.M.; Colette, A.; Fernandez, J.; Gobiet, A.; et al. Regional climate hindcast simulations within EURO-CORDEX: Evaluation of a WRF multi-physics ensemble. *Geosci. Model Dev.* **2015**, *8*, 603–618. [CrossRef]
15. Lavin-Gullon, A.; Fernandez, J.; Bastin, S.; Cardoso, R.M.; Fita, L.; Giannaros, T.M.; Goergen, K.; Gutiérrez, J.M.; Kartsios, S.; Katragkou, E.; et al. Internal variability versus multi-physics uncertainty in a regional climate model. *Int. J. Climatol.* **2021**, *41*, E656–E671. [CrossRef]
16. Lucas, D.D.; Simpson, M.; Cameron-Smith, P.; Baskett, R.L. Bayesian inverse modeling of the atmospheric transport and emissions of a controlled tracer release from a nuclear power plant. *Atmos. Chem. Phys.* **2017**, *17*, 13521–13543. [CrossRef]
17. Jensen, D.D.; Lucas, D.D.; Lundquist, K.A.; Glascoe, L.G. Sensitivity of a Bayesian source-term estimation model to spatiotemporal sensor resolution. *Atmos. Environ. X* **2019**, *3*, 100045. [CrossRef]
18. Watson, P.A. Applying machine learning to improve simulations of a chaotic dynamical system using empirical error correction. *J. Adv. Model. Earth Syst.* **2019**, *11*, 1402–1417. [CrossRef]
19. Calbó, J.; Pan, W.; Webster, M.; Prinn, R.G.; McRae, G.J. Parameterization of urban subgrid scale processes in global atmospheric chemistry models. *J. Geophys. Res. Atmos.* **1998**, *103*, 3437–3451. [CrossRef]
20. Mayer, M.; Wang, C.; Webster, M.; Prinn, R.G. Linking local air pollution to global chemistry and climate. *J. Geophys. Res. Atmos.* **2000**, *105*, 22869–22896. [CrossRef]
21. Beddows, A.V.; Kitwiroon, N.; Williams, M.L.; Beevers, S.D. Emulation and Sensitivity Analysis of the Community Multiscale Air Quality Model for a UK Ozone Pollution Episode. *Environ. Sci. Technol.* **2017**, *51*, 6229–6236. [CrossRef]

22. Wang, J.; Balaprakash, P.; Kotamarthi, R. Fast domain-aware neural network emulation of a planetary boundary layer parameterization in a numerical weather forecast model. *Geosci. Model Dev.* **2019**, *12*, 4261–4274. [CrossRef]
23. Krasnopolsky, V.M.; Fox-Rabinovitz, M.S.; Chalikov, D.V. New approach to calculation of atmospheric model physics: Accurate and fast neural network emulation of longwave radiation in a climate model. *Mon. Weather Rev.* **2005**, *133*, 1370–1383. [CrossRef]
24. Pal, A.; Mahajan, S.; Norman, M.R. Using deep neural networks as cost-effective surrogate models for super-parameterized E3SM radiative transfer. *Geophys. Res. Lett.* **2019**, *46*, 6069–6079. [CrossRef]
25. Lucas, D.; Prinn, R. Parametric sensitivity and uncertainty analysis of dimethylsulfide oxidation in the clear-sky remote marine boundary layer. *Atmos. Chem. Phys.* **2005**, *5*, 1505–1525. [CrossRef]
26. Kelp, M.M.; Tessum, C.W.; Marshall, J.D. Orders-of-magnitude speedup in atmospheric chemistry modeling through neural network-based emulation. *arXiv* **2018**, arXiv:1808.03874.
27. Ivatt, P.D.; Evans, M.J. Improving the prediction of an atmospheric chemistry transport model using gradient-boosted regression trees. *Atmos. Chem. Phys.* **2020**, *20*, 8063–8082. [CrossRef]
28. Murphy, K.P. *Machine Learning: A Probabilistic Perspective*; MIT Press: Cambridge, MA, USA, 2012.
29. Gelman, A.; Hill, J. *Data Analysis Using Regression and Multilevel/Hierarchical Models*; Cambridge University Press: Cambridge, UK, 2006.
30. Norment, H.G. *DELFIC: Department of Defense Fallout Prediction System. Volume I-Fundamentals*; Final Report 16 Jan–31 Dec 79; Atmospheric Science Associates: Bedford, MA, USA, 1979.
31. Lucas, D.D.; Pallotta, G.; Simpson, M.D. Using Machine Learning to Intelligently Select Members of Large Atmospheric Model Ensembles. In Proceedings of the AGU Fall Meeting Abstracts, Washington, DC, USA, 10–14 December 2018; Volume 2018, p. GC43J-1663.
32. Lucas, D.D.; Simpson, M.; Pallotta, G. Probabilistic Predictions and Uncertainty Estimation Using Adaptively Designed Ensembles for Radiological Plume Modeling. In Proceedings of the CTBT Science and Technology 2019 Conference, Vienna, Austria, 24–28 June 2019.
33. Aoyama, M. Long-range transport of radiocaesium derived from global fallout and the Fukushima accident in the Pacific Ocean since 1953 through 2017—Part I: Source term and surface transport. *J. Radioanal. Nucl. Chem.* **2018**, *318*, 1519–1542. [CrossRef]
34. Hong, S.Y.; Noh, Y.; Dudhia, J. A new vertical diffusion package with an explicit treatment of entrainment processes. *Mon. Weather Rev.* **2006**, *134*, 2318–2341. [CrossRef]
35. Dudhia, J. A multi-layer soil temperature model for MM5. In Proceedings of the Sixth PSU/NCAR Mesoscale Model Users' Workshop, Boulder, CO, USA, 22–24 July 1996; pp. 22–24.
36. Kain, J.S. The Kain–Fritsch convective parameterization: An update. *J. Appl. Meteorol.* **2004**, *43*, 170–181. [CrossRef]
37. Chen, S.H.; Sun, W.Y. A One-dimensional Time Dependent Cloud Model. *J. Meteorol. Soc. Jpn. Ser. II* **2002**, *80*, 99–118. [CrossRef]
38. Dudhia, J. Numerical study of convection observed during the winter monsoon experiment using a mesoscale two-dimensional model. *J. Atmos. Sci.* **1989**, *46*, 3077–3107. [CrossRef]
39. Janjić, Z.I. The step-mountain eta coordinate model: Further developments of the convection, viscous sublayer, and turbulence closure schemes. *Mon. Weather Rev.* **1994**, *122*, 927–945. [CrossRef]
40. Tewari, M.; Chen, F.; Wang, W.; Dudhia, J.; LeMone, M.; Mitchell, K.; Ek, M.; Gayno, G.; Wegiel, J.; Cuenca, R. Implementation and verification of the unified NOAH land surface model in the WRF model. In *20th Conference on Weather Analysis and Forecasting/16th Conference on Numerical Weather Prediction*; American Meteorological Society: Seattle, WA, USA, 2004; Volume 1115, pp. 2165–2170.
41. Hong, S.Y.; Dudhia, J.; Chen, S.H. A revised approach to ice microphysical processes for the bulk parameterization of clouds and precipitation. *Mon. Weather Rev.* **2004**, *132*, 103–120. [CrossRef]
42. Collins, W.D.; Rasch, P.J.; Boville, B.A.; Hack, J.J.; McCaa, J.R.; Williamson, D.L.; Kiehl, J.T.; Briegleb, B.; Bitz, C.; Lin, S.J.; et al. Description of the NCAR community atmosphere model (CAM 3.0). *NCAR Tech. Note NCAR/TN-464+ STR* **2004**, *226*, 1326–1334.
43. Sukoriansky, S.; Galperin, B.; Perov, V. Application of a new spectral theory of stably stratified turbulence to the atmospheric boundary layer over sea ice. *Bound.-Layer Meteorol.* **2005**, *117*, 231–257. [CrossRef]
44. Benjamin, S.G.; Grell, G.A.; Brown, J.M.; Smirnova, T.G.; Bleck, R. Mesoscale weather prediction with the RUC hybrid isentropic–terrain-following coordinate model. *Mon. Weather Rev.* **2004**, *132*, 473–494. [CrossRef]
45. Grell, G.A.; Freitas, S.R. A scale and aerosol aware stochastic convective parameterization for weather and air quality modeling. *Atmos. Chem. Phys.* **2014**, *14*, 5233–5250. [CrossRef]
46. Iacono, M.J.; Delamere, J.S.; Mlawer, E.J.; Shephard, M.W.; Clough, S.A.; Collins, W.D. Radiative forcing by long-lived greenhouse gases: Calculations with the AER radiative transfer models. *J. Geophys. Res. Atmos.* **2008**, *113*. [CrossRef]
47. Nakanishi, M.; Niino, H. An improved Mellor–Yamada level-3 model: Its numerical stability and application to a regional prediction of advection fog. *Bound.-Layer Meteorol.* **2006**, *119*, 397–407. [CrossRef]
48. Gilliam, R.C.; Pleim, J.E. Performance assessment of new land surface and planetary boundary layer physics in the WRF-ARW. *J. Appl. Meteorol. Climatol.* **2010**, *49*, 760–774. [CrossRef]
49. Grell, G.A.; Dévényi, D. A generalized approach to parameterizing convection combining ensemble and data assimilation techniques. *Geophys. Res. Lett.* **2002**, *29*, 38-1–38-4. [CrossRef]
50. Rogers, E.; Black, T.; Ferrier, B.; Lin, Y.; Parrish, D.; DiMego, G. National Oceanic and Atmospheric Administration Changes to the NCEP Meso Eta Analysis and Forecast System: Increase in resolution, new cloud microphysics, modified precipitation assimilation, modified 3DVAR analysis. *NWS Tech. Proced. Bull.* **2001**, *488*, 15.

51. Pleim, J.E. A combined local and nonlocal closure model for the atmospheric boundary layer. Part I: Model description and testing. *J. Appl. Meteorol. Climatol.* **2007**, *46*, 1383–1395. [CrossRef]
52. Berg, L.K.; Gustafson Jr, W.I.; Kassianov, E.I.; Deng, L. Evaluation of a modified scheme for shallow convection: Implementation of CuP and case studies. *Mon. Weather Rev.* **2013**, *141*, 134–147. [CrossRef]
53. Virtanen, P.; Gommers, R.; Oliphant, T.E.; Haberland, M.; Reddy, T.; Cournapeau, D.; Burovski, E.; Peterson, P.; Weckesser, W.; Bright, J.; et al. SciPy 1.0–Fundamental Algorithms for Scientific Computing in Python. *arXiv* **2019**, arXiv:1907.10121.
54. Pedregosa, F.; Varoquaux, G.; Gramfort, A.; Michel, V.; Thirion, B.; Grisel, O.; Blondel, M.; Prettenhofer, P.; Weiss, R.; Dubourg, V.; et al. Scikit-learn: Machine Learning in Python. *J. Mach. Learn. Res.* **2011**, *12*, 2825–2830.
55. Lam, S.K.; Pitrou, A.; Seibert, S. Numba: A LLVM-based Python JIT Compiler. In Proceedings of the Second Workshop on the LLVM Compiler Infrastructure in HPC, Austin, TX, USA, 15 November 2015; ACM: New York, NY, USA, 2015; pp. 1–6. [CrossRef]
56. Kasim, M.; Watson-Parris, D.; Deaconu, L.; Oliver, S.; Hatfield, P.; Froula, D.; Gregori, G.; Jarvis, M.; Khatiwala, S.; Korenaga, J.; et al. Building high accuracy emulators for scientific simulations with deep neural architecture search. *arXiv* **2020**, arXiv:2001.08055.
57. Chang, J.C.; Hanna, S.R. Air quality model performance evaluation. *Meteorol. Atmos. Phys.* **2004**, *87*, 167–196. [CrossRef]
58. Maurer, C.; Baré, J.; Kusmierczyk-Michulec, J.; Crawford, A.; Eslinger, P.W.; Seibert, P.; Orr, B.; Philipp, A.; Ross, O.; Generoso, S.; et al. International challenge to model the long-range transport of radioxenon released from medical isotope production to six Comprehensive Nuclear-Test-Ban Treaty monitoring stations. *J. Environ. Radioact.* **2018**, *192*, 667–686. [CrossRef]
59. Swets, J.A. Measuring the accuracy of diagnostic systems. *Science* **1988**, *240*, 1285–1293. [CrossRef] [PubMed]

atmosphere

MDPI

Article

Using Task Farming to Optimise a Street-Scale Resolution Air Quality Model of the West Midlands (UK)

Jian Zhong [1], Christina Hood [2], Kate Johnson [2], Jenny Stocker [2], Jonathan Handley [2], Mark Wolstencroft [3], Andrea Mazzeo [1], Xiaoming Cai [1] and William James Bloss [1,*]

1 School of Geography, Earth & Environmental Sciences, University of Birmingham, Edgbaston, Birmingham B15 2TT, UK; j.zhong.1@bham.ac.uk (J.Z.); a.mazzeo@bham.ac.uk (A.M.); x.cai@bham.ac.uk (X.C.)
2 Cambridge Environmental Research Consultants, Cambridge CB2 1SJ, UK; chood@cerc.co.uk (C.H.); kjohnson@cerc.co.uk (K.J.); jstocker@cerc.co.uk (J.S.); jonathan.handley@cerc.co.uk (J.H.)
3 Birmingham City Council, Birmingham B5 5BD, UK; mark.wolstencroft@birmingham.gov.uk
* Correspondence: w.j.bloss@bham.ac.uk

Abstract: High resolution air quality models combining emissions, chemical processes, dispersion and dynamical treatments are necessary to develop effective policies for clean air in urban environments, but can have high computational demand. We demonstrate the application of task farming to reduce runtime for ADMS-Urban, a quasi-Gaussian plume air dispersion model. The model represents the full range of source types (point, road and grid sources) occurring in an urban area at high resolution. Here, we implement and evaluate the option to automatically split up a large model domain into smaller sub-regions, each of which can then be executed concurrently on multiple cores of a HPC or across a PC network, a technique known as task farming. The approach has been tested for a large model domain covering the West Midlands, UK (902 km^2), as part of modelling work in the WM-Air (West Midlands Air Quality Improvement Programme) project. Compared to the measurement data, overall, the model performs well. Air quality maps for annual/subset averages and percentiles are generated. For this air quality modelling application of task farming, the optimisation process has reduced weeks of model execution time to approximately 35 h for a single model configuration of annual calculations.

Keywords: air pollution; air quality modelling; ADMS-Urban; high performance computing; HPC; West Midlands

Citation: Zhong, J.; Hood, C.; Johnson, K.; Stocker, J.; Handley, J.; Wolstencroft, M.; Mazzeo, A.; Cai, X.; Bloss, W.J. Using Task Farming to Optimise a Street-Scale Resolution Air Quality Model of the West Midlands (UK). *Atmosphere* **2021**, *12*, 983. https://doi.org/10.3390/atmos12080983

Academic Editor: Patrick Armand

Received: 28 May 2021
Accepted: 26 July 2021
Published: 30 July 2021

Publisher's Note: MDPI stays neutral with regard to jurisdictional claims in published maps and institutional affiliations.

1. Introduction

Air pollution has become the biggest environmental risk for public health both globally and locally [1–4]. Air pollution can cause adverse health effects, e.g., diseases associated with respiratory, circulatory, nervous, digestive and urinary systems [5]. In 2016, the World Health Organisation (WHO) estimated [6,7] premature deaths attributed to ambient air pollution as about 4.2 million per year and that about 91% of the world's population dwelt in areas with air pollution levels higher than WHO guidelines [8]. The mortality burden associated with ambient air pollution is about 28–36,000 per year in the UK [9]. The availability of air quality information is of vital importance to improve the understanding of the associated health effects [10,11], and to develop effective and equitable air pollution control policies.

Air quality measurements can provide direct information about the levels of air pollutants in the atmosphere. The UK Automatic Urban and Rural Network (AURN) [12] is the largest automatic air quality monitoring network across the UK. The quality-assured stationary sites in AURN can normally provide continuous measurements of air pollution concentrations at high temporal resolution (e.g., hourly air quality data), but with coarse spatial resolution due to the limited number of sites [12], and at significant capital and operational cost. Owing to the advanced development of Internet of Things, low-cost sensors [13] are also increasingly used for air quality measurements, as indicative measures. These techniques can enable the dense network of air quality monitoring required for building smart cities. Other monitoring approaches, such as mobile measurements using bicycles [14,15] and vehicles [16], generate air quality information at both high temporal and spatial resolutions within relatively small domains, while satellite measurements can provide a globally consistent air quality monitoring service at a coarse spatial resolution [17]. However, these measurement approaches are unable to provide the high-resolution spatial and temporal air pollutant concentration data required for some detailed population exposure calculations, or to evaluate potential policy options.

To complement the information obtainable from air quality monitoring services, the use of air quality modelling has rapidly increased over recent decades. These tools play a key role in environmental science because of their capability to quantify the deterministic relationships between emission sources, dispersion, mixing, concentrations, advection and deposition over different distance and time scales [18]. Their use has been promoted by the 2008 European Directive on Ambient Air Quality and Cleaner Air for Europe that explicitly encourages the adoption of modelling for air quality management such as forecasting and emission reduction plans [19]. Air quality models use mathematical equations to simulate physical and chemical processes affecting air pollution in the atmosphere using different approaches depending on the degree of meteorological and chemical detail required for a given application [20,21]. Dispersion, transport and chemical processes are modelled from local to regional scales using different types of models. Local-scale models can represent explicit source properties, such as geometry and efflux conditions, incorporating a simplified chemical scheme and using representative meteorological and emission data [22,23]. Regional-scale models use diffusion equations (e.g., Eulerian models) [24,25] or instantaneous flow approaches (e.g., Lagrangian models) [26,27] to simulate full chemistry and physical mechanisms acting in the atmosphere, accounting for the interaction of the emissions, homogeneously mixed on each grid, with meteorology. Other models adopt a simpler and less data-demanding approach to estimate air pollutant concentrations using a statistical or empirical approach [28–31]. The simplification of these models is achieved by ignoring the time-varying processes affecting air pollutant concentrations connected with variations in emissions, processing and meteorological conditions. Models that represent physical and chemical processes are the most suitable for air quality assessment planning for a number of reasons including: the capability to simulate at different spatial scales (from hemispheric simulations to regional and local scale) [32] and temporal scales (from short time period or event analysis to annual and inter-annual simulations) and the possibility to conduct several types of analysis, from the dispersion processes of inert and/or trace pollutants at a particular ground-level receptor influenced by an emission source, to simulations of the full chemistry acting in the atmosphere on pollutants from all available emission sources, and their interactions with meteorology. This also allows the models to be used to assess the likely changes in air pollutant concentrations resulting from differing scenarios of emissions reductions and/or forecasts of future climatic conditions [33].

Different mesoscale meteorological and chemistry-transport models (CTMs) (e.g., WRF [34], CMAQ [35] or WRF-Chem [36]) are commonly used worldwide by governments and researchers to study air pollution exposure [37,38], plan emissions reductions [39] and create scenarios [40] to reduce air pollution in urban areas. These systems require extensive computational time and resources in comparison to statistical and empirical models. This is due to the necessity to account for atmospheric dynamics and complex chemical dispersion and deposition processes of potentially thousands of primary and secondary pollutants over urban areas at different spatial resolution (from a few hundreds of meters to several km) [28]. Parallel HPC (High Performance Computer) clusters are generally used to supply the computational resources for this type of dispersion modelling. HPC clusters are able to run calculations from simulations of three-dimensional domains with different cell dimensions and sizes in parallel with a reasonable computational time [41].

In contrast, for most applications, local-scale dispersion models execute sequentially in terms of spatial and temporal calculations, leading to extended runtimes for the simulation of large urban areas.

Computer models representing physical or chemical processes such as pollutant dispersion are commonly initially written in a sequential form, as this is simple to develop and it is easily portable across different types of computational architecture. However, the computational burden associated with modelling complex atmospheric processes often requires runtime optimisation, such as code parallelisation whereby calculations are distributed over multiple cores on HPC clusters [42]. Sequential code can be converted into parallel code using parallelisation algorithms such as OpenMP [43], Parallel Virtual Machine (PVM) [44] and Message Passing Interface (MPI) [45], but a simpler approach, not requiring changes to code architecture, is to use task farming. Task farming involves running the same, possibly sequential, code on multiple processors using differing model configuration parameters and data inputs [46]. The application of task farming to modelling of physical processes with differing configurations relating to different spatial areas is sometimes known as spatial parallelisation. For some applications, task farming may be wasteful in terms of computational resources compared to full code parallelisation; for instance, the same code will be executed separately on each core and, for spatial parallelisation, there may be a need for additional calculations at the edge of each computational sub-domain. However, spatial parallelisation is relatively simple to implement from a code development perspective and can lead to runtime optimisations that broadly scale with the number of processors available. A task farming approach has previously been applied to the AERMOD Gaussian plume dispersion model with preliminary testing [47], alongside a qualitative assessment of the possibility of code parallelisation, which concluded that it would require significant development effort.

ADMS-Urban [48,49] is a quasi-Gaussian plume air dispersion model that represents the structure of the atmospheric boundary layer using two governing parameters: the boundary layer depth and the Monin–Obukhov length. It uses a physics-based approach, so it requires a range of data inputs (meteorological, emissions and long-range pollutant transport data). ADMS-Urban explicitly represents the full range of source types occurring in an urban area at high resolution (industry, transport and diffuse sources); the model is able to account for the influence of complex urban morphology (building density, street canyons [50,51]) on dispersion and generates street-scale resolution maps that highlight both pollution hotspots and areas of better air quality. The model has been used to quantify urban air pollution levels in many cities worldwide [49,52], but model run times can be extensive when run sequentially, with city-scale calculations taking weeks to execute on standard Windows PCs. Multiple model runs can be required in order to assess different policy or emissions scenarios, and to perform sensitivity analyses, so improving model run times is a key requirement for enabling the analysis of a broad range of scenarios.

This paper presents the results of a novel approach to running ADMS-Urban, where task farming has been used to spatially parallelise the model configuration, and each run component has been executed on an HPC—Bluebear at the University of Birmingham. The approach has been tested for a large model domain covering the West Midlands (WM), UK (902 km^2), as part of modelling work in the WM-Air (the West Midlands Air Quality Improvement Programme) project [53]. WM-Air is a five-year impact-focussed programme to support the improvement of air quality and associated health, environmental and economic benefits in the West Midlands. Section 2 describes the methodology of task farming in the ADMS-Urban model and presents the modelling configuration for the WM case study. Section 3 reports the model evaluation from the receptor run and several types of air quality maps from the contour run. Section 4 discusses the results and Section 5 gives a summary.

2. Methodology

2.1. ADMS-Urban Model

ADMS-Urban can model pollution sources with explicit point, line, area or volume geometry, using quasi-Gaussian plume dispersion expressions, with skewed vertical profiles used in convective conditions [54]. Road sources are modelled as a special case of line sources, where traffic-induced turbulence effects are included based on user-defined emission rates and/or traffic flow and speed data [55]. Point sources are modelled as elevated sources for large industry sources and stack parameters (e.g., stack height and diameter, efflux temperature and exit velocity) are needed. A regular grid of volume sources with uniform source depth is also used to represent total emissions of both the explicit sources and other sources where less detailed source characteristics are available, such as domestic heating or minor industrial processes.

Dispersion calculations for each explicit source and a single volume source (forming one cell of the uniform regular grid) are initially carried out along an "internal grid" of calculation locations following the downwind plume centreline, with along-wind interpolation, lateral and vertical profile factors used to obtain concentrations at the required output locations. The internal calculation grid resolution is finest at the source location and increases in geometric sequence with increasing distance from the source, as plume properties are expected to vary more slowly further from the source. The single volume source grid cell dispersion patterns are spatially translated to the location of each cell, scaled by the individual cell emissions and applied to the final output locations.

For pollutants which are considered inert on local scales, concentrations from all included sources are summed at each output location to form the total output concentration. For pollutants where local chemistry processes are significant, such as NO_X and NO_2, a concentration-weighted average of dispersion time is also calculated at each output location and used in an implementation of the Generic Reaction Set (GRS) chemistry scheme [56,57].

2.2. Run-Time Optimisation Using Task Farming

ADMS-Urban is a serial program designed to run on a single processor. However, the latest version of the model includes the ability to split up a large modelling region into smaller sub-regions, each of which can then be executed concurrently on multiple cores of a HPC or across a PC network, a technique known as task farming.

Only those output points that fall within a given sub-region are included in that sub-region run. Conversely, since the concentration at any output point can be affected by any upwind source within the modelling region, it is important that all source emissions are included in each sub-region run. A study of agricultural non-point source dispersion modelling, for sources with a maximum horizontal dimension of 60–80 m, showed that the exact geometry of neutrally buoyant non-point source types made little difference to predicted downwind concentrations beyond approximately 100 m [58]. Large efficiency gains can therefore be achieved, for appropriate source types, by only explicitly modelling those sources that fall within the sub-region (plus an additional "buffer" zone), while the emissions from more distant sources can be modelled via the (computationally much cheaper) grid source. This also provides justification for regional-scale chemical transport models typically only requiring gridded input emissions.

In ADMS-Urban, road sources are modelled as neutrally buoyant sources (with an initial mixing depth to account for vertical spread in the wake of vehicles) and can therefore be spatially truncated to sub-regions in this way. Figure 1 below shows an example of which road sources are explicitly modelled for a particular sub-region run. Run times can be optimised by ensuring a similar number of explicitly modelled sources and associated output points are included within each sub-region, hence, smaller sub-regions are used in areas with a higher density of explicit road sources.

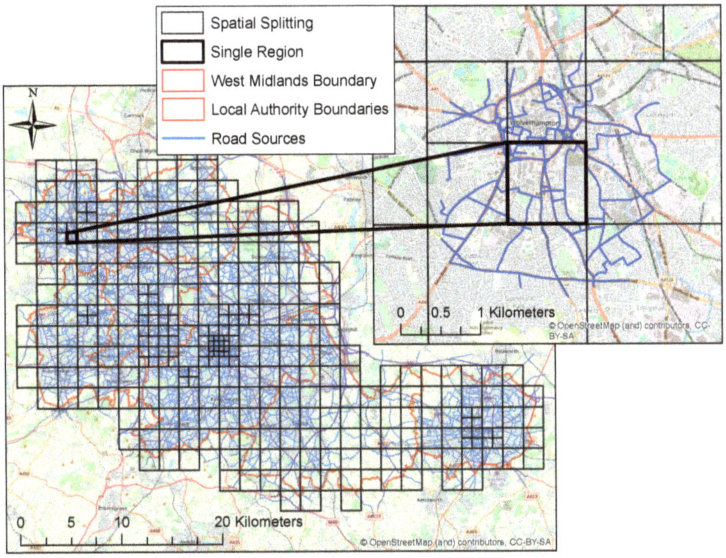

Figure 1. Example of explicit road source truncation for a particular sub-region. A buffer zone of 750 m has been used. Note that spatial sub-regions outside the WM boundary are run in the model but not included in contour output, and thus are not shown here.

Conversely, point sources with high pollutant emission rates are always modelled explicitly due to their non-negligible buoyancy and elevated source height, which affect their dispersion over a long distance. Point sources are generally selected for explicit modelling if they have annual average emissions greater than 1 g/s of a pollutant of interest, or are subject to specific national regulation ("'Part A" sources). The number of point sources of this type is often much smaller than the number of modelled road sources and so the run-time cost of including all point sources in each sub-region run, with extents normally of the order of 1 km, is comparatively small.

2.3. *Case Study*

The West Midlands Combined Authority (WMCA) in the UK covers seven constituent local authorities (Birmingham, Coventry, Dudley, Sandwell, Solihull, Walsall and Wolverhampton). Geographically, the West Midlands (WM) is an area of around 902 km^2 roughly centred on Birmingham. Air quality modelling is an important tool for the investigation of air quality within the WM region and for the assessment of the impact of specific intervention scenarios on air quality within the region.

2.3.1. Emissions

Emission sources in the model included explicit point sources, explicit road sources and 1 km $\times$ 1 km horizontal resolution grid sources for the baseline year of 2016 (shown as Figure 2). The EMIT Atmospheric Emissions Inventory Toolkit (developed by CERC) was used to pre-process the emission data before import into the ADMS-Urban model. Table 1 shows an overview of total emissions for different source types over WM computational domain.

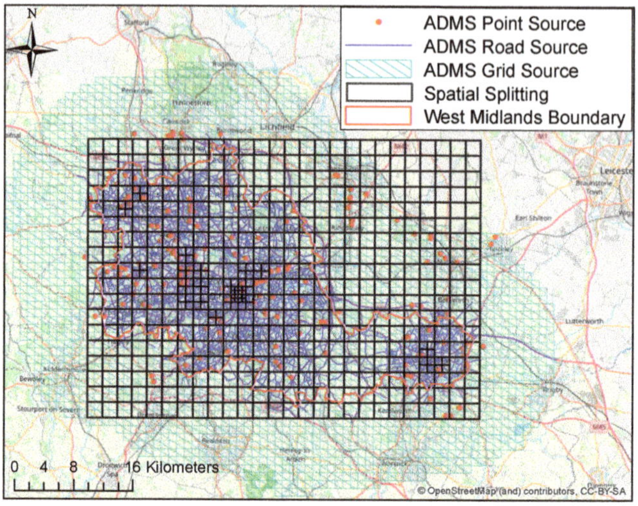

Figure 2. Emission sources in the model and spatial splitting for the modelling domain over West Midlands.

Table 1. Overview of total emissions (in tonnes/year) for different sources over the WM computational domain.

Group [1]	NO_2	NO_x	VOC	PM_{10}	$PM_{2.5}$	SO_2
Major Roads	3165	16,740	502	985	460	24
Point Sources	148	2951	1441	318	230	153
SNAP01 energyprod	20	400	4	6	6	106
SNAP02 domcom	150	3000	1801	1826	1775	994
SNAP03 indcom	165	3294	46	248	243	660
SNAP04 indproc	1	24	411	1773	197	20
SNAP05 offshore	0	0	1746	0	0	0
SNAP06 solvents	0	3	18,140	552	233	2
SNAP07 Minor Roads	2025	10,830	1362	586	314	50
SNAP08 othertrans	530	6620	2135	368	403	59
SNAP09 waste	3	57	251	129	119	3
SNAP10 agric	4	84	1049	194	40	0
SNAP11 nature	0	4	721	89	82	0

[1] For SNAP sectors, see Section Grid Sources.

Point Sources

Point source emission rates were taken from the UK National Atmospheric Emissions Inventory (NAEI) [59], which collected detailed emission data from large individual sources. Other smaller emission sources in the industrial and commercial sector were included as grid sources (Section Grid Sources). Large industrial point sources were considered explicitly as elevated point sources in the dispersion model. The emission inventory for these point sources combined the NAEI 2016 data (for emission rates) and Birmingham City Council (BCC) Airviro [60] model data (for stack parameters, e.g., stack height and diameter, efflux temperature and exit velocity). Representative typical stack characteristics by sector were used for the point sources where the stack characteristics are not known. The emission rates from the point sources were given for a wide range of pollutants; those of interest are NO_x as NO_2, PM_{10}, $PM_{2.5}$, Non-Methane Volatile Organic Compounds (NMVOC) and SO_2. The location of the point sources was given in the British National Grid Coordinate System (OSGB) and has been converted to the modelling coordinate system in Lambert Conformal Conic Projection (LCC). The use of modelling coordinate was consistent with that in the regional Community Multiscale Air Quality (CMAQ) modelling system, to prepare for the development of a coupling system between CMAQ and ADMS-Urban model under WM-Air.

Road Sources

Road sources in the current baseline model combined the traffic maps from Transport for West Midlands (TfWM) PRISM model [61] and BCC's SATURN model [62]. The SATURN model has more road links within the forthcoming Clean Air Zone of Birmingham [63]. The traffic map covers major roads, e.g., motorways and "A" roads. Minor roads not represented by the current traffic map are modelled as grid sources. The traffic data for AM peak, PM peak and inter-peak time periods have been combined and converted into Annual Average Daily Traffic (AADT). The traffic flows were categorised into heavy and light vehicles. These traffic model output data were evaluated against the TfWM's traffic count data. The light vehicle from the traffic model agrees well with traffic counts, while the heavy vehicle is consistently underestimated compared to traffic counts and an adjustment was made. Bus timetable data from Remix [64] were also processed and included in the model input. Representative fleet composition data (Euro classification for each sub-type of heavy and light vehicles) were taken from ANPR data in a recent Birmingham Clean Air Zone (CAZ) document [62] and has been incorporated into the EMIT calculations. The UK NAEI 2014 road traffic emission factors, with real-world adjustments following the approach described in Hood et al. 2018 [49], were used for the calculation of emission rates.

Grid Sources

Grid sources for 2016 were defined at 1 km $\times$ 1 km resolution with a typical depth of 10 m. The base gridded emissions were downloaded from the NAEI website [59] in the OSGB coordinate system, and have been converted to the LCC modelling coordinates. NAEI emissions are available for all SNAP (Selected Nomenclature for Air Pollution) sectors, i.e.,:

- SNAP 01—Combustion in Energy Production and Transformation (energyprod);
- SNAP02—Combustion in Commercial, Industrial, Residential and Agriculture (domcom);
- SNAP03—Combustion in Industry (indcom);
- SNAP04—Production Processes (indproc);
- SNAP05—Extraction and Distribution of Fossil Fuels (offshore);
- SNAP06—Solvent Use (solvents);
- SNAP07—Road Transport (roadtrans);
- SNAP08—Other Transport and Mobile Machinery (othertrans);
- SNAP09—Waste Treatment and Disposal (waste);
- SNAP10—Agriculture, Forestry and Landuse Change (agric);
- SNAP11—Nature (nature).

The pollutants of interest are NO_x as NO_2, NMVOC, PM_{10}, $PM_{2.5}$ and SO_2. SNAP07 has been reduced by subtracting the emission contribution from the explicit major road sources. EMIT also aggregates the explicit major road emissions into the same 1 km $\times$ 1 km grid. The residual emission for this SNAP07 sector can be then derived and modelled as SNAP07_minor road.

2.3.2. Time Varying Factors

Time-varying factors from the EMEP model [65,66] were available for each hour of the day by SNAP sector and pollutant. An emissions inventory covering the area of interest was available, with total emission for each sector. These emission rates were used to calculate a combined set of weighted average monthly emission factors for each pollutant, which were applied to the total gridded emission rates. Separate time varying factors were applied to particulate and gaseous gridded emissions, reflecting different balances between sectors and source types for these pollutants. In additional to the gridded emission rates, time varying factors have been also applied to explicit road sources. The monthly factors used for explicit road source emissions were taken from Community Modelling and Analysis System (CAMS) regional emissions v3.1 [67]. Diurnal profiles for road traffic have been calculated using 24-h flow and speed data from automatic traffic count sites (data downloaded from TfWM), typically available for 1 week per site. The roads of interest were isolated, and the light and heavy vehicle hourly flows and speeds were processed through an Emissions Factor Toolkit (EFT, version 9.0) [68] spreadsheet to calculate hourly emission rates of the pollutant of interest. The emission rates were then normalised by the average emission rate on the road, to give a time varying profile for the road. The roads were classified into medium or high flow and average time-varying profiles were calculated for each type. The diurnal profile for medium roads was also applied to the grid source, representing both the significant contribution of minor roads to the residual gridded emissions and the representation of emissions from roads outside the current sub-region and buffer zones in the gridded emissions.

2.3.3. Background Data

Background concentration files were created using historic observation data from a variety of rural background sites surrounding the West Midlands modelling area, available from the Department for Environment, Food and Rural Affairs (Defra) UK-Air website [12]. Data were limited in the West Midlands area, so a suitable background file was created using the following sites for different pollutants: (1) NO_x, NO_2, O_3: Ladybower (Lat, Lon: 53.403370, −0.752006), Market Harbough (52.554444, −0.772222), Chilbolton (51.149617, −1.438228), Leominster (52.221740, −2.736665), (2) SO_2: Ladybower, Narberth (51.781784, −4.691462), Chilbolton and (3) PM_{10} and $PM_{2.5}$: Chilbolton (with large periods of missing data filled using data from Sheffield Devonshire Green). The direction of each monitoring site from the centre of the modelling region, and wind direction sectors which were appropriate for each site, were calculated. The monitored wind direction for each hour was used to identify upwind monitoring data for that hour. The use of Chilbolton for particulate background concentrations was due to the fact that appropriate background monitoring sites for PM were scarce around the West Midlands area. The monitored Chilbolton concentration was multiplied by the ratio of the annual average concentration at a rural area bordering the West Midlands to that at Chilbolton based on Defra's background concentration maps [69].

2.3.4. Meteorological Data

For the West Midlands, an appropriate synoptic meteorological measurement site is located at Birmingham Elmdon, within Birmingham Airport, with data obtained from Met Office MIDAS in CEDA Archive [70]. "UK Hourly weather data", "UK Mean Wind" and "UK Hourly rainfall data" have been combined to create the met data format required by the model. The generated met file included hourly data for wind direction, wind speed (converted from knots to m/s), total cloud fraction, air temperature, relative humidity and precipitation.

2.3.5. Advanced Canyon and Urban Canopy Files

The data required to carry out the advanced canyon [51] and urban canopy [50] calculations are (1) a road network shapefile and (2) a buildings shapefile, including a height field. The building data have been obtained from Digimap database [71] via the University. The ADMS-Urban software package included ArcGIS tools [72] which have been used to calculate an Advanced Canyon file. The building height and canyon width along each road link were derived. The gridded urban canopy parameters have also been calculated for use in representing urban wind flow variations. These will enable the ADMS-Urban model to account for the street canyon effect for road emissions and spatially varying urban canopy flow for all source types.

2.3.6. Spatial Splitting

The task farming approach was achieved by spatially splitting the domain within the ADMS model. The overall rectangular output grid domain extent covering WM was first divided into 468 smaller sub-domains (forming a grid of 26 by 18 sub-domains, shown as Figure 2), each with a size of 2 km × 2 km. For some 2 km × 2 km subdomains with denser road links in city centre areas, which also included increased numbers of output points to fully resolve the near-road concentrations, a further spatial splitting into 1 km × 1 km or 500 m × 500 m was adopted in order to reduce the overall computation time. The total number of sub-domains was 540 (although no output points have been specified beyond 1 km outside the WM boundaries). The maximum number of road sources in a single sub-domain was 598 and the maximum number of output points in a domain was 4725. A buffer zone [73] of 750 m for road sources (to exclude explicit road sources unlikely to contribute significantly to modelled concentrations in the sub-domain) was used for each sub-domain.

3. Results

3.1. Receptor Run: Model Evaluation

For the purpose of model evaluation, the model was first run in a "Receptor" Mode (a run with output for a limited number of specified receptors) for 32 air quality measurement sites within the WM over the whole year of 2016, with measured concentration data obtained from local authorities and Defra's AURN [12] (shown as Figure 3, mostly with available hourly air quality measurements). These sites included three types, i.e., 1 airport site, 19 roadside sites and 12 urban background sites. In order to reduce the model computational time, the source exclusion option [73] was used to not explicitly model road sources far away from specified receptors, and therefore unlikely to contribute significantly to modelled concentrations at receptors, with a specified exclusion distance of 750 m. The Receptor run was conducted in a Windows PC and it took about 12 h' computation time to get the hourly output of five air pollutants (NO_x, NO_2, O_3, PM_{10} and $PM_{2.5}$) across a whole year for all 32 receptors. The Model Evaluation Toolkit [74] was used to conduct the evaluation of the model by comparing to the measured air quality data using statistical and graphical methods.

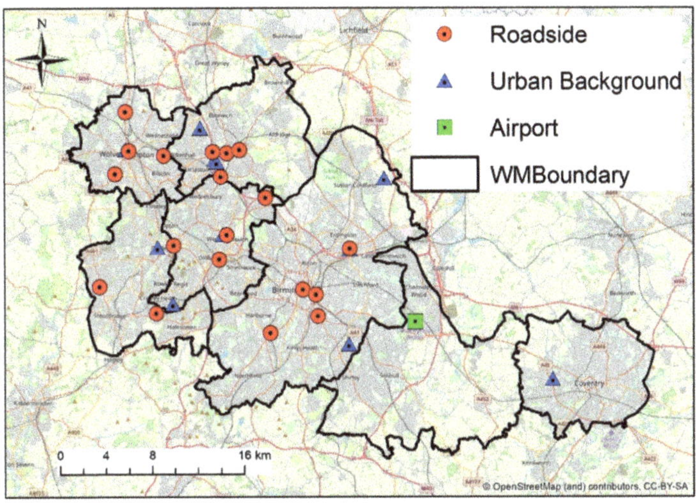

Figure 3. Monitoring sites within West Midland used for the model evaluation.

3.1.1. NO_x and Chemistry

Figure 4 shows the evaluation of modelled annual NO_x, NO_2 and O_3 against observations using scatter plots divided by site type. Overall, the model performed well in terms of NO_x and NO_2 for all site types. The good fits for O_3 further suggested good performance of the model chemistry. Table 2 shows the statistics (see definitions in [49]) for the model evaluation for the baseline year of 2016, calculated from the hourly modelled and measured concentrations. Fb (Fractional bias) measures the mean concentration difference between the model and measurement (ideal value is 0). Fac2 measures the fraction of modelled data within a factor of 2 of observations (ideal value is 1.0). NMSE (normalised mean square error) measures the mean concentration difference between matched pairs of modelled and measured data (ideal value is 0). R (correlation coefficient) measures a linear relationship of concentrations between the model and measurement (ideal value is 1.0). Fb varies between (−0.24, 0.03) for NO_x and (−0.08, 0.06) for NO_2 and O_3, while (0, 0.05) for NO_x and (−0.01, 0.06) for NO_2 and O_3 were derived for the ADMS-Urban run with adjusted road traffic in Hood et al. (2018) [49]. Fac2 indicates that more than 62% of NO_x, 76% of NO_2 and 73% of O_3 are within a factor of 2 of observations. NMSE varies between (1.03, 1.24) for NO_x, (0.12, 0.32) for NO_2 and O_3, which is similar to (0.62, 0.86) for NO_x and (0.21, 0.29) for NO_2 and O_3 in Hood et al. (2018) [49]. R varies between (0.58, 0.83) for NO_x, NO_2 and O_3, which is close to (0.61, 0.77) in Hood et al. (2018) [49].

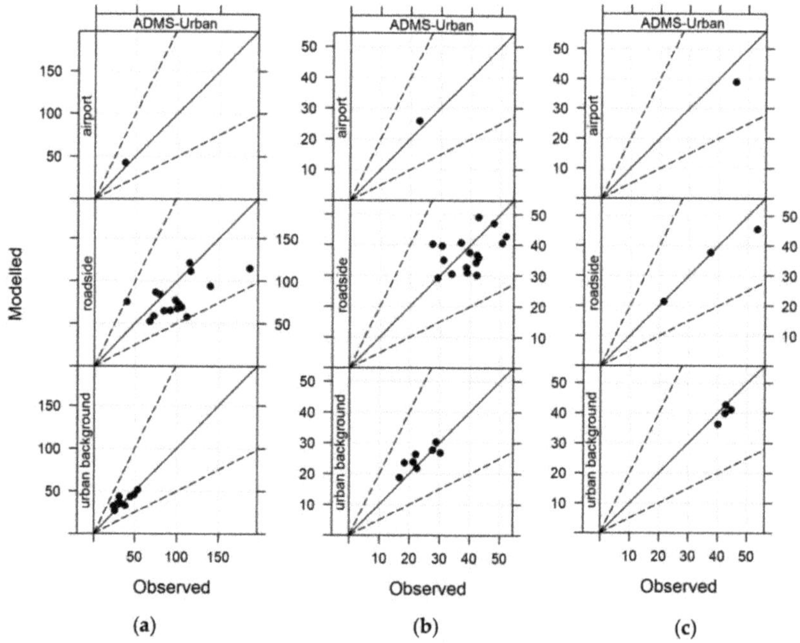

Figure 4. Comparison of annual averages between model output and measurement: (**a**) for NO_x (in µg m^{-3}), (**b**) for NO_2 (in µg m^{-3}) and (**c**) for O_3 (in µg m^{-3}).

Table 2. Statistics for model evaluation for the baseline year of 2016 calculated from the hourly modelled and measured concentrations. nSites: number of sites; Obs: observed concentration; Mod: modelled concentration; Fb: fraction bias (ideal value is 0); Fac2: fraction of modelled data within a factor of 2 of observations (ideal value is 1.0); NMSE: normalised mean square error (ideal value is 0); R: correlation coefficient (ideal value is 1.0).

Pollutant	Site Type	nSites	Obs (μg m^{-3})	Mod (μg m^{-3})	Fb	Fac2	NMSE	R
NOx	roadside	16	96.6	75.8	−0.24	0.62	1.03	0.58
Nox	urban background	9	38.2	39.4	0.03	0.74	1.24	0.63
NO2	roadside	17	39.0	36.0	−0.08	0.76	0.32	0.59
NO2	urban background	8	23.7	25.0	0.06	0.81	0.32	0.71
O3	roadside	3	34.3	33.8	−0.01	0.73	0.17	0.80
O3	urban background	4	42.5	40.0	−0.06	0.79	0.12	0.83
PM10	roadside	8	15.6	17.1	0.09	0.78	0.43	0.48
PM10	urban background	3	13.9	15.6	0.12	0.79	0.43	0.50
PM2.5	roadside	3	13.4	11.8	−0.13	0.76	0.53	0.61
PM2.5	urban background	3	10.7	10.3	−0.03	0.74	0.50	0.66

3.1.2. PM_{10} and $PM_{2.5}$

Figure 5 shows the evaluation of modelled annual average PM_{10} and $PM_{2.5}$ against observations using scatter plots divided by site types; note that there were no $PM_{2.5}$ measurements at the single airport site. PM_{10} had a very good fit for the airport and urban background sites. PM_{10} tended to slightly over-predict at roadside sites, possibly related to uncertainties in traffic non-exhaust emissions and background data. The model had good predictions for the small number of sites with available $PM_{2.5}$ measurement data. For PM_{10} and $PM_{2.5}$, Fb ranges between (−0.13, 0.12), slightly wider than (−0.07, 0.09) in Hood et al. (2018) [49]. Fac2 indicates that more than 74% of PM_{10} and $PM_{2.5}$ are within a factor of 2 of observations. NMSE varies between (0.43, 0.53) for PM_{10} and $PM_{2.5}$, slightly higher than (0.27, 0.37) in Hood et al. (2018) [49]. R varies between (0.48, 0.66) for PM_{10} and $PM_{2.5}$, slightly lower than (0.58, 0.77) in Hood et al. (2018) [49].

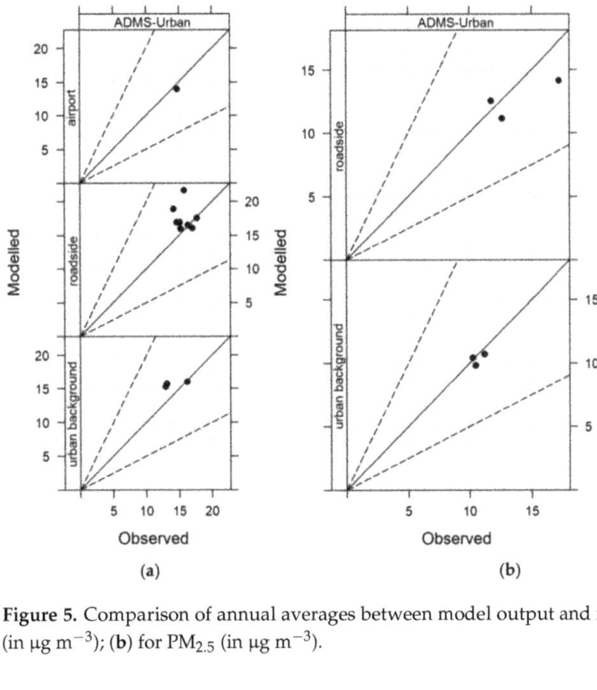

(a) (b)

Figure 5. Comparison of annual averages between model output and measurement: (**a**) for PM_{10} (in μg m^{-3}); (**b**) for $PM_{2.5}$ (in μg m^{-3}).

3.2. Contour Run: Air Quality Maps

For the generation of air quality maps, the model was then run in a "Contour" Mode (with the splitting option activated) to include output points covering the whole WM (and extending up to 1 km outside the WM boundary). An array job with 540 cores, each for a single sub-domain as shown in Figure 2, was submitted to the HPC at the University of Birmingham using the Linux version of the ADMS-Urban model. The overall elapsed time for the run (determined by the slowest core of 540 cores) for the typical whole year 2016 baseline case is about 35 h, with a median core run time of about 5 h and a minimum run time of 16 s for sub-domains without any emission sources. The total computational time (summing over all 540 cores) is about 169 days. Figure 6 shows a comparison of elapsed (clock) time of the slowest core and total computational time using task farming, (estimated by a typical one day simulation due to the substantial computational time requirement for a single core simulation). From 1 core to 4 core, the typical elapsed time can reduce by about 80%. From 156 cores, the typical elapsed time reduces more gradually. The total computational time profile has a slower decrease than the profile for the elapsed time of the slowest core, which is due to the increase in the buffer zone calculations for larger numbers of cores (especially for 540 cores). The choice of the number of cores can be dependent on the local HPC service, such as limitations in the number of available cores and the maximum allowed time for a single core (walltime).

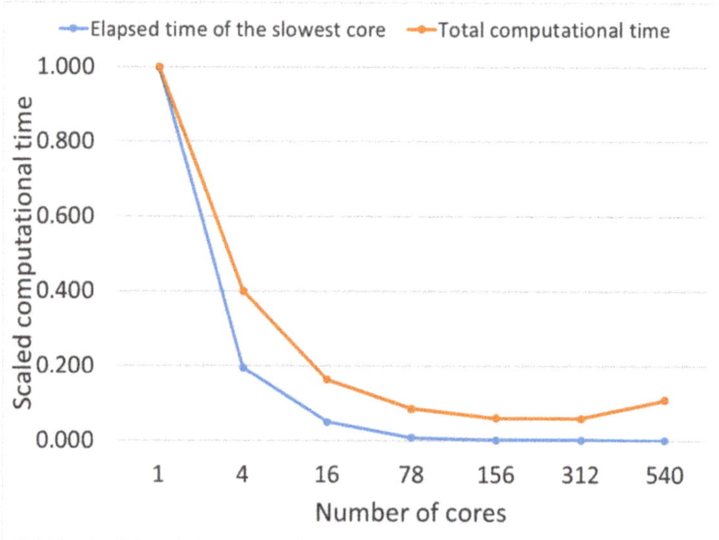

Figure 6. Comparison of elapsed (clock) time of the slowest core and total computational time using task farming; note that the horizontal axis scale is non-linear.

The output for each subdomain was in netcdf file format, which has been combined and interpolated using the CombineCOF and AddInterpIGP utilities developed by CERC. The re-combination and interpolation time was about 1 h. The recombined and interpolated outputs for the hourly output of the whole year over WM region contained ~0.61 million and ~1.26 million output locations, and had file sizes of about 120 GB and 247 GB, respectively. The final netcdf output was then processed to derive annual/subset averages and other statistical output (e.g., percentiles) using the "Process comprehensive output" tool in the model. This process took a couple of hours, dependent on the number of output air pollutants. The final contour plots over WM at a specified resolution (e.g., 10 m × 10 m) were created via GIS tools, in particular using interpolation in Surfer and display in ArcGIS.

3.2.1. Annual Air Quality Map

Figure 7 presents a map of the annual average NO_2 concentration which is a key air quality challenge for roadside locations in the UK, at 10 m × 10 m horizontal resolution for the baseline year of 2016. Other pollutants such as NO_x, O_3, PM_{10} and $PM_{2.5}$ are shown in Figure A1. The legend of Figure 7 indicates colour scales with annual mean NO_2 concentrations higher than the UK objective value of 40 μg m^{-3} [75] shown in orange and red. There were relatively higher concentrations of NO_2 near motorways and major roads in city centre areas, mostly due to the higher traffic-related emissions. Away from major roads and in rural areas, NO_2 concentrations were generally lower.

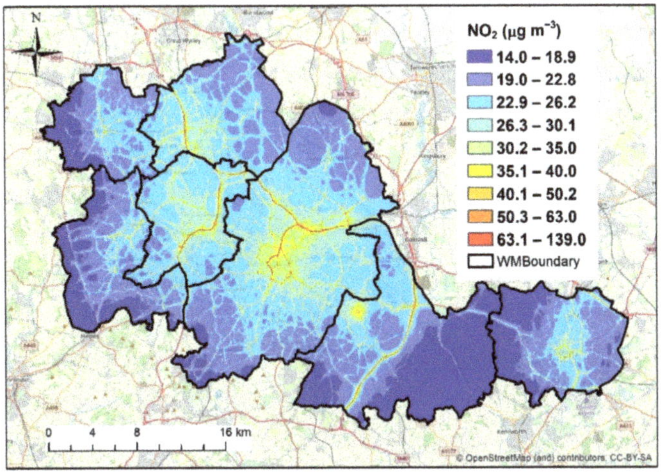

Figure 7. Annual air quality maps for NO_2 (in μg m^{-3}) at 10 m × 10 m resolution. Areas exceeding the UK objective value of 40 μg m^{-3} are shown in yellow, orange and red.

3.2.2. Projected Air Quality Map for Health

For the purposes of health-related research, including assessment of personal exposure and exploration of relationships between air pollution levels and socio-demographic characteristics typically available on different spatial scales, the 10 m × 10 m horizontal resolution annual air quality map may need to be further aggregated into other polygon layers, e.g., Lower Layer Super Output Areas (LSOA) and ward levels (averages over these layers). Figure 8 shows examples of projected annual air quality maps for NO_2 averaged over LSOA layer and ward level layer. There were clear patterns of higher concentration in city centre areas and lower concentration in rural areas. The projected air quality maps for other pollutants such as NO_x, O_3, PM_{10} and $PM_{2.5}$ are shown in Figure A2 (in LSOA layer) and Figure A3 (in the ward layer). These can then be linked to population and health data for the assessment of the health impacts of air pollution. Note that the spatial averaging process leads to a narrower range of concentrations across the whole area, as both the lowest concentrations in the most rural areas and the highest concentrations adjacent to road sources are no longer fully represented; this reduction in dynamic range is significant and dependent on spatial resolution.

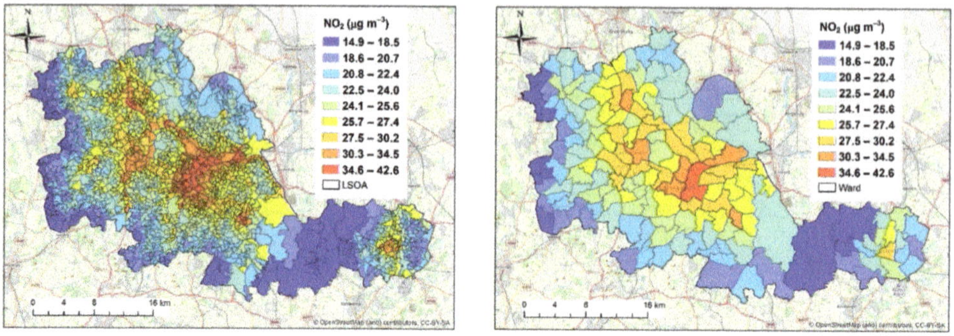

(a) (b)

Figure 8. Projected annual air quality maps for NO_2 (in μg m^{-3}) (**a**) in the Lower Layer Super Output Areas (LSOA) layer and (**b**) in the ward level.

3.2.3. Percentile Air Quality Map

Apart from annual average air quality targets, percentiles are also a useful indicator for the exceedance of the air quality objective value. For NO_2, the 99.8 percentile for the 1 h mean [75] is normally used, which represents the 18th highest concentration in hourly series over a whole calendar year. The UK objective for 1 h mean NO_2 concentration (where public exposure occurs) is "200 μg m^{-3} not to be exceeded more than 18 times a year" [75]. Figure 9 shows the 99.8 percentile for 1 h mean NO_2 concentration. The exceedance of air quality objective value (200 μg m^{-3}) for 1 h NO_2 concentration was found mostly along motorways and major roads linking to motorways, areas in which public exposure may be limited.

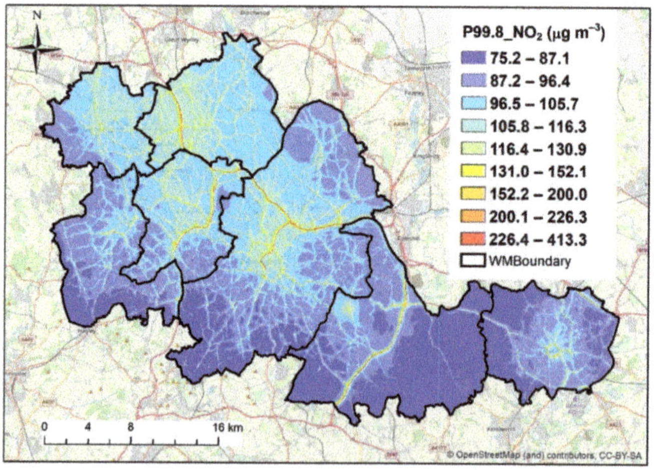

Figure 9. The 99.8 percentile annual air quality maps for 1 h mean NO_2 (in μg m^{-3}) at 10 m × 10 m resolution. Areas exceeding the UK objective value of 200 μg m^{-3} are shown in orange and red.

3.2.4. Air Quality Maps over Temporal Subsets

Post-processing tools provide the flexibility to obtain concentration averages over temporal subsets, which may be useful when mapped for health/exposure study. Figure 10 shows air quality maps of NO_2 over selected temporal subsets, i.e., AM (7 a.m.–9 a.m.) weekday, IP (inter-Peak) (9 a.m.–3 p.m.) weekday, PM (3 p.m.–7 p.m.) weekday, and IP (9 a.m.–3 p.m.) weekend. For AM weekday and PM weekday, the influence of major road emissions is clearly visible over the whole WM region, due to the region-wide increased traffic activity during these peak periods. For IP weekday and IP weekend, the major roads have a lesser contribution to concentrations over the WM region, compared with peak periods on weekdays. The NO_2 concentration for IP weekday shows higher overall concentration levels and a clearer pattern of influence from road emissions than that for IP weekend.

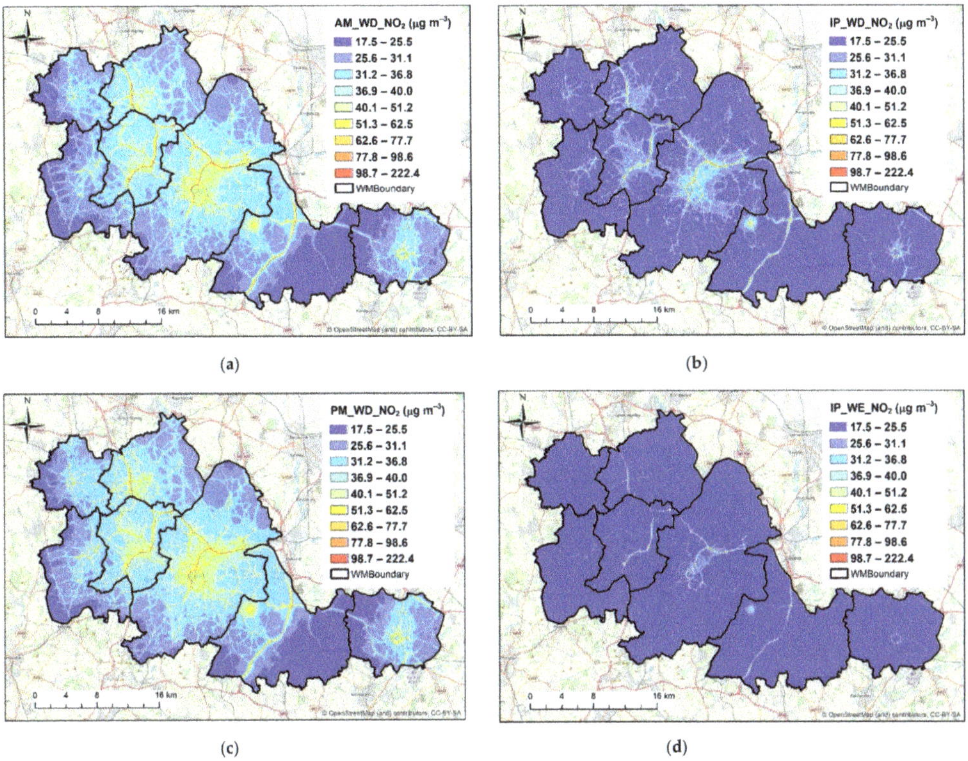

Figure 10. Air quality maps for NO_2 (in μg m^{-3}) at 10 m × 10 m resolution averaged over (**a**) AM weekday, (**b**) IP weekday, (**c**) PM weekday, (**d**) IP weekend.

4. Discussion

The air quality sites used to evaluate the model performance included three representative types (i.e., airport, roadside and urban background sites). Overall, the model performed well for all pollutants. For airport and urban background sites, the model results reproduce the measured values well since these are less influenced by local emissions and complex building geometry. For roadside sites, the concentrations of air pollutants are more influenced by local emission (e.g., traffic NO_x) and street canyon geometry, which may be reflected in higher uncertainty for some of the roadside sites. NO_x and NO_2 concentration levels are most closely related to local emissions and can be well predicted by the current model. PM concentrations are more related to the regional background, which may have some uncertainty. In order to reduce the model uncertainty, the model has been set up with the best available emissions, meteorology, building data, source locations and monitor locations for the WM region. From the model best practice and model evaluation, the model configuration is satisfactory for the wider WM contour run. It is of note that, for PM_{10} and $PM_{2.5}$, the background concentrations (constrained to observations) are greater than the increment associated with emissions in the model domain at all sites.

The contour run was performed by using the modelling capability of task farming (via spatial splitting) within the ADMS-Urban model. For this air quality modelling application in a large urban area, the optimisation process has reduced weeks of model execution time to only ~35 h and the model can generate high horizontal resolution "street-scale" air quality maps over WM. There are relatively higher annual concentrations of NO_2 in city centre areas (e.g., Birmingham), mainly due to the higher local traffic emissions. As NO_2 concentrations are closely related to local traffic emissions, the control of traffic in city centre areas would have a substantial effect on reducing proximate NO_2 levels. A Birmingham CAZ is proposed to be in place from June 2021 to reduce air pollution levels within Birmingham city centre [63].

The high horizontal resolution air quality maps generated from the model output can be further aggregated into other health-related layers (such as LSOA and ward layers) to study the relationship between air quality and health data. Apart from the annual averages, the model output and post-processing flexibility enables the calculation of other statistics, such as percentiles. Exceedances of the air quality objective value for the 99.8 percentile of the 1 h mean NO_2 concentration were generally only found for motorways and major roads directing to motorways, which is unsurprising due to intensive traffic activity. The reduction in traffic speed limits for some motorways (e.g., trials on speed limits on M6 and M5 motorways near Birmingham reduced from 70 mph to 60 mph by Highways England [76]) may thus help to reduce the NO_2 exceedance, although exposure at such locations may be limited.

Air quality maps over temporal subsets can be also derived for health/exposure study. As expected, AM weekday and PM weekday have clear patterns of region-wide traffic, while NO_2 concentrations over these periods are much higher than the annual averages. The influences of traffic over IP weekday and IP weekend are less significant compared with AM weekday and PM weekday. These findings can be useful for exposure and health studies over working periods. Reis et al. [77] also highlighted the importance of workday population mobility on the exposure to air pollutants.

5. Conclusions

A WM-Air ADMS-Urban baseline model configuration has been developed and model predictions for NO_x, NO_2, O_3, PM_{10} and $PM_{2.5}$ have been evaluated using measurement data. Overall, the model performed well and run times are manageable using the task farming approach. A regional (e.g., CMAQ) modelling system can provide spatially varying regional background predictions, which can be coupled with the ADMS-Urban model in future, but may also have its own uncertainties in terms of model configuration, compared with observational constraints. The post-processing flexibility enables the creation of air quality maps for annual/subset averages and other statistical output (e.g., percentiles). The model outputs can be useful for the study of health impacts of air pollutants.

Future work will draw upon the demonstrated efficient execution of multiple air quality modelling scenarios on HPC. It is important to ensure that the model configuration includes model inputs that are sufficiently detailed that they allow different scenarios to be represented. There are a range of possible air quality modelling scenarios: local and national, short-term and long-term, transport related and non-transport related. The combination of detailed inventory and spatially defined emissions, high resolution dispersion simulation, efficient parallelisation and flexible post-processing will allow the exploration of multiple scenarios. These may investigate concentration responses to interventions such as Clean Air Zones, the influence of solid fuel combustion, agricultural emissions, air quality–climate interactions and the relationships between air pollution exposure and population distribution. These in turn enable optimisation of combined benefits and equity of possible future limit value and exposure reduction based on air quality targets.

Author Contributions: Conceptualisation, X.C.; methodology, C.H. and K.J.; software, J.S. and J.H.; validation, K.J. and C.H.; formal analysis, J.Z.; investigation, J.Z.; resources, X.C., M.W. and A.M.; data curation, K.J., C.H. and J.Z.; writing—original draft preparation, J.Z. and C.H.; writing—review and editing, A.M., J.S. and W.J.B.; visualisation, J.Z.; supervision, J.S., J.H., M.W. and X.C.; project administration, J.S. and X.C.; funding acquisition, W.J.B. All authors have read and agreed to the published version of the manuscript.

Funding: This research was funded by the UK Natural Environment Research Council (NERC) project WM-Air, grant number NE/S003487/1.

Institutional Review Board Statement: Not applicable.

Data Availability Statement: AURN data are available via Defra website, https://uk-air.defra.gov.uk/networks/network-info?view=aurn (accessed on 18 June 2019). The NAEI emission data are available via http://naei.beis.gov.uk/data (accessed on 18 July 2019). The traffic data may be accessed subject to a license with TfWM/BCC. MIDAS met data are available via http://data.ceda.ac.uk/badc (accessed on 14 January 2019). EMEP time-variation data are available via https://www.emep.int/. The building data are available via https://digimap.edina.ac.uk (accessed on 28 May 2019). Other data (e.g., modelling output) may be made available upon request.

Acknowledgments: The authors appreciate the University of Birmingham's BlueBEAR HPC service (http://www.bear.bham.ac.uk, accessed on 23 April 2021) for providing the computational resource. The authors would like to acknowledge local authorities within WM for provision of local air quality measurement and modelling data and for their review of local emissions data, with special thanks to John Grant and Curtis Dean in Walsall council. The authors also thank Transport for West Midlands (TfWM) and Birmingham City Council for provision of traffic data, previous modelling and reports.

Conflicts of Interest: The authors declare no conflict of interest.

Appendix A

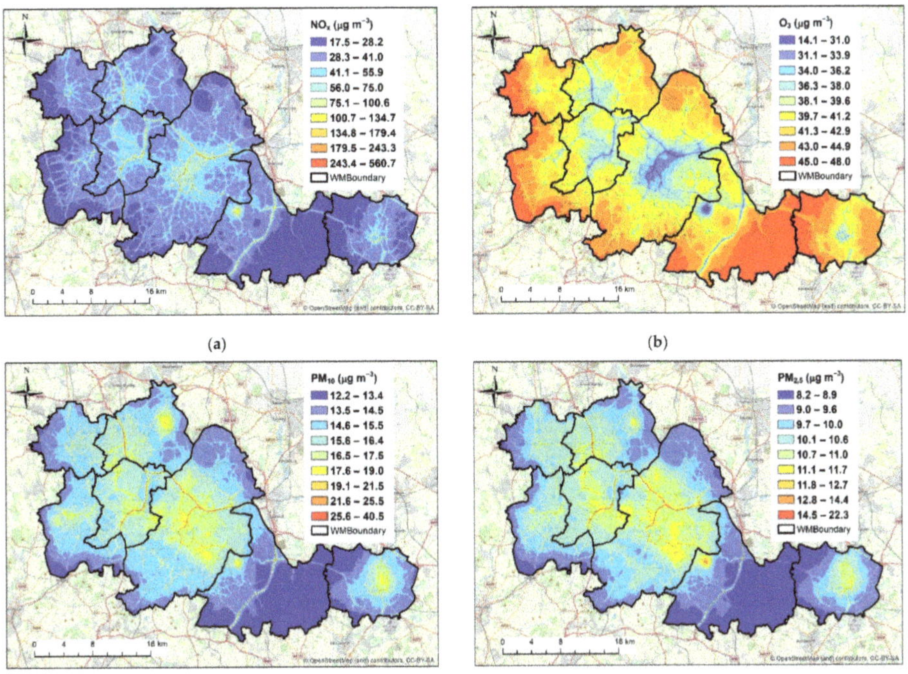

Figure A1. Annual air quality maps for (**a**) NO_x (in $\mu g\ m^{-3}$), (**b**) O_3 (in $\mu g\ m^{-3}$), (**c**) PM_{10} (in $\mu g\ m^{-3}$) and (**d**) $PM_{2.5}$ (in $\mu g\ m^{-3}$) at 10 m × 10 m resolution.

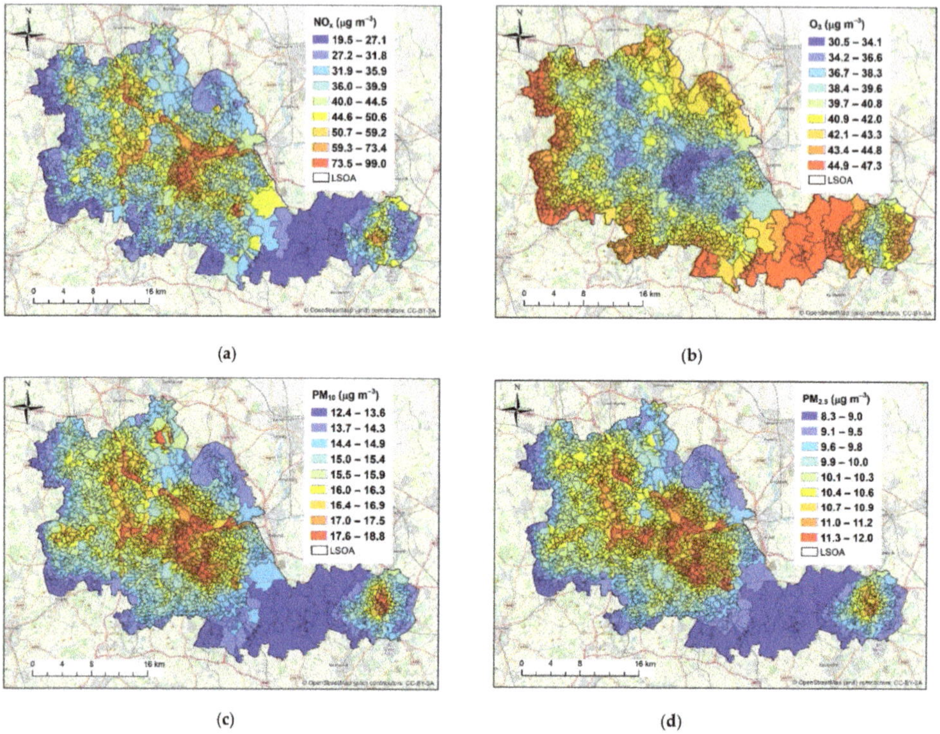

Figure A2. Projected annual air quality maps for (**a**) NO_x (in μg m^{-3}), (**b**) O_3 (in μg m^{-3}), (**c**) PM_{10} (in μg m^{-3}) and (**d**) $PM_{2.5}$ (in μg m^{-3}) in the Lower Layer Super Output Areas (LSOA) layer.

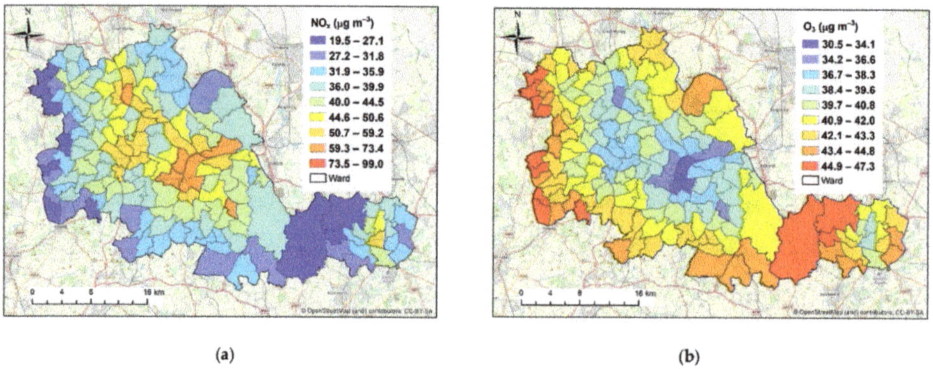

Figure A3. *Cont.*

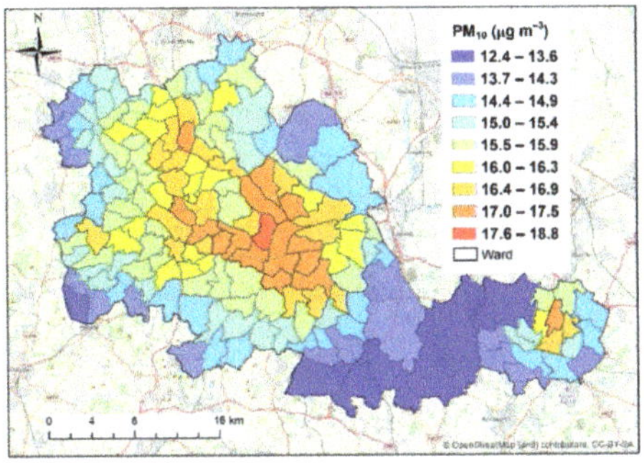

(c)

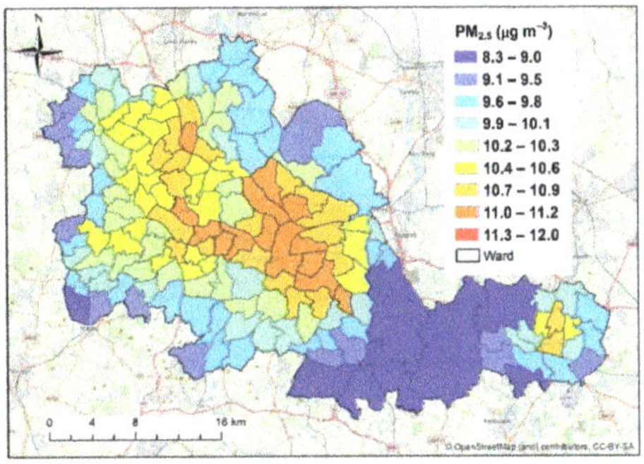

(d)

Figure A3. Projected annual air quality maps for (**a**) NO_x (in µg m^{-3}), (**b**) O_3 (in µg m^{-3}), (**c**) PM_{10} (in µg m^{-3}) and (**d**) $PM_{2.5}$ (in µg m^{-3}) in the ward level.

References

1. WHO. Available online: https://www.who.int/news-room/fact-sheets/detail/ambient-(outdoor)-air-quality-and-health (accessed on 11 January 2021).
2. Xu, J.; Srivastava, D.; Wu, X.; Hou, S.; Vu, T.V.; Liu, D.; Sun, Y.; Vlachou, A.; Moschos, V.; Salazar, V.; et al. An evaluation of source apportionment of fine OC and PM 2.5 by multiple methods: APHH-Beijing campaigns as a case study. *Faraday Discuss.* **2020**, *226*, 290–313. [CrossRef]
3. Gordon, T.; Balakrishnan, K.; Dey, S.; Rajagopalan, S.; Thornburg, J.; Thurston, G.; Agrawal, A.; Collman, G.; Guleria, R.; Limaye, S.; et al. Air pollution health research priorities for India: Perspectives of the Indo-US Communities of Researchers. *Environ. Int.* **2018**, *119*, 100–108. [CrossRef]
4. UK_gov. Available online: https://assets.publishing.service.gov.uk/government/uploads/system/uploads/attachment_data/file/633270/air-quality-plan-detail.pdf (accessed on 11 January 2021).
5. Hu, F.P.; Guo, Y.M. Health impacts of air pollution in China. *Front. Environ. Sci. Eng.* **2021**, *15*, 1–18. [CrossRef]
6. WHO. Available online: https://www.who.int/phe/publications/air-pollution-global-assessment/en/ (accessed on 11 January 2021).
7. Shaddick, G.; Thomas, M.L.; Mudu, P.; Ruggeri, G.; Gumy, S. Half the world's population are exposed to increasing air pollution. *Npj Clim. Atmos. Sci.* **2020**, *3*, 1–5. [CrossRef]
8. WHO. Available online: https://www.who.int/teams/environment-climate-change-and-health/air-quality-and-health/ambient-air-pollution (accessed on 11 January 2021).
9. COMEAP. Available online: https://www.gov.uk/government/publications/nitrogen-dioxide-effects-on-mortality/associations-of-long-term-average-concentrations-of-nitrogen-dioxide-with-mortality-2018-comeap-summary (accessed on 24 May 2021).
10. McLaren, J.; Williams, I.D. The impact of communicating information about air pollution events on public health. *Sci. Total Environ.* **2015**, *538*, 478–491. [CrossRef] [PubMed]
11. Delmas, M.A.; Kohli, A. Can Apps Make Air Pollution Visible? Learning About Health Impacts Through Engagement with Air Quality Information. *J. Bus. Ethics* **2020**, *161*, 279–302. [CrossRef]
12. Defra. Available online: https://uk-air.defra.gov.uk/networks/network-info?view=aurn (accessed on 18 June 2019).
13. Wesseling, J.; de Ruiter, H.; Blokhuis, C.; Drukker, D.; Weijers, E.; Volten, H.; Vonk, J.; Gast, L.; Voogt, M.; Zandveld, P.; et al. Development and Implementation of a Platform for Public Information on Air Quality, Sensor Measurements, and Citizen Science. *Atmosphere* **2019**, *10*, 445. [CrossRef]
14. Samad, A.; Vogt, U. Mobile air quality measurements using bicycle to obtain spatial distribution and high temporal resolution in and around the city center of Stuttgart. *Atmos. Environ.* **2021**, *244*, 117915. [CrossRef]
15. Elen, B.; Peters, J.; Van Poppel, M.; Bleux, N.; Theunis, J.; Reggente, M.; Standaert, A. The Aeroflex: A Bicycle for Mobile Air Quality Measurements. *Sensors* **2013**, *13*, 221–240. [CrossRef]
16. Hagemann, R.; Corsmeier, U.; Kottmeier, C.; Rinke, R.; Wieser, A.; Vogel, B. Spatial variability of particle number concentrations and NOx in the Karlsruhe (Germany) area obtained with the. mobile laboratory 'AERO-TRAM'. *Atmos. Environ.* **2014**, *94*, 341–352. [CrossRef]
17. de Vries, J.; Loenen, E.; Mijling, B.; van der Meulen, W.; van der Meer, A. Satellite and Local Measurements Based Services for Air Quality Improvement. *Asian J. Atmos. Environ.* **2019**, *13*, 39–44. [CrossRef]
18. EEA. Available online: https://www.eea.europa.eu/publications/fairmode (accessed on 25 February 2021).
19. EUR-Lex. Available online: https://eur-lex.europa.eu/legal-content/EN/TXT/?uri=celex%3A32008L0050 (accessed on 25 February 2021).

20. AIR-EIA. Available online: http://www.ess.co.at/AIR-EIA/LECTURES/L001.html (accessed on 25 February 2021).
21. Lamb, R.G.; Seinfeld, J.H. Mathematical modeling of urban air pollution. General theory. *Environ. Sci. Technol.* **1973**, *7*, 253–261. [CrossRef] [PubMed]
22. Pasquill, F. The Estimation of The Dispersion of Windborne Material. *Meteorol. Mag.* **1961**, *90*, 33–40.
23. Pasquill, F. Some observed properties of medium-scale diffusion in the atmosphere. *Q. J. R. Meteorol. Soc.* **1962**, *88*, 70–79. [CrossRef]
24. SCOPE. Available online: https://storage.googleapis.com/carnegie-dge-scope-mirror/SCOPE_30/SCOPE_30_2.28_Piver_635-650.pdf (accessed on 25 February 2021).
25. Liu, M.-K.; Seinfeld, J.H. On the validity of grid and trajectory models of urban air pollution. *Atmos. Environ.* **1975**, *9*, 555–574. [CrossRef]
26. Friedlander, S.K.; Seinfeld, J.H. Dynamic model of photochemical smog. *Environ. Sci. Technol.* **1969**, *3*, 1175–1181. [CrossRef]
27. Fisher, B.E.A. The long range transport of sulphur dioxide. *Atmos. Environ.* **1975**, *9*, 1063–1070. [CrossRef]
28. El-Harbawi, M. Air quality modelling, simulation, and computational methods: A review. *Environ. Rev.* **2013**, *21*, 149–179. [CrossRef]
29. Russo, A.; Raischel, F.; Lind, P.G. Air quality prediction using optimal neural networks with stochastic variables. *Atmos. Environ.* **2013**, *79*, 822–830. [CrossRef]
30. Voukantsis, D.; Karatzas, K.; Kukkonen, J.; Räsänen, T.; Karppinen, A.; Kolehmainen, M. Intercomparison of air quality data using principal component analysis, and forecasting of PM_{10} and $PM_{2.5}$ concentrations using artificial neural networks, in Thessaloniki and Helsinki. *Sci. Total Environ.* **2011**, *409*, 1266–1276. [CrossRef]
31. Gardner, M.W.; Dorling, S.R. Artificial neural networks (the multilayer perceptron)—A review of applications in the atmospheric sciences. *Atmos. Environ.* **1998**, *32*, 2627–2636. [CrossRef]
32. Thunis, P.; Miranda, A.; Baldasano, J.M.; Blond, N.; Douros, J.; Graff, A.; Janssen, S.; Juda-Rezler, K.; Karvosenoja, N.; Maffeis, G.; et al. Overview of current regional and local scale air quality modelling practices: Assessment and planning tools in the EU. *Environ. Sci. Policy* **2016**, *65*, 13–21. [CrossRef]
33. Carnevale, C.; Finzi, G.; Guariso, G.; Pisoni, E.; Volta, M. Surrogate models to compute optimal air quality planning policies at a regional scale. *Environ. Modell. Softw.* **2012**, *34*, 44–50. [CrossRef]
34. Skamarock, W.C.; Klemp, J.B. A time-split nonhydrostatic atmospheric model for weather research and forecasting applications. *J. Comput. Phys.* **2008**, *227*, 3465–3485. [CrossRef]
35. Chemel, C.; Fisher, B.E.A.; Kong, X.; Francis, X.V.; Sokhi, R.S.; Good, N.; Collins, W.J.; Folberth, G.A. Application of chemical transport model CMAQ to policy decisions regarding PM2.5 in the UK. *Atmos. Environ.* **2014**, *82*, 410–417. [CrossRef]
36. Grell, G.A.; Peckham, S.E.; Schmitz, R.; McKeen, S.A.; Frost, G.; Skamarock, W.C.; Eder, B. Fully coupled "online" chemistry within the WRF model. *Atmos. Environ.* **2005**, *39*, 6957–6975. [CrossRef]
37. Hu, J.L.; Li, X.; Huang, L.; Ying, Q.; Zhang, Q.; Zhao, B.; Wang, S.X.; Zhang, H.L. Ensemble prediction of air quality using the WRF/CMAQ model system for health effect studies in China. *Atmos. Chem. Phys.* **2017**, *17*, 13103–13118. [CrossRef]
38. San Jose, R.; Perez, J.L.; Perez, L.; Barras, R.M.G. Effects of climate change on the health of citizens modelling urban weather and air pollution. *Energy* **2018**, *165*, 53–62. [CrossRef]
39. Liu, H.; Deng, Y.; Lu, C.; Zhang, X.; Yang, X.; Song, D.; Tan, Q. A CMAQ-based Study on Evaluation of Achievements of Measures on Ozone Pollution Prevention and Control in Summer 2017 in Chengdu. *IOP Conf. Ser. Earth Environ. Sci.* **2020**, *610*, 012015. [CrossRef]
40. Cheng, F.Y.; Feng, C.Y.; Yang, Z.M.; Hsu, C.H.; Chan, K.W.; Lee, C.Y.; Chang, S.C. Evaluation of real-time PM2.5 forecasts with the WRF-CMAQ modeling system and weather-pattern-dependent bias-adjusted PM2.5 forecasts in Taiwan. *Atmos. Environ.* **2021**, *244*, 117909. [CrossRef]
41. Zlatev, Z.; Brandt, J.; Builtjes, P.J.H.; Carmichael, G.; Dimov, I.; Dongarra, J.; Dop, H.V.; Georgiev, K.; Hass, H.; José, R.S. *Large Scale Computations in Air Pollution Modelling*, 1st ed.; Springer: Dordrecht, The Netherlands, 1999; pp. 1173–1184.
42. Dabdub, D.; Seinfeld, J.H. Air quality modeling on massively parallel computers. *Atmos. Environ.* **1994**, *28*, 1679–1687. [CrossRef]
43. Dagum, L.; Menon, R. OpenMP: An industry standard API for shared-memory programming. *IEEE Comput. Sci. Eng.* **1998**, *5*, 46–55. [CrossRef]
44. Sunderam, V.S. PVM: A framework for parallel distributed computing. *Concurr. Pract. Exper.* **1990**, *2*, 315–339. [CrossRef]
45. Aoyama, Y.; Nakano, J. *RS/6000 SP: Practical MPI Programming*; International Technical Support Organisation, IBM Corporation: Austin, TX, USA, 1999; pp. 11–37.
46. Kolodziej, J.; Pop, F.; Dobre, C. *Modeling and Simulation in HPC and Cloud Systems*; Springer International Publishing: Cham, Switzerland, 2018; pp. 45–57.
47. Delic, G. Exploring parallel processing opportunities in AERMOD. In Proceedings of the 15th Annual CMAS Conference, Chapel Hill, NC, USA, 24–26 October 2016.
48. Owen, B.; Edmunds, H.A.; Carruthers, D.J.; Singles, R.J. Prediction of total oxides of nitrogen and nitrogen dioxide concentrations in a large urban area using a new generation urban scale dispersion model with integral chemistry model. *Atmos. Environ.* **2000**, *34*, 397–406. [CrossRef]
49. Hood, C.; MacKenzie, I.; Stocker, J.; Johnson, K.; Carruthers, D.; Vieno, M.; Doherty, R. Air quality simulations for London using a coupled regional-to-local modelling system. *Atmos. Chem. Phys.* **2018**, *18*, 11221–11245. [CrossRef]

50. Hood, C.; Carruthers, D.; Seaton, M.; Stocker, J.; Johnson, K. Urban canopy flow field and advanced street canyon modelling in ADMS-Urban. In Proceedings of the 16th International Conference on Harmonisation, Varna, Bulgaria, 8–11 September 2014; pp. 8–11.
51. Hood, C.; Stocker, J.; Seaton, M.; Johnson, K.; O'Neill, J.; Thorne, L.; Carruthers, D. Comprehensive evaluation of an advanced street canyon air pollution model. *J. Air Waste Manag. Assoc.* **2021**, *71*, 247–267. [CrossRef] [PubMed]
52. Biggart, M.; Stocker, J.; Doherty, R.M.; Wild, O.; Hollaway, M.; Carruthers, D.; Li, J.; Zhang, Q.; Wu, R.L.; Kotthaus, S.; et al. Street-scale air quality modelling for Beijing during a winter 2016 measurement campaign. *Atmos. Chem. Phys.* **2020**, *20*, 2755–2780. [CrossRef]
53. WM-Air. Available online: https://wm-air.org.uk/ (accessed on 10 May 2021).
54. CERC. Available online: http://www.cerc.co.uk/environmental-software/assets/data/doc_techspec/P10_01.P12_01.pdf (accessed on 26 March 2021).
55. CERC. Available online: http://www.cerc.co.uk/environmental-software/assets/data/doc_techspec/P31_01.pdf (accessed on 26 March 2021).
56. CERC. Available online: http://www.cerc.co.uk/environmental-software/assets/data/doc_techspec/P18_03.pdf (accessed on 26 March 2021).
57. Venkatram, A.; Karamchandani, P.; Pai, P.; Goldstein, R. The development and application of a simplified ozone modeling system (SOMS). *Atmos. Environ.* **1994**, *28*, 3665–3678. [CrossRef]
58. Stocker, J.; Ellis, A.; Smith, S.; Carruthers, D.; Venkatram, A.; Dale, W.; Attree, M. A review of dispersion modelling of agricultural emissions with non-point sources. *Int. J. Environ. Pollut.* **2017**, *62*, 247–263. [CrossRef]
59. NAEI. 2019. Available online: http://naei.beis.gov.uk/data (accessed on 18 July 2019).
60. Airviro. Available online: http://80.229.174.184/iairviro/ (accessed on 31 October 2018).
61. PRISM. 2019. Available online: https://www.tfwm.org.uk/strategy/data-insight/transport-modelling/about-prism (accessed on 22 July 2019).
62. BCC_Report. 2018. Available online: https://www.birmingham.gov.uk/download/downloads/id/11353/aq3_-_birmingham_caz_fbc_report-_air_quality_v3_4-12-18.pdf+&cd=1&hl=zh-CN&ct=clnk&gl=uk (accessed on 1 October 2019).
63. BCC. Available online: https://www.birmingham.gov.uk/info/20076/pollution/1763/a_clean_air_zone_for_birmingham (accessed on 2 May 2021).
64. Remix. Available online: platform.remix.com (accessed on 9 April 2019).
65. Simpson, D.; Benedictow, A.; Berge, H.; Bergström, R.; Emberson, L.D.; Fagerli, H.; Flechard, C.R.; Hayman, G.D.; Gauss, M.; Jonson, J.E.; et al. The EMEP MSC-W chemical transport model—Technical description. *Atmos. Chem. Phys.* **2012**, *12*, 7825–7865. [CrossRef]
66. EMEP. Available online: https://www.emep.int/ (accessed on 17 May 2021).
67. ECCAD. Available online: https://eccad.aeris-data.fr/ (accessed on 19 May 2021).
68. EFT. Available online: https://laqm.defra.gov.uk/review-and-assessment/tools/emissions-factors-toolkit.html (accessed on 12 August 2019).
69. Defra. Available online: https://uk-air.defra.gov.uk/data/laqm-background-home (accessed on 18 June 2019).
70. CEDA. Available online: http://data.ceda.ac.uk/badc (accessed on 14 January 2019).
71. Digimap. Available online: https://digimap.edina.ac.uk (accessed on 28 May 2019).
72. Jackson, M.; Hood, C.; Johnson, C.; Johnson, K. Calculation of Urban Morphology Parameterisations for London for use with the ADMS-Urban Dispersion Model. *Int. J. Adv. Remote Sens. GIS* **2016**, *5*, 1678–1687. [CrossRef]
73. CERC. ADMS-Urban User Guide Version 5.0. Available online: https://www.cerc.co.uk/environmental-software/assets/data/doc_userguides/CERC_ADMS-Urban5.0_User_Guide.pdf (accessed on 26 March 2021).
74. Stidworthy, A.; Jackson, M.; Johnson, K.; Carruthers, D.; Stocker, J. Evaluation of local and regional air quality forecasts for London. *Int. J. Environ. Pollut.* **2018**, *64*, 178–191. [CrossRef]
75. Defra. Available online: https://uk-air.defra.gov.uk/assets/documents/Air_Quality_Objectives_Update.pdf (accessed on 10 May 2021).
76. HIghwaysEngland. Available online: https://highwaysengland.co.uk/our-work/air-quality/air-quality-speed-limit-trials/ (accessed on 10 May 2021).
77. Reis, S.; Liška, T.; Vieno, M.; Carnell, E.J.; Beck, R.; Clemens, T.; Dragosits, U.; Tomlinson, S.J.; Leaver, D.; Heal, M.R. The influence of residential and workday population mobility on exposure to air pollution in the UK. *Environ. Int.* **2018**, *121*, 803–813. [CrossRef] [PubMed]

atmosphere

Article

Accelerated Time and High-Resolution 3D Modeling of the Flow and Dispersion of Noxious Substances over a Gigantic Urban Area—The EMERGENCIES Project

Olivier Oldrini [1,*], Patrick Armand [2], Christophe Duchenne [2], Sylvie Perdriel [3] and Maxime Nibart [4]

1 MOKILI, 75014 Paris, France
2 CEA, DAM, DIF, 91297 Arpajon, France; patrick.armand@cea.fr (P.A.); christophe.duchenne@cea.fr (C.D.)
3 AmpliSIM, 75014 Paris, France; info@amplisim.com
4 ARIA Technologies, 92100 Boulogne-Billancourt, France; mnibart@aria.fr
* Correspondence: info@mokili.fr

Abstract: Accidental or malicious releases in the atmosphere are more likely to occur in built-up areas, where flow and dispersion are complex. The EMERGENCIES project aims to demonstrate the operational feasibility of three-dimensional simulation as a support tool for emergency teams and first responders. The simulation domain covers a gigantic urban area around Paris, France, and uses high-resolution metric grids. It relies on the PMSS modeling system to model the flow and dispersion over this gigantic domain and on the Code_Saturne model to simulate both the close vicinity and the inside of several buildings of interest. The accelerated time is achieved through the parallel algorithms of the models. Calculations rely on a two-step approach: the flow is computed in advance using meteorological forecasts, and then on-demand release scenarios are performed. Results obtained with actual meteorological mesoscale data and realistic releases occurring both inside and outside of buildings are presented and discussed. They prove the feasibility of operational use by emergency teams in cases of atmospheric release of hazardous materials.

Keywords: operational emergency modeling; atmospheric release; high-resolution metric grid; 3D; PMSS modeling system; Code_Saturne; EMERGENCIES project

Citation: Oldrini, O.; Armand, P.; Duchenne, C.; Perdriel, S.; Nibart, M. Accelerated Time and High-Resolution 3D Modeling of the Flow and Dispersion of Noxious Substances over a Gigantic Urban Area—The EMERGENCIES Project. *Atmosphere* **2021**, *12*, 640. https://doi.org/10.3390/atmos12050640

Academic Editors: Ashok Luhar and Enrico Ferrero

Received: 28 February 2021
Accepted: 14 May 2021
Published: 18 May 2021

1. Introduction

Many types of accidents, malicious actions, or terrorist attacks lead to the releases of noxious gases into the atmosphere. Densely built-up and highly populated areas, such as industrial sites or urban districts, are the critical locations for such events. The health effects are most severe close to the source of emission. On a local scale, the buildings significantly influence the airflow and consequently the dispersion. Thus, atmospheric dispersion models need to account for the influence on the flow of the three-dimensional (3D) geometry of the buildings.

Up to now, first responders and emergency teams essentially relied on modified Gaussian modeling. Most of the fast response modeling tools indeed use simplified flow formulations in the urban canopy and concentration analytical solutions with assumptions about the initial size of the plume and increased turbulence due to the urban environment: for examples, see ADMS-Urban [1], PRIME [2], SCIPUFF [3], and the dispersion model tested by Hanna and Baja [4]. However, Gaussian approaches do not apply to the complex flow patterns in the vicinity of the release in built-up areas, where the most serious consequences of the release are happening. Further away from the source, they cannot model specific effects due to buildings, such as the entrapment and subsequent releases of noxious substances due to canyon streets, cavity zones behind buildings, etc. Field experiments, such as the Mock Urban Setting Test (MUST) field experiment [5], clearly demonstrated the importance of high-resolution modeling in complex built-up areas: critical effects, such as change in the plume centerline direction compared to the inflow condition, were observed and studied [6,7]. Models developed for scales too large to handle individual buildings have difficulties taking into account such effects.

At the same time, more precise modeling tools are becoming increasingly available for emergency situations at the mesoscale [8] and local scale [9,10]. Computational fluid dynamics (CFD) models provide reference solutions by solving the Navier–Stokes equations. In particular, the Reynolds-averaged Navier–Stokes (RANS) approach can be used on small setups by relying on heavy parallel computing and optimization [11] or precomputation [12,13].

The computational cost of CFD calculations, such as RANS and especially large eddy simulation (LES), indeed makes their usability complex for real-time emergency situations. To overcome this, a common strategy is to precompute a database of cases [12,13]. Nonetheless, the capability of the database to represent the variability of actual flows is an important issue, especially for complex 3D flows: it requires discretizing not only many inflow parameters, such as wind speed and wind direction, but also turbulence parameters and requires consideration of multiple inflow locations, usually at least along a vertical distribution. This leads to very large databases still lacking representativeness. The representativeness is particularly critical in built-up environments where an actual flow can be drastically different from a database case with similar but just slightly different inflow conditions.

Other approaches strike a balance between model speed and accuracy. Röckle [14] suggested such an approach by using mass consistency in combination with local wind observations to solve the mean flow. Kaplan et al. [15] extended the approach by coupling the flow modeling with Lagrangian particle dispersion modeling (LPDM, see [16]). This approach has been completed to provide modeling solutions (see [17,18]).

Since first responders and emergency teams still essentially rely on modified Gaussian modeling, the EMERGENCIES project was designed to prove the operational feasibility of 3D high-resolution modeling as a support tool for first responders or emergency teams. As such, the modeling approach was defined to apply to a realistic chain of events and applied to a geographic area compatible with the area of responsibility of an emergency team. The approach selected was to precompute a 3D and high-resolution flow over this huge domain by using mesoscale forecasts from one day to the next and then to compute the dispersion on an on-demand basis.

While source term estimation is a critical issue, it is not covered in this paper, which is dedicated to the modeling of flow and dispersion. Source term estimation requires dedicated modeling to deal both with uncertainties regarding the situation and with the particular physics that may be involved, such as fire or explosions.

The paper is organized in five parts: Section 2 contains a brief description of the modeling tools, both the numerical models and the computing infrastructure, and Section 3 summarizes the EMERGENCIES project modeling set-up. Section 4 presents and comments on the results of the simulations, including the operational aspects. Section 5 draws conclusions on the potential use of this approach for emergency preparedness and response in the case of an atmospheric release.

2. Modeling Tools

After describing the numerical models, we introduce the computing cluster used for the simulations.

2.1. Presentation of the PMSS Modeling System

Operational modeling systems must provide reliable results in a limited amount of time. This is all the more of a challenge when one needs to consider large simulation domains, e.g., covering a whole urban area, at a high spatial resolution. The challenge can be tackled by combining appropriate models with efficient parallelization. This was the basic motivation for the development of the Parallel Micro SWFT and SPRAY (PMSS) modeling system.

2.1.1. The Modeling System

The PMSS modeling system [17,19–21] is a parallelized flow and dispersion modeling system. It is constituted of the Parallel-SWIFT (PSWIFT) flow model and the Parallel-SPRAY (PSPRAY), an LPDM, both used in small-scale urban mode.

The PSWIFT model [22–24] is a parallelized mass-consistent 3D diagnostic model able to handle complex terrain using terrain-following coordinates. It uses analytical relationships for the flow velocity around buildings and produces diagnostic wind velocity, pressure, turbulence, temperature, and humidity fields. A calculation is performed using the three sequential steps:

- First guess computation starting from heterogeneous meteorological input data, mixing surface and profile measurements with mesoscale model outputs;
- Modification of the first guess using analytical zones defined around isolated buildings or within groups of buildings [14,15];
- Mass consistency with impermeability on ground and buildings.

The PSWIFT model also incorporates a RANS solver [24,25]. It can be used as an alternate in the third step above. The RANS solver provides a more accurate pressure field: this is essential to derive the pressure on facades of buildings and model the infiltration inside.

The PSPRAY model [23,26,27] is a parallelized LPDM [16] that takes into account obstacles. The PSPRAY model simulates the dispersion of an airborne contaminant by following the trajectories of numerous numerical particles. The velocity of each virtual particle is the sum of a transport component and a turbulence component. The transport component is derived from the local average wind vector, while the turbulence component is derived from the stochastic scheme developed by Thomson [28] that solves a 3D form of the Langevin equation. This equation comprises a deterministic term that depends on the Eulerian probability density function of the turbulent velocity, and is determined from the Fokker–Planck equation, and a stochastic diffusion term that is obtained from a Lagrangian structure function. The PSPRAY model treats elevated and ground-level emissions, instantaneous and continuous emissions, or time-varying sources. Additionally, it is able to deal with plumes with initial arbitrarily oriented momentum, negative or positive buoyancy, and cloud spread at the ground due to gravity.

2.1.2. Scalability

In order to allow for very large calculations and near-real-time calculations, the PMSS modeling system integrates parallelism using both weak and strong scalability [17].

The weak scalability relies on domain decomposition (DD). The domain is divided into tiles that are compatible in size with the memory available to a single core. The strong scalability is implemented differently within the PSWIFT and PSPRAY models. Within the PSWIFT model, strong scalability relies on the diagnostic property of the code: time frames are computed in parallel. Within the PSPRAY model, strong scalability is achieved by distributing the virtual particles among the computing cores. An example of a combination of strong and weak scaling is illustrated for the PSWIFT model and for the PSPRAY model in Figure 1.

(**a**)

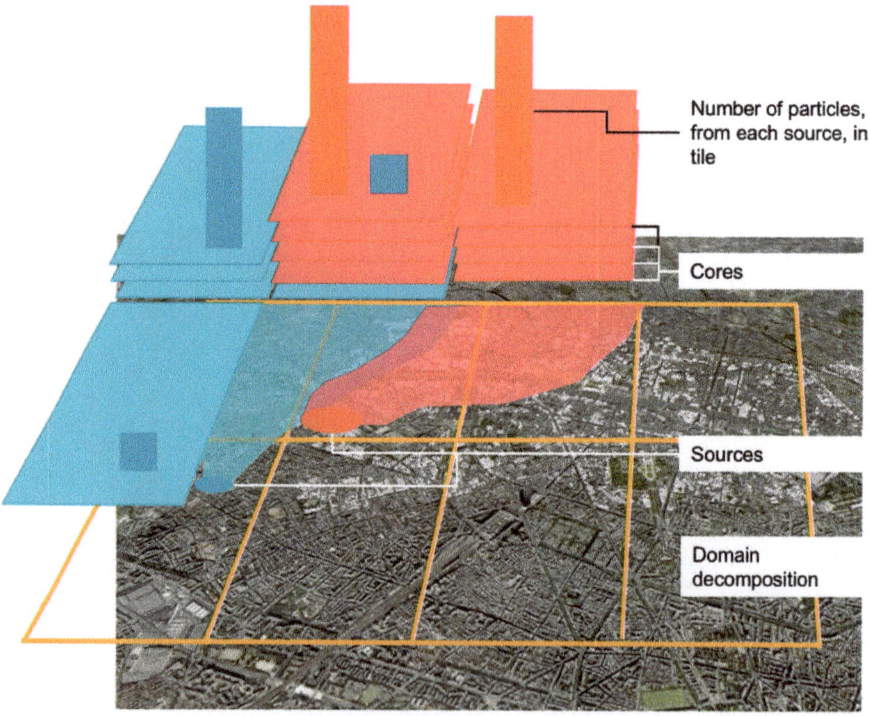

(**b**)

Figure 1. Example of parallel settings: (**a**) domain decomposition and time frame parallel treatment for the PSWIFT model using 17 cores on a domain decomposed into 8 tiles; (**b**) domain decomposition and particle distribution for the PSPRAY model.

The PMSS modeling system has also the ability to handle multiple nested domains. The PSPRAY model can hence use a nested domain computed by the PWIFT flow model or computed by another flow model such as, in our case, Code_Saturne (see Section 2.2). The results of this flow model on the nested domain must be stored in the same binary format as a PSWIFT calculation, and they must contain at least wind velocity and turbulence fields.

The PMSS modeling system has been validated both in scalar mode [22,26,27] and regarding parallel algorithms [17,20]. The parallel testing was performed in computational environments ranging from a multicore laptop to several hundred cores in a high-performance computing center.

2.2. Presentation of the Code_Saturne Model

In an urban area, one cannot exclude transfers of pollutants from outside to the inner part of buildings or, vice versa, pollution originating in a building and being transferred outside. Neither the diagnostic nor the momentum version of PSWIFT is appropriate to model the flow inside a building. Thus, computational fluid dynamics (CFD) is needed; more specifically, we used the Code_Saturne model [29], an open-source general-purpose and environmental application-oriented CFD model.

The Code_Saturne model (http://code-saturne.org, accessed on 17 May 2021) is an open-source CFD model developed at EDF R&D. Based on a finite volume method, it simulates incompressible or compressible laminar and turbulent flows in complex 2D and 3D geometries. Code_Saturne solves the RANS equations for continuity, momentum, energy, and turbulence. The turbulence modeling uses eddy-viscosity model or second-moment closure, such as k-epsilon. The time-marching scheme is based on a prediction of velocity followed by a pressure correction step. Additional details are provided in [29].

The Code_Saturne model also has an atmospheric option [7]. The model has been used extensively, not only for atmospheric flow but also to model the flow within buildings, including the complicated setup of an intensive care hospital room [30].

2.3. The Computing Infrastructure

Modeling the flow and dispersion at high resolution in very large domains of several tens of kilometers of extent requires parallel domain decomposition using several hundreds or even thousands of cores of a supercomputer.

The simulations were carried out on a supercomputer consisting of 5040 B510 bull-X nodes, each with two eight-core Intel Sandy Bridge EP (E5-2680) processors at 2.7 GHz and with 64 GB of memory. The network is an InfiniBand QDR Full Fat Tree network. The file system offers 5 PB of disk storage.

3. The EMERGENCIES Project

The EMERGENCIES project aims to demonstrate the operational feasibility of accelerated time tracking of toxic atmospheric releases, be they accidental or intentional, in a large city and its buildings through 3D numerical simulation.

After describing the domain setup, we present the flow modeling and then the release scenarios.

3.1. Domain Setup

The modeling domain covers Greater Paris: it includes the City of Paris, the Hauts-de-Seine, the Seine-Saint-Denis, and the Val-de-Marne (see Figure 2). It extends to the airports of Orly, to the south, and of Roissy Charles de Gaulle, to the north. This geographic area is under the authority of the Paris Fire Brigade.

Figure 2. View of the domain boundaries (green line) and the department boundaries (blue line) (© OpenStreetMap Contributors).

The simulation domain has a uniform horizontal resolution of 3 m and dimensions of 38.4 km × 40.8 km, leading to 12,668 × 13,335 points in a horizontal plane. The vertical grid has 39 points from the ground up to 1000 m, the resolution near the ground being 1.5 m. The grid, therefore, has more than 6 billion points.

The static data used for the modeling consists of the following:

- The topography at an original resolution of 25 m taken from the BD Topo of the French National Geographical Institute (IGN);
- The land use data with the same gridded resolution but obtained from the CORINE Land Cover data [31];
- The buildings data, taken also from the BD Topo of IGN, consisting of 1.15 million polygons with height attributes.

These data have a disk footprint ranging from 1 GB for the topography or land use to 3 GB for the building data. Figure 3 presents a view of the topography and the building data.

Figure 3. 3D view of the topography and buildings.

This very large domain is treated using domain decomposition: the domain is divided into horizontal tiles of 401 × 401 horizontal grid points (see Figure 4). In total, 1088 tiles are used to cover the domain, 32 along the west–east axis, and 34 along the south–north axis.

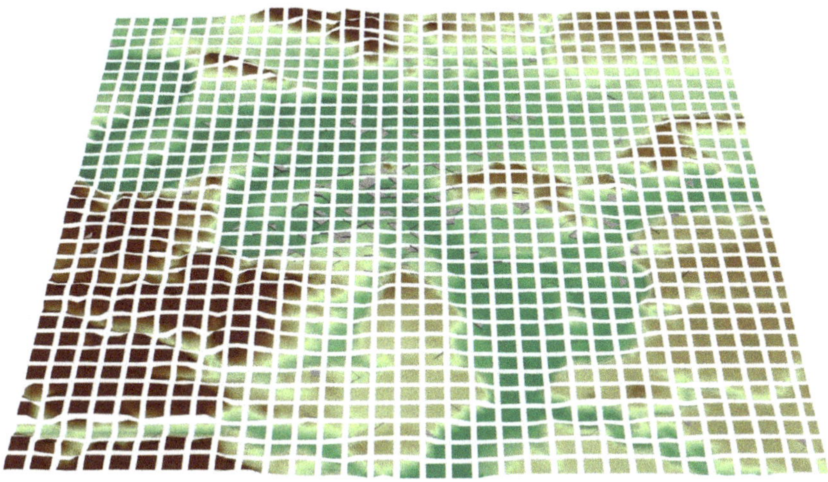

Figure 4. Domain decomposition of the domain in 1088 tiles.

Three buildings of particular interest have been chosen: a museum (M), a train station (TS), and an administration building (A). Their locations are presented in Figure 5. They have been selected to demonstrate the capability of the system to provide relevant information regarding releases either propagating from inner parts of buildings to the outside or occurring in the urban environment and then penetrating buildings. Three nested domains around M, TS, and A have been defined using a grid with a very high horizontal and vertical resolution both inside the buildings and in their vicinity. The characteristics of these nested domains are summarized in Table 1. The grid size is about 1 m horizontally and vertically.

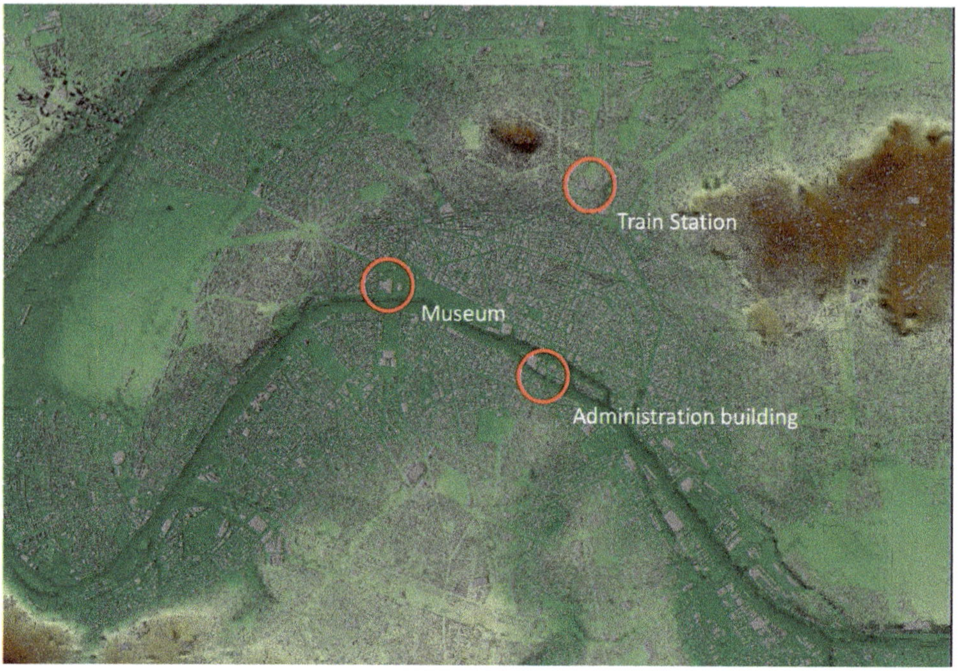

Figure 5. Aerial view zoom of the city of Paris with the specific buildings of interest (red circles).

Table 1. Nested domain main characteristics.

Domain	M	TS	A
Domain extension (m)	390 × 350 × 100	385 × 290 × 100	230 × 170 × 100
Number of vertical grid levels	42	43	43
Number of cells (million)	5.6	4.6	2.6

3.2. Flow Simulations

The flow accounting for the buildings in the huge EMERGENCIES domain is downscaled from the forecasts of the mesoscale Weather Research and Forecasting (WRF) model [32]. These simulations run every day and require 100 computing cores for 2 h 10 min for 72 h simulated.

The WRF forecasts are extracted every hour on vertical profiles inside the huge urban simulation domain and provided as inputs to the PMSS modeling system. Thus, 24 timeframes of the microscale 3D flow around buildings are computed by PSWIFT. Finally, the inflow conditions for flow and turbulence are extracted on the boundaries of the very high resolution nested domain to perform the Code_Saturne simulations.

These calculations can be performed routinely every day provided the computing resource is available. The flow and turbulence fields are then available every day in advance both for the full domain and the chosen nests. Accidental scenarios can then be simulated on an on-demand basis.

3.3. Dispersion Simulations

In the framework of the project, fictitious multiple attacks consisting of the releases of substances with potential health consequences were considered. These substances could be radionuclides, chemical products, or pathogenic biological agents. Substances were assumed to be released inside or near the public buildings introduced in the previous section. All sources are fictitious point releases of short durations occurring on the same day (taken as an example), in real meteorological conditions.

The day for the release was chosen arbitrarily to be 23 October 2014. The releases were carried out in the morning, starting at 10 a.m. for the first one. The simulated period for the dispersion is a 5-h duration period between 10 a.m. and 3 p.m.

The flow conditions for 23 October 2014 show an average wind speed of 3 m/s at 20 m above ground with winds coming from the southwest.

Each release has a 10 min duration, and the material emitted is considered to be a gas, without density effect, or fine particulate matter, i.e., with aerodynamic diameters below 2.5 µm. One kilogram of the material is emitted during each release. The first release occurs at 10 a.m. local time in the museum, the second one occurs at 11 a.m. inside the administration building, and the last one starts at 12 a.m. in front of the train station.

All these releases are purely fictional, but the intention is to try to apply the modeling system to a scenario that could be faced by emergency teams and first responders.

The nested modeling is handled in a two-way mode by the LPDM PSPRAY model within the PMSS modeling system: numerical particles move from the nested domains around the buildings where the attacks take place to the large domain encompassing the Paris area or, vice versa, from this large domain back to a nested domain.

4. Results and Discussion

In this section, the results obtained in the framework of the EMERGENCIES project are presented and discussed, taking into account the characteristics of the simulations and the associated computational resources (duration, CPU required, and storage). After describing the results for the flow simulations, we then present the results obtained for the dispersion before briefly discussing the constraints in rapidly and efficiently visualizing very large results.

4.1. Flow Simulations

After presenting the flow results for the large domain and the nested domains, we discuss the operating applicability of the system developed in the frame of the EMERGENCIES project.

The calculation for the flow on the whole domain using the PMSS modeling system requires a minimum of 1089 computing cores, i.e., one core per tile plus a master core. Time frames can be simulated in parallel. Several parallel configurations have been tested and are presented in Table 2. An example of the level of precision of the flow obtained throughout the whole domain is presented in Figure 6.

Table 2. Duration of simulations for the flow on the large domain.

Number of Time Frames in Parallel	Number of Cores	Simulation Duration (min)
1	1089	161
4	4353	92
8	8705	82

The duration ranges from around 2 h 40 min using the minimum number of cores down to around 1 h 20 min when computing eight timeframes concurrently. Regarding the storage, a single time frame requires 200 GB for the whole domain, leading to around 8 TB for 24 timeframes.

When considering the simulation for the nested domains, the Code_Saturne CFD model has been set up using around 200 cores per nested domain and per time frame. It requires 14,400 cores to handle the 24 timeframes for each of the three domains. The simulation duration is then 100 min for the domain M, 69 min for the domain TS, and 64 min for the domain A. The storage per time frame is 281, 232, and 80 MB, respectively. The storage for the detailed computations in and around the buildings is low compared to the large urban domain covering Greater Paris, contrarily to the computational costs. Indeed, the CFD model is much more computationally intensive than the diagnostic approach used in the PMSS modeling system. An example of the flow both in the close vicinity of and inside the administration building is presented in Figure 7.

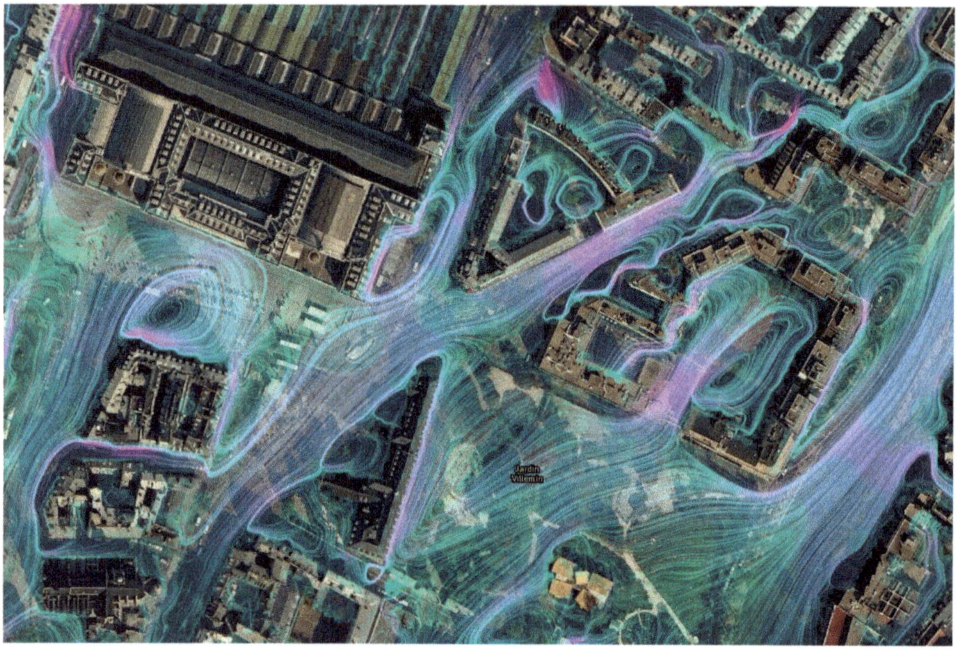

Figure 6. Closeup view of the flow streamlines computed in the streets south and east of the train station. The wind speed ranges from 0 m/s in cyan to 2 m/s in magenta.

The flow modeling is intended for use on a daily basis: each day, a high-resolution flow modeling over the whole Paris area is simulated to be made available should any dispersion modeling of accidental or malevolent airborne releases be necessary.

After the WRF forecast modeling, around 3 h down to no more than 1 h 20 min is required to downscale the flow on the large Parisian domain at 3 m resolution, depending on the number of cores dedicated, respectively 1089 and 8705.

Then, around 5000 cores are required to handle the 24 timeframes of each nest in less than 1 h 40 min. If the three nested domains at very high resolution are treated, around 10,000 cores need to be available for 2 h: 5000 cores available each hour to handle the domain TS and then the domain A and 5000 cores to handle the domain M for 1 h 40 min.

Hence, the whole microscale flow modeling could be performed in 3 h 20 min using 10,000 cores.

4.2. Dispersion Simulations

After presenting the dispersion results for the large domain and the nested domains, we discuss the operating applicability of the system developed in the framework of the EMERGENCIES project.

Figure 7. 3D view of the flow streamlines for the nested domain both in the vicinity of and inside the administration building. The wind speed ranges from 0 m/s in deep blue to 3.5 m/s in orange.

The dispersion modeling benefits from two-way nested computations: numerical particles can move from the inner-most 1 m resolution domains to the large domain over Paris and its vicinity, and vice versa. At each 5 s emission step, 40,000 numerical particles are emitted. Since there are three 10 min-long releases, 14.4 million particles are emitted. A parametric study regarding the number of cores is presented in Table 3. Concentrations are calculated every 10 min using 10 min averages.

Table 3. Influence of the number of cores on the duration of the dispersion simulation.

Number of Cores	Simulation Duration (min)
250	150
500	88
750	96
1000	120

Five hundred cores were retained for operational use. The simulation results produce 90 GB for 5 simulated hours. Eighty tiles of the large domain were reached by the plumes during the calculation, plus the three nested domains.

Figure 8 illustrates the evolution of the plume inside, in the vicinity of, and further away from the museum.

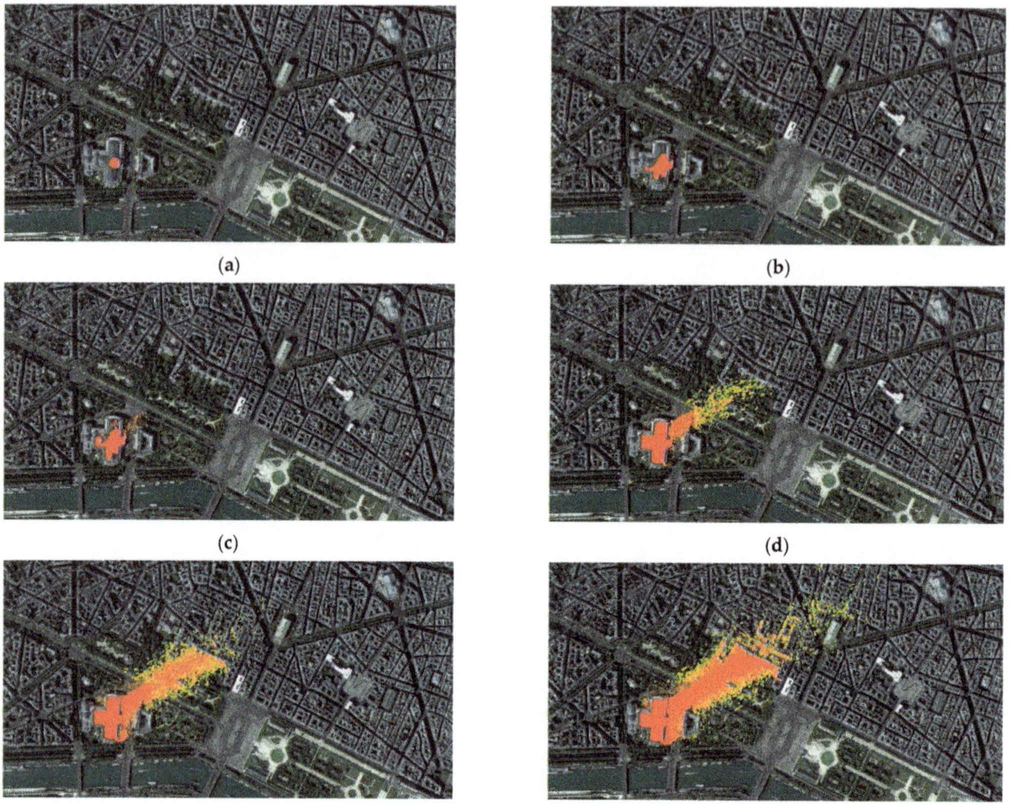

(a) (b) (c) (d) (e) (f)

Figure 8. View of the plume inside and in the vicinity of the museum: (**a**) 1 min after the release; (**b**) 3 min; (**c**) 4 min; (**d**) 6 min (**e**) 8 min; (**f**) 10 min. The change from the 1 m resolution of the inner domain to the 3 m resolution of the large domain can be noticed. The threshold for 500 $\mu g/m^3$ of concentration is in red, while that for 1 $\mu g/m^3$ is in orange and that for 0.001 $\mu g/m^3$ is in green.

In an operational context, the dispersion simulation would be activated on demand. Using 500 cores, the duration of the calculation, one and a half hours, has to be compared to the physical duration of five hours simulated here. The simulation is thus 3.3 times faster than the reality. Moreover, the results can be analyzed on the fly, without requiring the whole simulation to be completed.

4.3. Visualization of the Simulation Results

Due to the large amount of data generated, the efficient visualization of data for operational use has proved to be a challenge in itself. After introducing a first attempt using a traditional scientific visualization application, we present a short overview of a more operative approach [33].

3D modeling results can be explored using 3D scientific visualization applications. Due to the large quantity of data, the viewer chosen has to handle heavy parallelization. We chose the open-source software ParaView and developed a dedicated plugin. Nonetheless, interactive visualization using several hundred cores required several seconds or tens of seconds for refreshing: this is too ineffective for a user to navigate the results. Non-interactive image rendering has also been used (see Figure 1), but it would be ineffective for the user in an emergency situation.

Hence, we relied finally on a tiled web map approach. The results of the simulation are processed on the fly as results are produced [33]. Multiple vertical levels can be treated to provide views not solely on the ground but also in the volume at multiple heights above the ground. Processing takes a few minutes for each time frame since it consists only of a change of file format. Then, results can be consulted in real time through a tile map service using any web browser. Figures 6, 8 and 9 use such an approach.

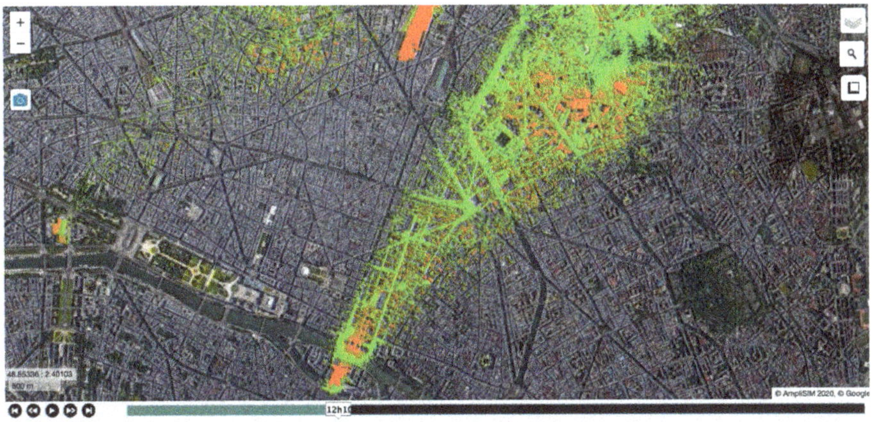

Figure 9. View of the three release locations and of the plumes, at the ground level, using tiled navigation. The threshold for 0.01 μg/m^3 of concentration is in green, while that for 1 μg/m^3 is in red (background satellite image © TerraMetrics, LLC—www.terrametrics.com © Google 2021, accessed on 30 April 2021).

5. Conclusions

The EMERGENCIES project proved it is feasible to provide emergency teams and first responders with high-resolution simulation results to support field decisions. The project showed that the areas of responsibility of several tens of kilometers of length can be managed operationally. The domain, at a 3 m resolution, includes very high resolution modeling for specific buildings of interest where both the outside and the inside of the building have been modeled at 1 m resolution.

The modeling is carried out in two steps: First, the modeling of the flow is computed in advance starting from mesoscale forecasts provided each day for the following day. Then, these flows are used for on-demand modeling of dispersion that can occur everywhere in the domain and at multiple locations.

The modeling of the flow for the 24 timeframes can be achieved in 3 h 20 min using 10,000 cores both for the large domain and the three nested domains. If 5000 cores are available, the large domain requires 1 h 30 min and each nested domain requires an additional duration ranging between 1 h and 1 h 40 min. Five hundred cores are then required to be ready for on-demand modeling of the dispersion: the dispersion is modeled 3.3 times faster than it occurs in real life.

This amount of computer power is large but not inconsistent with the means usually dedicated to crisis management.

The dispersion requires a lower amount of computing cores than the modeling of the flow. This allows the modeler to perform several dispersion calculations concurrently, particularly to create alternate scenarios for the source terms. This is of particular interest considering the uncertainties related to the source term estimation. Successive scenarios may also be considered as information is updated and the comprehension of the situation by the emergency teams improves.

A large amount of 3D data was produced: 200 GB for the flow and 90 GB for the dispersion. While data produced near the ground are of the uttermost importance, the vertical distribution of concentration from the ground to a moderate elevation is also particularly relevant. Firstly, calculations of inflow and outflow exchanges for specific buildings of interest require such information. Then, the vertical distribution of concentration can support decisions to design evacuation strategies for specific buildings, such as moving people toward the building rooftop rather than evacuating them at the ground level into the neighboring streets.

One key aspect also put forward during the project is the difficulty of efficiently visualizing this huge amount of data produced. After focusing initially on a parallel scientific visualization application, we relied on on-the-fly data processing to create tiled web maps for the flow and the dispersion. These data can then be accessed in real time by the emergency team. This approach to visualizing data will be the focus of a dedicated paper.

Author Contributions: Conceptualization, O.O. and P.A.; Data curation, C.D., S.P. and M.N.; Investigation, O.O., C.D., S.P. and M.N.; Methodology, O.O., P.A. and S.P.; Resources, C.D., S.P. and M.N.; Supervision, O.O. and P.A.; Visualization, S.P.; Writing—original draft, O.O.; Writing—review & editing, O.O. and P.A. All authors have read and agreed to the published version of the manuscript.

Funding: This research received no external funding.

Conflicts of Interest: The authors declare no conflict of interest.

References

1. McHugh, C.A.; Carruthers, D.J.; Edmunds, H.A. ADMS-Urban: An air quality management system for traffic, domestic and industrial pollution. *Int. J. Environ Pollut.* **1997**, *8*, 666–674.
2. Schulman, L.L.; Strimaitis, D.G.; Scire, J.S. Development and evaluation of the PRIME plume rise and building down-wash model. *J. Air Waste Manag. Assoc.* **2000**, *5*, 378–390. [CrossRef] [PubMed]
3. Sykes, R.I.; Parker, S.F.; Henn, D.S.; Cerasoli, C.P.; Santos, L.P. *PC-SCIPUFF Version 1.3—Technical Documentation*; Technical Report 725; Titan Corporation, ARAP Group: Princeton, NJ, USA, 2000.
4. Hanna, S.; Baja, E. A simple urban dispersion model tested with tracer data from Oklahoma City and Manhattan. *Atmos. Environ.* **2009**, *43*, 778–786. [CrossRef]
5. Biltoft, C.A. *Customer Report for Mock Urban Setting Test*; DPG Document Number 8-CO-160-000-052; Defense Threat Reduction Agency: Alexandria, VA, USA, 2001.
6. Warner, S.; Platt, N.; Heagy, J.F.; Jordan, J.E.; Bieberbach, G. Comparisons of Transport and Dispersion Model Predictions of the Mock Urban Setting Test Field Experiment. *J. Appl. Meteorol. Climatol.* **2006**, *45*, 1414–1428. [CrossRef]
7. Milliez, M.; Carissimo, B. Numerical simulations of pollutant dispersion in an idealized urban area, for different meteorological conditions. *Bound. Layer Meteorol.* **2006**, *122*, 321–342. [CrossRef]
8. Rolph, G.; Stein, A.; Stunder, B. Real-time Environmental Applications and Display System: READY. *Environ. Modell. Softw.* **2017**, *95*, 210–228. [CrossRef]
9. Leitl, B.; Trini Castelli, S.; Baumann-Stanzer, K.; Reisin, T.G.; Barmpas, P.; Balczo, M.; Andronopoulos, S.; Armand, P.; Jurcakova, K.; Milliez, M. Evaluation of air pollution models for their use in emergency response tools in built environments: The 'Michelstadt' Case study in COST ES1006 action. In *Air Pollution Modelling and Its Application XXIII*; Springer: Cham, Switzerland, 2014; pp. 395–399.
10. Armand, P.; Bartzis, J.; Baumann-Stanzer, K.; Bemporad, E.; Evertz, S.; Gariazzo, C.; Gerbec, M.; Herring, S.; Karppinen, A.; Lacome, J.M.; et al. *Best Practice Guidelines for the Use of the Atmospheric Dispersion Models in Emergency Response Tools at Local-Scale in Case of Hazmat Releases into the Air*; Technical Report COST Action ES 1006; University of Hamburg: Hamburg, Germany, 2015.
11. Gowardhan, A.A.; Pardyjak, E.R.; Senocak, I.; Brown, M.J. A CFD-based wind solver for urban response transport and dispersion model. *Environ. Fluid Mech.* **2011**, *11*, 439–464. [CrossRef]
12. Yee, E.; Lien, F.S.; Ji, H. *A Building-Resolved Wind Field Library for Vancouver: Facilitating CBRN Emergency Response for the 2010 Winter Olympic Games*; Defense Research and Development: Suffield, AB, Canada, 2010.
13. Patnaik, G.; Moses, A.; Boris, J.P. Fast, Accurate Defense for Homeland Security: Bringing High-Performance Computing to First Responders. *J. Aerosp. Comput. Inf. Commun.* **2010**, *7*, 210–222. [CrossRef]
14. Röckle, R. Bestimmung der Strömungsverhaltnisse im Bereich Komplexer Bebauungs-Strukturen. Ph.D. Thesis, Technical University Darmstadt, Darmstadt, Germany, 1990.
15. Kaplan, H.; Dinar, N. A Lagrangian dispersion model for calculating concentration distribution within a built-up domain. *Atmos. Environ.* **1996**, *30*, 4197–4207. [CrossRef]
16. Rodean, H.C. *Stochastic Lagrangian Models of Turbulent Diffusion*; American Meteorological Society: Boston, MA, USA, 1996; Volume 45.
17. Oldrini, O.; Armand, P.; Duchenne, C.; Olry, C.; Tinarelli, G. Description and preliminary validation of the PMSS fast response parallel atmospheric flow and dispersion solver in complex built-up areas. *J. Environ. Fluid Mech.* **2017**, *17*, 1–18. [CrossRef]

18. Singh, B.; Pardyjak, E.R.; Norgren, A.; Willemsen, P. Accelerating urban fast response Lagrangian dispersion simulations using inexpensive graphics processor parallelism. *Environ. Modell. Softw.* **2011**, *26*, 739–750. [CrossRef]
19. Oldrini, O.; Olry, C.; Moussafir, J.; Armand, P.; Duchenne, C. Development of PMSS, the Parallel Version of Micro SWIFT SPRAY. In Proceedings of the 14th International Conference on Harmonisation within Atmospheric Dispersion Modelling for Regulatory Purposes, Kos Island, Greece, 2–6 October 2011; pp. 443–447.
20. Oldrini, O.; Armand, P.; Duchenne, C.; Perdriel, S. Parallelization Performances of PMSS Flow and Dispersion Modeling System over a Huge Urban Area. *Atmosphere* **2019**, *10*, 404. [CrossRef]
21. Oldrini, O.; Armand, P. Validation and Sensitivity Study of the PMSS Modelling System for Puff Releases in the Joint Urban 2003 Field Experiment. *Bound. Layer Meteorol.* **2019**, *171*, 513–535. [CrossRef]
22. Moussafir, J.; Oldrini, O.; Tinarelli, G.; Sontowski, J.; Dougherty, C. A new operational approach to deal with dispersion around obstacles: The MSS (Micro-Swift-Spray) software suite. In Proceedings of the 9th International Conference on Harmonisation within Atmospheric Dispersion Modelling for Regulatory Purposes, Garmisch-Partenkirchen, Germany, 1–4 June 2004; Volume 2, pp. 114–118.
23. Tinarelli, G.; Brusasca, G.; Oldrini, O.; Anfossi, D.; Trini Castelli, S.; Moussafir, J. Micro-Swift-Spray (MSS) a new modelling system for the simulation of dispersion at microscale. General description and validation. In *Air Pollution Modelling and Its Applications XVII*; Borrego, C., Norman, A.N., Eds.; Springer: Cham, Switzerland, 2007; pp. 449–458.
24. Oldrini, O.; Nibart, M.; Armand, P.; Olry, C.; Moussafir, J.; Albergel, A. Introduction of Momentum Equations in Micro-SWIFT. In Proceedings of the 15th International Conference on Harmonisation within Atmospheric Dispersion Modelling for Regulatory Purposes, Madrid, Spain, 6–9 May 2013.
25. Oldrini, O.; Nibart, M.; Duchenne, C.; Armand, P.; Moussafir, J. Development of the parallel version of a CFD—RANS flow model adapted to the fast response in built-up environments. In Proceedings of the 17th International Conference on Harmonisation within Atmospheric Dispersion Modelling for Regulatory Purposes, Budapest, Hungary, 9–12 May 2016.
26. Anfossi, D.; Desiato, F.; Tinarelli, G.; Brusasca, G.; Ferrero, E.; Sacchetti, D. TRANSALP 1989 experimental campaign—II. Simulation of a tracer experiment with Lagrangian particle models. *Atmos. Environ.* **1998**, *32*, 1157–1166. [CrossRef]
27. Tinarelli, G.; Mortarini, L.; Trini Castelli, S.; Carlino, G.; Moussafir, J.; Olry, C.; Armand, P.; Anfossi, D. Review and validation of Micro-Spray, a Lagrangian particle model of turbulent dispersion. In *Lagrangian Modelling of the Atmosphere, Geophysical Monograph*; American Geophysical Union (AGU): Washington, DC, USA, 2013; Volume 200, pp. 311–327.
28. Thomson, D.J. Criteria for the selection of stochastic models of particle trajectories in turbulent flows. *J. Fluid Mech.* **1987**, *180*, 529–556. [CrossRef]
29. Archambeau, F.; Méchitoua, N.; Sakiz, M. Code Saturne: A finite volume code for the computation of turbulent incompressible flows-Industrial applications—Industrial Applications. *Int. J. Finite Vol.* **2004**, *1*, hal-01115371.
30. Beauchêne, C.; Laudinet, N.; Choukri, F.; Rousset, J.L.; Benhamadouche, S.; Larbre, J.; Chaouat, M.; Benbunan, M.; Mimoun, M.; Lajonchère, J.; et al. Accumulation and transport of microbial-size particles in a pressure protected model burn unit: CFD simulations and experimental evidence. *BMC Infect. Dis.* **2011**, *11*, 58. [CrossRef]
31. Bossard, M.; Feranec, J.; Otahel, J. *CORINE Land Cover Technical Guide—Addendum 2000*; Technical Report No. 40; European Environment Agency: Copenhagen, Denmark, 2000.
32. Skamarock, W.C.; Klemp, J.B.; Dudhia, J.; Gill, D.O.; Barker, D.; Duda, M.G.; Huang, X.-Y.; Wang, W.; Powers, J.G. *A Description of the Advanced Research WRF Version 3*; No. NCAR/TN-475+STR; University Corporation for Atmospheric Research: Boulder, CO, USA, 2008; p. 113.
33. Oldrini, O.; Perdriel, S.; Armand, P.; Duchenne, C. Web visualization of atmospheric dispersion modelling applied to very large calculations. In Proceedings of the 18th International Conference on Harmonisation within Atmospheric Dispersion Modelling for Regulatory Purposes HARMO, Bologna, Italy, 9–12 October 2017; Volume 18, pp. 9–12.

atmosphere

MDPI

Article

High-Speed Visualization of Very Large High-Resolution Simulations for Air Hazard Transport and Dispersion

Olivier Oldrini [1,*], Sylvie Perdriel [2], Patrick Armand [3] and Christophe Duchenne [3]

1 MOKILI, F-75014 Paris, France
2 AmpliSIM, F-75014 Paris, France; info@amplisim.com
3 CEA, DAM, DIF, F-91297 Arpajon, France; patrick.armand@cea.fr (P.A.); christophe.duchenne@cea.fr (C.D.)
* Correspondence: info@mokili.fr

Abstract: In the case of an atmospheric release of a noxious substance, modeling remains an essential tool to assess and forecast the impact of the release. The impact of such situations on populated, and hence built-up, areas is of the uttermost importance. However, modeling on such areas requires specific high-resolution approaches, which are complex to set up in emergency situations. Various approaches have been tried and evaluated: The EMERGENCIES and EMED project demonstrated an effective strategy using intensive parallel computing. Large amounts of data were produced that proved initially to be difficult to visualize, especially in a crisis management framework. A dedicated processing has been set up to allow for rapid and effective visualization of the modeling results. This processing relies on a multi-level tiled approach initiated in web cartography. The processing is using a parallel approach whose performances were evaluated using the large amounts of data produced in the EMERGENCIES and EMED projects. The processing proved to be very effective and compatible with the requirements of emergency situations.

Keywords: operational emergency modeling; atmospheric release; high-resolution metric grid; web visualization; web mapping; emergencies project

Citation: Oldrini, O.; Perdriel, S.; Armand, P.; Duchenne, C. High-Speed Visualization of Very Large High-Resolution Simulations for Air Hazard Transport and Dispersion. *Atmosphere* **2021**, *12*, 920. https://doi.org/10.3390/atmos12070920

Academic Editor: Ashok Luhar

Received: 7 June 2021
Accepted: 13 July 2021
Published: 17 July 2021

Publisher's Note: MDPI stays neutral with regard to jurisdictional claims in published maps and institutional affiliations.

1. Introduction

In the context of crisis management in case of atmospheric release, such as accidental or malevolent releases, numerical simulation can be an important asset [1]. Locations with high population densities are where the simulation capability is the most critical, in particular due to the number of potential casualties. Such locations are usually built-up areas. Nonetheless, built-up areas require specific and precise modeling due to complex flow and dispersion patterns [2–6].

Modified Gaussian approaches are already largely in use in decisions support systems in the case of emergency situations, see for instance ALOHA [7], or SCIPUFF [8] within HPAC [9]. They offer a capability to handle in tens of minutes the global impact of built-up areas on the flow and turbulence, and hence for phenomena at a minimum spatial scale of several tens of meters. They offer indeed limited capabilities regarding the complexity of the flow and dispersion, especially in the vicinity of the release, where the effects are the most acute. They also lack, further away from the release, important effects of the buildings such as entrapment of the pollutants, for instance in the canyon streets, or change in the plume direction due to the building pattern [10]. On the contrary, the computational fluid dynamic (CFD) often succeeds in accurately describing complicated patterns of flow and dispersion. Still, Reynolds-averaged Navier–Stokes (RANS) models and, more importantly, large eddy simulation (LES) models may require very large computational times compared to Gaussian plume approaches. Such difficulties may be solved on small setups by relying on heavy parallel computing and optimization [11] or precomputation [12].

Nonetheless, responsibility areas of emergency teams are usually quite large. Relying on shared expertise with the emergency practitioners during exercises [13], we developed a capability to use complex 3-dimensional (3D) modeling that takes buildings explicitly into account in such large areas. This capability was demonstrated during the projects EMERGENCIES [14] and Emergencies-MEDiterranean (EMED [15]. The responsibility areas of the Paris Fire Brigade and of the Marseille area Fire Brigade (Bouches-du-Rhône) used in the EMERGENCIES and EMED projects, respectively, are very large areas of several tens of kilometers of extension from north to south and west to east. Using a high-resolution 3D approach, the projects generated large amount of data. However, at the same time, in the framework of crisis management, the projects also illustrated the necessity to provide the emergency teams with rapid visualizations able to describe the simulation results both at the local and global scale. Indeed, global views of the simulation results on the whole domain proved to be impossible to obtain using on the shelf traditional scientific viewing. This was all the more an issue in the framework of crisis management where rapidity and efficiency are of the essence.

Initial attempts relied on parallel scientific viewer usage with specific developments to handle the modeling system specifics. However, while being able to visualize the results as a whole, it was not applicable in practice due to the lengthy and non-interactive process. In order to tackle this challenge, we introduced a web-based multi-zoom tiled approach relying on parallel treatment of massive simulations results. This kind of approach has been introduced in the field of web mapping services to handle maps with various levels of details and very large dimensions, see for instance [16].

The paper is organized as follows: Section 2 contains a brief description of the EMERGENCIES and EMED projects and the modeling data to visualize. Section 3 introduces the approach, including the parallel treatments of the results to support the subsequent web visualization, while Section 4 presents and comments on the results. Section 5 draws conclusions on the potential use of this approach for emergency preparedness and response in case of an atmospheric release.

2. The EMERGENCIES and EMED Project

After offering an overview of the projects and the associated modeling, we describe the data produced.

2.1. Overview of the Projects

The EMERGENCIES and EMED projects are dedicated at demonstrating the operational capability, for crisis management, of high-resolution modeling required in built-up areas. As such, the modeling domain has been chosen according to the responsibility areas of actual emergency teams, namely the Fire Brigades of Paris, Marseille, Toulon, and Nice.

These urban areas have geographical extension of up to several tens of kilometers. To be able to model such areas at almost metric resolution, specific modeling tools and computing clusters were used. After giving an overview of the modeling and computing capabilities, we summarize the setup of each project.

2.1.1. Modeling and Computing Capabilities

The PMSS modeling system (see [17–20]) has been used to model the flow and dispersion on domain with extension of several tens of kilometers and horizontal grid resolution of 3 m. The modeling system consists of the individual PSWIFT and PSPRAY models. The PSWIFT model [21–24] is a flow and turbulence parallel, 3D, mass-consistent, and terrain-following diagnostic model that uses a Röckle-type [25] approach to take into account buildings. It also incorporates an optional fast momentum solver. The dispersion is handled by the PSPRAY model [22,26,27] using the flow and turbulence computed by the PSWIFT model. The PSRPAY model is a parallel Lagrangian particle dispersion model (LPDM) [28] that takes into account obstacles. It simulates the dispersion of an airborne contaminant by following the trajectories of numerous numerical particles. The velocity of each virtual particle is the sum of a transport and turbulent component, the latest being derived from the stochastic scheme developed by Thomson [29] that solves a 3D form of the Langevin equation. While not being a source term model, the PSPRAY model treats complex releases, including elevated and ground-level emissions, instantaneous and continuous emissions, or time-varying sources. Additionally, it is able to deal with plumes with initial arbitrarily oriented momentum, negative or positive buoyancy, radioactive decay, and cloud spread at the ground due to gravity.

The parallel scheme of the PMSS modeling system allows domain decomposition to be performed. It is particularly adapted to large domains while taking into account explicitly buildings.

Very high-resolution modeling at 1-m resolution inside and in close vicinity of specific buildings of interest was performed using the Reynolds-averaged Navier–Stokes (RANS) computational fluid dynamic (CFD) model Code_Saturne [30]. Based on a finite volume method, it simulates incompressible or compressible laminar and turbulent flows in complex 2D and 3D geometries. Code_Saturne solves the RANS equations for continuity, momentum, energy and turbulence.

Calculations were performed on a supercomputer consisting of 5040 B510 bull-X nodes, each with 2 eight cores Intel Sandy Bridge EP (E5-2680) processors at 2.7 GHz and with 64nGB of memory. The network is an InfiniBand QDR full-fat tree network. The file system offers 5 PB of disk storage.

The approach is a two-step approach:

- Flow and turbulence are computed in advance each day for the next, and for the whole domain starting from meso-scale meteorological forecasts computed using the WRF model [31];
- Dispersion is computed on-demand when a situation occurs.

2.1.2. Experimental Setting for the EMERGENCIES Project

The domain covered the Greater Paris area, with an extension of roughly 38 × 41 km^2 (see Figure 1). The horizontal grid resolution was 3 m. The vertical grid had 39 grid points from the ground up to 1000 m with 1.5-m grid resolution near the ground. The grid contained more than 6 billion points.

Figure 1. 3D view of the topography and buildings. North is at the top.

The domain contained three very high-resolution nested domains around specific buildings, a museum, a train station, and an administrative building, with 1-m grid size. The dispersion scenario consisted of three malevolent releases, close or inside the buildings of interest, with a duration of several minutes and occurring during a period of two hours.

The domain was divided into more than 1000 tiles distributed among the computing cores (see Figure 2).

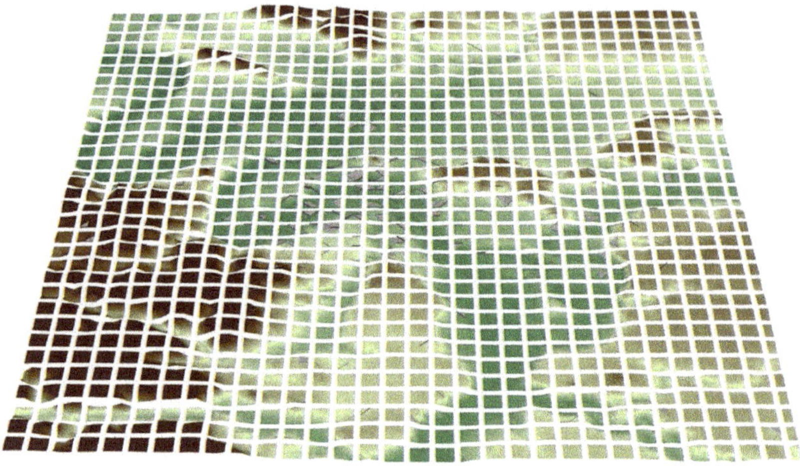

Figure 2. Domain decomposition of the domain in 1088 tiles.

The release scenarios consisted of several 10 min duration releases of one kilogram each. The release could be a gas, without density effects, or fine particulate matter, i.e., with aerodynamic diameters below 2.5 μm.

2.1.3. Experimental Setting for the EMED Project

The EMED project can be considered as a follow-up of the EMERGENCIES project and dedicated as a first step toward an industrialization of the approach.

In this project, three high-resolution domains around three major cities along the French Riviera were modeled: Marseille, Toulon, and Nice. The domain size ranged from 20×16 km^2 around Nice to 58×50 km^2 around Marseille (see Figure 3), each with a horizontal grid resolution of 3 m. The grid for the Marseille domain contained more than 10 billion points.

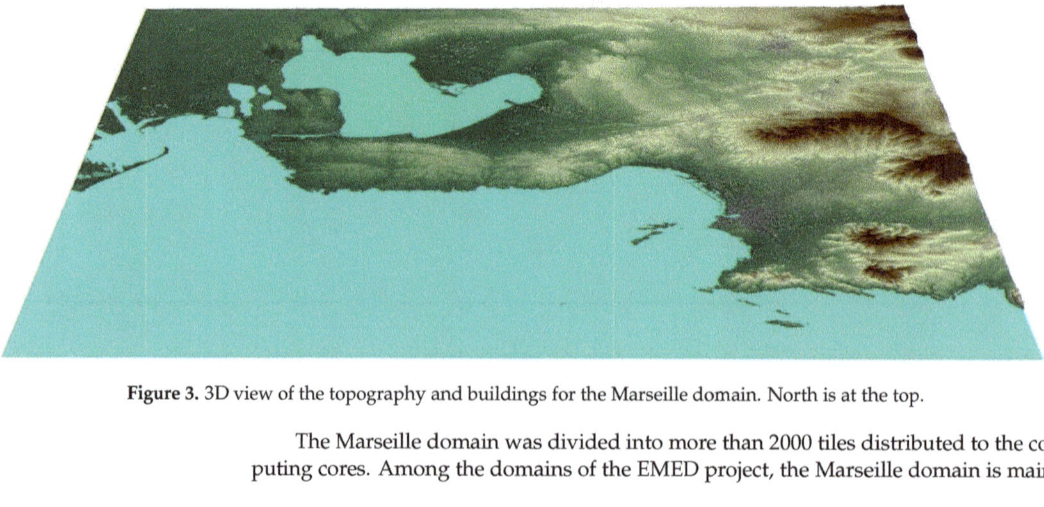

Figure 3. 3D view of the topography and buildings for the Marseille domain. North is at the top.

The Marseille domain was divided into more than 2000 tiles distributed to the computing cores. Among the domains of the EMED project, the Marseille domain is mainly referred to in the following sections since the volume of data generated was larger, and the difficulty to handle them was more acute.

The release scenarios consisted of several 20-min duration releases of one kilogram each. The release could be a gas, without density effects, or fine particulate matter, i.e., with aerodynamic diameters below 2.5 μm.

2.2. Data Produced

For each project, and each domain, the data produced were twofold:

- First the flow and turbulence data (FT data), on each and every tile of the domain;
- Then concentration data (C data), only on tiles reached by the plumes generated by the releases.

For the EMERGENCIES project, three additional nested domains are also available around and inside the buildings of interest.

FT data are produced for a 24-h duration. In the EMERGENCIES project, they are produced every hour, while in the EMED project, they are produced every 15 min. Since the grid contains roughly 13,000 × 13,000 × 39 point in EMERGENCIES, and 19,000 × 16,000 × 39 in EMED for the domain covering Marseille, the data produced for each timeframe are quite large:

- 200 GB per timeframe for the domain covering Paris;
- 668 GB per timeframe for the domain covering Marseille.

These data do not interest directly the team handling the crisis management, with the exception maybe of the flow for the firefighters in the case of a fire associated with a release. Still, the modelers are required to inspect and verify the modeling of the flow and turbulence.

Regarding the transport and dispersion of the release, the size of the C data produced is directly proportional to:

- The area covered by the plume: the larger the plume, the larger the number of tiles that contain C data;
- The persistency duration of the plume in the domain;
- The averaging period used.

The area and persistency are strongly related to the transport and diffusion due to the flow.

Regarding the averaging period, since the transport and dispersion model is an LPDM, to compute a concentration field, particles have to be projected onto the computation grid. This is done during the averaging period. If the averaging period is 10 min, 6 values per grid point are obtained for one hour simulated. If this averaging period is 1 min, 60 values are obtained for one hour simulated. An averaging period of 10 min is sufficient for the needs of a crisis management team. As a modeler, it is nonetheless interesting to produce 1-min averages to obtain a more detailed view of the plume evolution, especially close to the source.

For the EMERGENCIES project, with 10 min averages, 90 GB of concentration data were produced to simulate a period of 5 h. This reaches 900 GB if 1-min averages are used.

For the EMED project, using 10 min averages, 85 GB of concentration data were produced to simulate a period of 4 h. This reaches 850 GB for 1-min averages.

The amount of data produced by the simulation ranges between around 100 GB and several TB. The fact that they were decomposed onto calculation tiles, and also the necessity for them to be rapidly and efficiently explored and manipulated in the context of crisis management led us to develop the approach presented in the following section.

3. Treatment and Visualization of the Data

After presenting the initial attempts for visualization, we introduce the methodology chosen based on multilevel tiled images. Then, we describe its implementation, with a particular focus on the treatment of vector field and the parallel distribution of calculations.

3.1. *Initial Attemtps*

With this large amount of data, the requirement is to be able to zoom in specific locations with a high level of details while at the same time being able to get a good understanding of what is occurring globally, with all this under the constraint of keeping a good level of interactivity.

In the EMERGENCIES project, the ability to produce any visualization of the results over the whole domain was not possible out of the box.

At first, parallel 3D scientific viewing was tried. The open-source data analysis and visualization application PARAVIEW [32] was selected due to its parallel capabilities. A dedicated plugin was developed. The plugin is relying on the computation tiles, i.e., the way the domain is decomposed by the PMSS modeling system: each tile is loaded by a PARAVIEW visualization server, and images can be generated in batch or through interactive views. Hence, the Paris domain required more than 1000 cores, one per PARAVIEW server, and the Marseille domain more than 2000 to operate the flow visualization. While the batch permitted to generate several views of interest (such as Figures 1–3 above), through a lengthy iterative process of blind view setup then view production, it is not interactive. The interactive viewing was tested up to around 120 computation tiles, but it required several tens of second to change a point of view, which restricted its actual usability. It also required a large amount of core to be available at the exact time of the visualization, and during the whole visualization session.

3.2. *Introduction to the Methodology*

The final methodology retained was taken from online cartography and very large high-resolution maps. The idea was to provide different levels of details at different zoom levels and to slice the data into visualization tiles at each level, as illustrated in Figure 4. The aim was to reduce the memory footprint of the data and limit it to the data being actually displayed in the view.

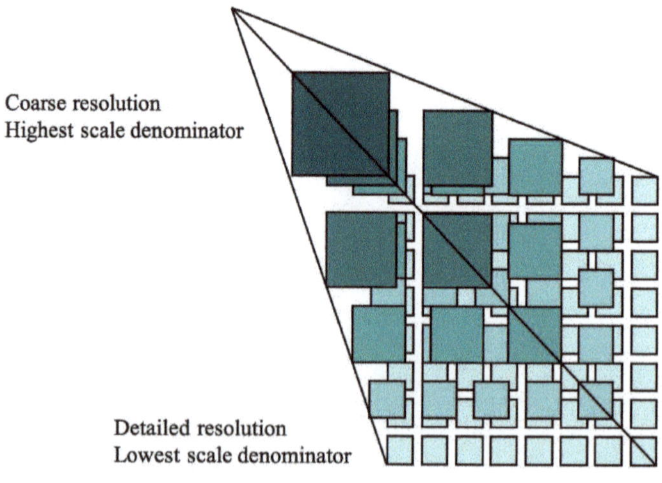

Figure 4. Multi-level tiled, or pyramidal, image (Copyright 2019 Open Geospatial Consortium).

At each zoom level z, the global map for the Earth surface is divided in 2^z visualization tiles, and each tile has a size of 256 × 256 pixels, since each tile is an image, traditionally a portable network graphics (PNG) type. At each zoom level, the tiles are numbered according to their location along the west–east, the x coordinate, and north–south axis, the y coordinate.

From now on, each time we refer to a tile, it is a visualization tile. A tile related to the domain decomposition of the parallel scheme of the model is referred to as a computational tile.

We retained such approach since it enabled fast browsing through a cartographic-type client, while being able to display high level of details when required.

The treatment was performed directly on the computing cluster as a post-processing step performed on the FT or C simulation data. The tiles were then made available through a web server and displayed by a JavaScript client on any web browser.

The post-processing was performed on a set of FT or C data, for a set of time frames and vertical levels. Accessing the actual values of each field, instead of producing a colored image, was required to be able to change the coloring scale or explore vector outputs such as the flow field. Hence, floats were encoded in 4 bytes and stored directly in the PNG files by using the 4 bytes normally distributed to the red, green, blue, and alpha channels.

3.3. Details on the Multilevel Tiling Implementation

The post-processing was performed in parallel using the message passing interface (MPI) standard. After describing the algorithm, we discuss the particular of vector fields such as the wind flow.

3.3.1. Parallel Scheme

The scheme relied on the Tiff file standard [33] and the Geospatial Data Abstraction Library (GDAL) [34]. It used the Tiff virtual stack approach to handle the domain decomposition inherited from the parallel calculation and see the group of Tiff files generated from each calculation tile as multiple parts of a single large Tiff file. It went through the following steps:

- First step, generation of the Tiff files:
    - Distribution of the analysis of available bin files to available cores to retrieve available fields, domain coordinates, available time steps;
    - Calculation by the master core of the large domain footprint, the tiles coordinates required for each zoom levels and the time steps to extract;
    - Generation of Tiff files for each FT or C data file, and for each vertical level and time step selected. The files are generated using the Google Mercator projection;
    - Creation of a Tiff virtual stack encompassing the whole calculation domain, the domain being, or not, decomposed in multiple computation tiles of arbitrary dimension;
- Second step, generation of the tiles from the Tiff files:
    - Loop on zoom levels being treated starting from the larger zoom level;
    - Distribution of each tile, from the total pool of tiles combining field name, time step and vertical level, to a core for generation.

3.3.2. Specificity for Vector Fields

For the vector fields visualization, additional treatment was performed but only on the client side. Indeed, the methodology for the post-processing was similar to the one used for a scalar field, with each component of the vector field being generated as a scalar field.

The visualization client then treated specifically the vector field by proposing either to view each component individually or by treating them as a vector field and offering streamline visualization. Streamlines were generated by using visualization particles which trajectories of are drawn on the screen. Visualization particles were exchanged between tiles to prevent streamline boundaries to appear at each tile boundary.

A dedicated presentation of the client will be available in a future publication.

4. Results

After presenting the results obtained for the EMERGENCIES and EMED data, we describe the performances of this approach then discuss its usability in crisis management.

4.1. Visualization

4.1.1. Flow and Turbulence Data

As described above, the flow was computed by the PMSS modeling system with a grid step of 3 m: A detailed view of the flow is presented in Figure 5 close to the train station and the museum of the project EMERGENCIES. The streamlines reveal the details of the flow and its importance for the dispersion pattern close to the source, especially regarding the entrapment of the concentration obviously in the buildings, but also in the recirculation zones and some narrow canyon streets nearby.

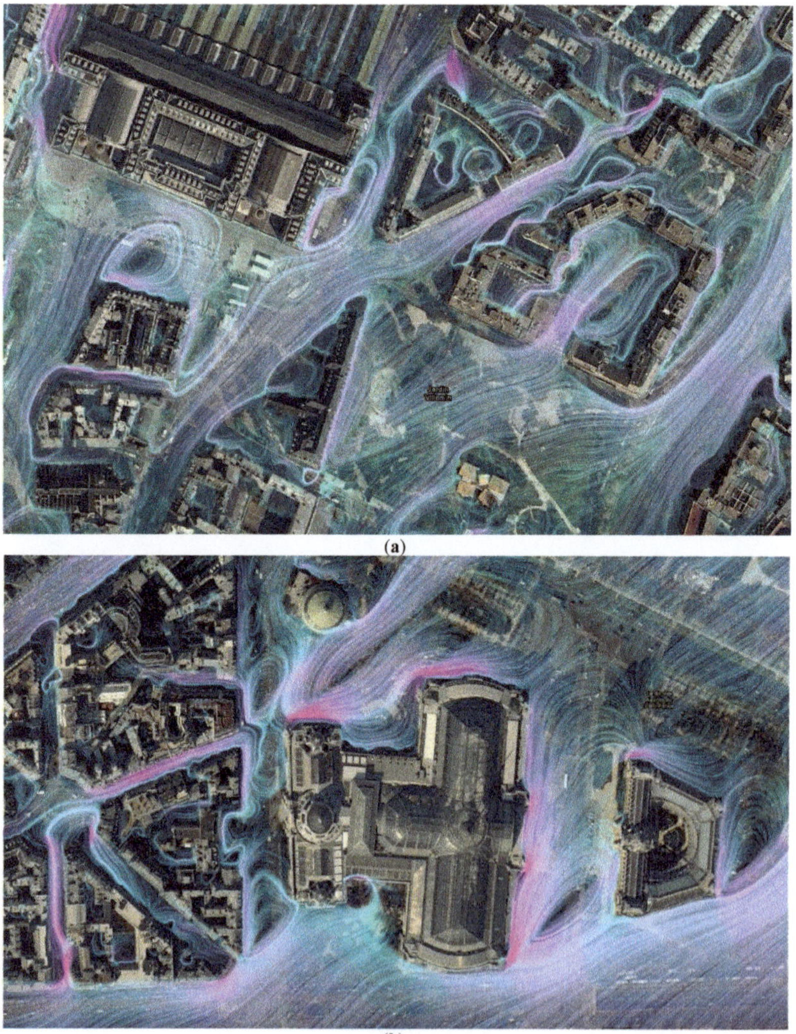

(**a**) (**b**)

Figure 5. Closeup view of the flow streamlines computed in the streets near (**a**) the train station and (**b**) the museum. North is at the top.

This type of analysis is particularly important for the modeler to evaluate the quality of the modeling and attain an understanding on the concentration patterns computed by the model.

The capability to visualize the flow streamlines on a large area is displayed in Figure 6. The flow can hence be visualized on a large area to understand the global flow pattern, while specific analysis at the scale of the road or the building is also available in any locations of these very large domains.

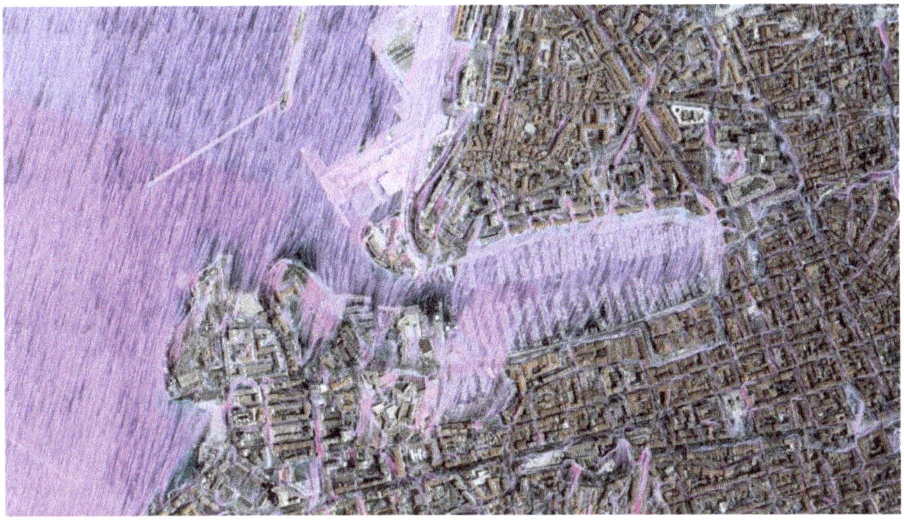

Figure 6. View of the flow streamlines computed in the Vieux Port area in Marseille. North is at the top.

Obviously, the wind speed can also be presented. Figure 7 displays the wind speed in the Marseille city center and on the topography south of the city.

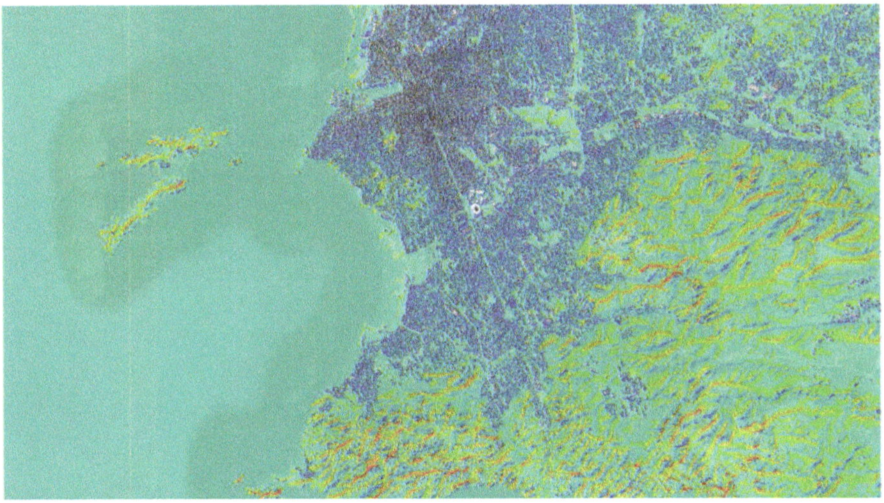

Figure 7. View of the wind speed in the Marseille city center and on the topography south of the city. North is at the top.

4.1.2. Concentration Data

The concentration field of the EMERGENCIES project for the first minutes after the release inside the museum is presented on Figure 8. It shows the dispersion pattern first within the building (Figure 8a–c), then in the rather open field close to the museum (Figure 8d,e), and finally in the streets to the north (Figure 8f).

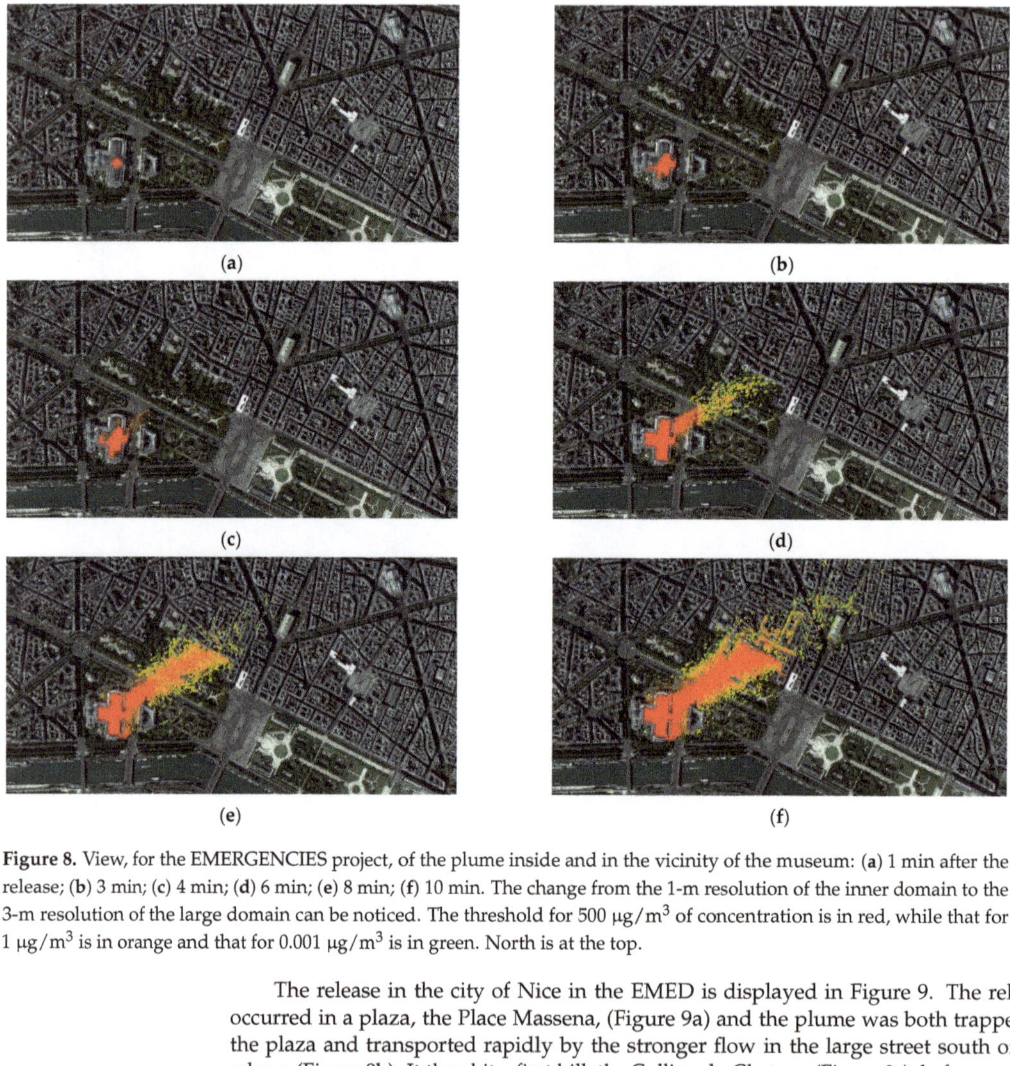

Figure 8. View, for the EMERGENCIES project, of the plume inside and in the vicinity of the museum: (**a**) 1 min after the release; (**b**) 3 min; (**c**) 4 min; (**d**) 6 min; (**e**) 8 min; (**f**) 10 min. The change from the 1-m resolution of the inner domain to the 3-m resolution of the large domain can be noticed. The threshold for 500 $\mu g/m^3$ of concentration is in red, while that for 1 $\mu g/m^3$ is in orange and that for 0.001 $\mu g/m^3$ is in green. North is at the top.

The release in the city of Nice in the EMED is displayed in Figure 9. The release occurred in a plaza, the Place Massena, (Figure 9a) and the plume was both trapped in the plaza and transported rapidly by the stronger flow in the large street south of the release (Figure 9b). It then hit a first hill, the Colline du Chateau (Figure 9c), before moving away and reaching the strong mountainous topography northeast of Nice, while remaining trapped in some narrow streets of the city center (Figure 9d).

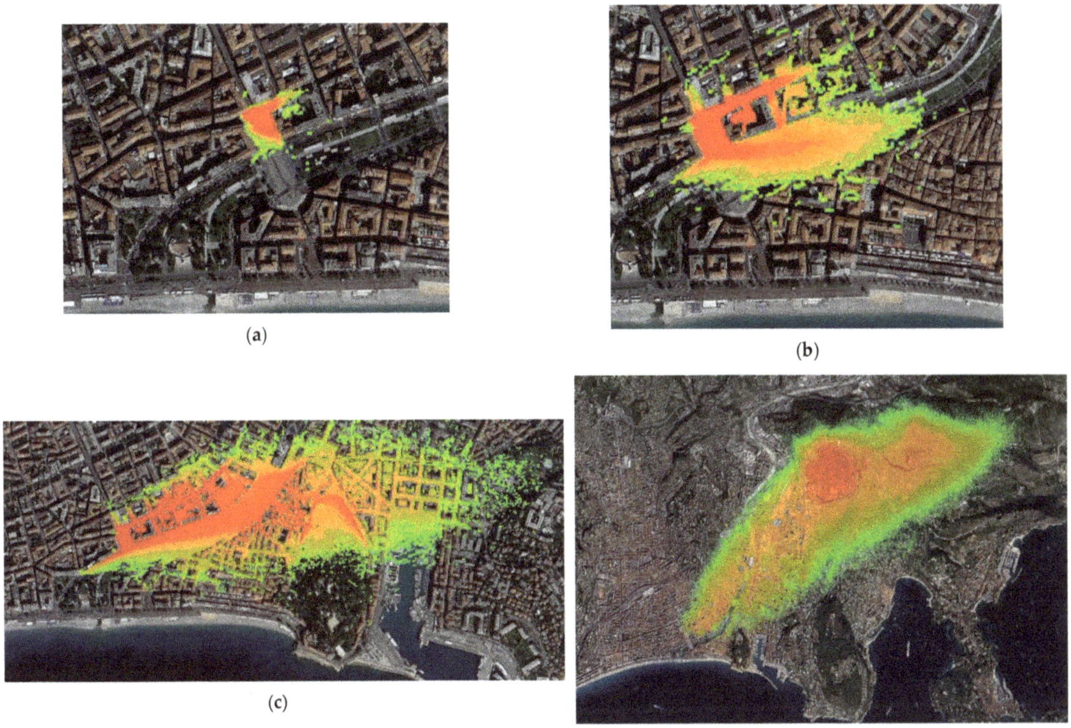

(**a**) (**b**) (**c**) (**d**)

Figure 9. View, for the EMED project and the release in the Nice domain, of the plume in the vicinity and further away from the release: (**a**) 2 min after the release; (**b**) 5 min; (**c**) 15 min; (**d**) 55 min. The scale, and the zoom level, vary between the views to cope with the evolution of the plume. The threshold for 500 $\mu g/m^3$ of concentration is in red, while that for 1 $\mu g/m^3$ is in orange and that for 0.01 $\mu g/m^3$ is in green. North is at the top.

In Figure 10, for the multiple releases in the Marseille domain, various zoom levels allowed the global pattern of the plume to be displayed while being able to also focus on pollutant entrapment inside canyon streets.

4.2. *Performances*

Performances of the methodology were evaluated both regarding the post-processing prior to the navigation, but also obviously during the navigation.

4.2.1. Post-processing Step

Performances of the creation of the multilevel tiled data were also relevant since they had to be included in the duration of the simulation, especially for a modeling system aimed at crisis management.

The parallel scheme had very limited communications between cores but was limited by the number of tiles that could be distributed among cores. Still, the fewer the tiles, the smaller the domain size and the lower the computational cost of the post-processing step.

The post-processing step had been performed on the same infrastructure as the simulations for the projects.

Regarding the flow and turbulence treatment, the zoom level was constructed between the levels 10 and 16. The levels above 16 were useless since the grid size of the mesh for the modeling system was 3 m, and at zoom level 16, the pixel size on the equator was below 3 m.

For the wind flow field, either in the Paris or Marseille domains, the treatment per timeframe using 100 cores required less than 10 min.

For the concentration data at 1 min, and since the data were stored in a single file per tile for the whole duration of the simulation, the treatment for the whole period of time used 50 cores and required roughly 2 h either for Marseille or Paris. For the Marseille domain, the simulation duration was 4 h and the binaries contained 240 timeframes; hence, the post-processing required 30 s on average per timeframe.

Regarding the disk size, the data produced for exploration had sizes similar to the raw binary output of the model, being binary too, with a maximum zoom level, 16, which led to a pixel size being of the same order of magnitude as the grid size of the model. A small overhead was due to the size of the additional zoom level above 16. On the other side, not all the vertical levels were extracted but mainly the levels near the ground.

As an illustration, the EMED concentration data, using 10-min averages for the concentration, had a size of 85 GB, while the multilevel tiles limited to the first two vertical levels had a size of 3.2 GB.

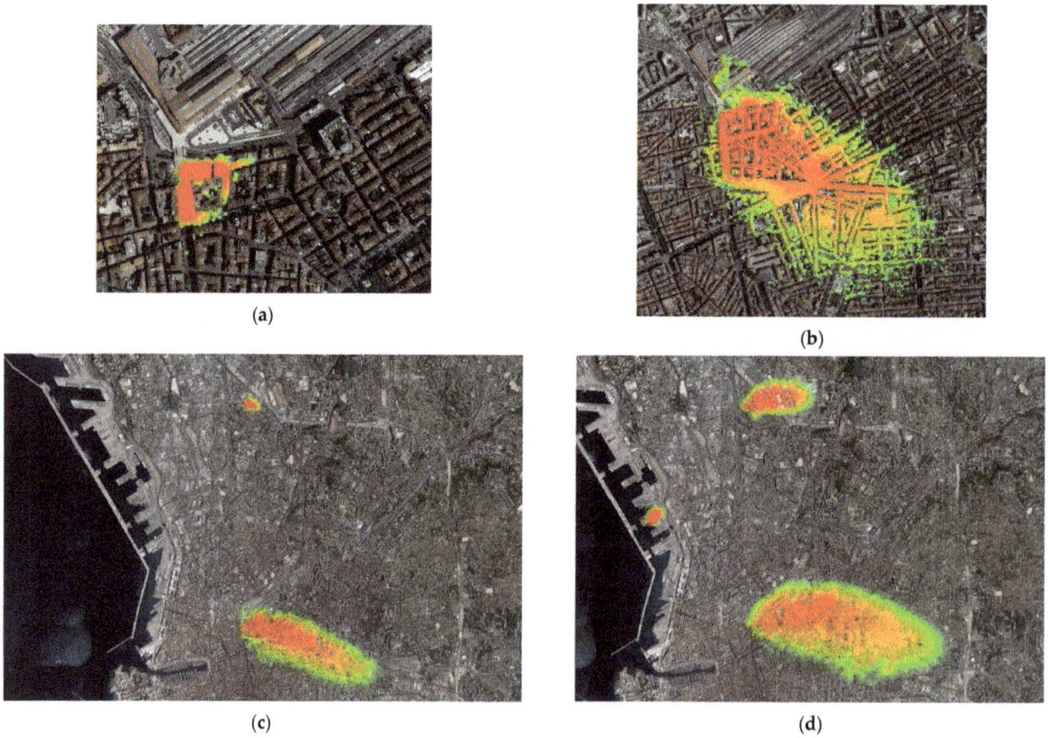

Figure 10. View, for the EMED project and the release in the Marseille domain, of the three plumes in the vicinity and further away from the releases: (**a**) 3 min after the first release; (**b**) 15 min; (**c**) 38 min; (**d**) 75 min. The scale, and the zoom level, vary between the views to cope with the evolution of the plumes. The threshold for 500 $\mu g/m^3$ of concentration is in red, while that for 1 $\mu g/m^3$ is in orange and that for 0.01 $\mu g/m^3$ is in green. North is at the top.

4.2.2. Navigation

The navigation through the data was limited by the capability of the server serving the tiles to support a defined level of workload. This being a problem of server scalability, which is rarely an issue in the context of crisis management where the number of team members accessing the data is very limited, the visualization on the client side was similar with the experience of a user browsing through any web mapping service:

- Data are accessible in less than several seconds, whatever the global size of the domain;
- Change in locations used the tiling capability, with additional tiles being loaded when required;
- Change in zoom allows both large scale and small-scale features to be monitored.

4.3. Discussion for Crisis Management

The approach taken here allowed us to:

- Take the most out of the computing infrastructure during the modeling phase by preparing the output;
- Enable a very efficient navigation in and consultation of the result during the exploitation of the simulation, even for non-expert users of modeling due to the very large diffusion of web mapping tools.

The web mapping approach allows a better integration of the simulation results in third parties integration tools, especially geographical information systems (GIS), which have increasing availability during crisis management.

It also allows the modeler to explore the results even for very large simulation results like the flow data presented above.

5. Conclusions

The EMERGENCIES and EMED project demonstrated the operational capability of high-resolution modeling in built-up areas for crisis management and supported by high-performance computing. These projects pointed out the vital importance of a capability to visualize easily and in time compatible with emergency situations the simulation results.

Since these projects were applied on responsibility areas of actual emergency teams, they generated very large amounts of data that made it all the more difficult to explore the data generated.

While an initial approach relying on traditional 3D scientific viewing permitted us to create visualizations in batch mode for such large results, this approach proved to be difficult to use as an actual operational capability. This was mainly related to limits in interactivity.

An alternative approach was introduced taken from the field of web mapping by using multilevel tiled images. The approach was implemented in a parallel library to be used as a post-processing step after the simulation of the flow and the dispersion and on the same infrastructure. The approach was also modified to allow us to display values rather than images, time-dependent and multi-height results, and scalar or vector fields.

The performance of the parallel post-processing step proved to be largely compatible with the requirement of the emergency situation, since the intensity in calculation and time required were significantly lower than those of the simulation in itself.

The actual navigation in the result either by the emergency team or the modeler was similar with the one experienced by any web mapping tool user, and hence very satisfactory to the requirement in terms of rapidity and capability do visualize both the larger scale but also the fine details.

This kind of tool is also widely known among the users, and it has natural integration capabilities in third party tools such as GIS.

This approach is hence of paramount importance for the adoption of high-resolution modeling in built-up areas by the crisis management teams, either during preparation of exercises or during an actual situation management.

In a context of increasing adoption of virtual or augmented reality (AR/VR) tools in the crisis management community, proving again the vital importance of very efficient visualization tools for the simulation community, additional perspectives are under studies to allow the integration of modeling results in the framework of AR/VR.

Author Contributions: Conceptualization, O.O., S.P. and P.A.; Data curation, S.P. and C.D.; Investigation, S.P.; Methodology, O.O., S.P. and P.A.; Resources, C.D.; Software, O.O. and S.P.; Supervision, O.O. and P.A.; Visualization, S.P.; Writing—original draft, O.O.; Writing—review & editing, O.O., S.P. and P.A. All authors have read and agreed to the published version of the manuscript.

Funding: This research received no external funding.

Conflicts of Interest: The authors declare no conflict of interest.

References

1. Armand, P.; Bartzis, J.; Baumann-Stanzer, K.; Bemporad, E.; Evertz, S.; Gariazzo, C.; Gerbec, M.; Herring, S.; Karppinen, A.; Lacome, J.M.; et al. Trini Castelli S (2015) Best practice guidelines for the use of the atmospheric dispersion models in emergency response tools at local-scale in case of hazmat releases into the air. Tech Rep, COST Action ES 1006. Available online: http://elizas.eu/images/Documents/Best%20Practice%20Guidelines_web.pdf (accessed on 16 July 2021).
2. Biltoft, C.A. *Customer Report for Mock Urban Setting Test*; DPG Document Number 8-CO-160-000-052; Defense Threat Reduction Agency: Alexandria, VA, USA, 2001.
3. Warner, S.; Platt, N.; Heagy, J.F.; Jordan, J.E.; Bieberbach, G. Comparisons of transport and dispersion model predictions of the mock urban setting test field experiment. *J. Appl. Meteorol. Climatol.* **2006**, *45*, 1414–1428. [CrossRef]
4. Allwine, K.J.; Flaherty, J. *Joint Urban 2003: Study Overview and Instrument Locations*; PNNL-15967; Pacific Northwest National Lab.: Richland, WA, USA, 2006.
5. Hernández-Ceballos, M.A.; Hanna, S.; Bianconi, R.; Bellasio, R.; Mazzola, T.; Chang, J.; Andronopoulos, S.; Armand, P.; Benbouta, N.; Čarný, P.; et al. UDINEE: Evaluation of multiple models with data from the JU2003 puff releases in Oklahoma City. Part I: Comparison of observed and predicted concentrations. *Bound. Layer Meteorol.* **2019**, *171*, 323–349. [CrossRef]
6. Hernández-Ceballos, M.A.; Hanna, S.; Bianconi, R.; Bellasio, R.; Chang, J.; Mazzola, T.; Andronopoulos, S.; Armand, P.; Benbouta, N.; Čarný, P.; et al. UDINEE: Evaluation of multiple models with data from the JU2003 puff releases in Oklahoma City. Part II: Simulation of puff parameters. *Bound. Layer Meteorol.* **2019**, *171*, 351–376. [CrossRef]
7. EPA; NOAA. *Area Locations of Hazardous Atmospheres (ALOHA)*; User's Manual; US Environmental Protection Agency (USEPA) and the National Oceanic and Atmospheric Administration (NOAA): Washington, DC, USA, 1999.
8. Sykes, R.I.; Parker, S.F.; Henn, D.S.; Cerasoli, C.P.; Santos, L.P. *PC-SCIPUFF Version 1.3 Technical Documentation*; ARAP Report No. 725; Titan Corporation, ARAP Group: Princeton, NJ, USA, 2000.
9. Chang, J.C.; Hanna, S.R.; Boybeyi, Z.; Franz, P. Use of Salt Lake City URBAN 2000 field data to evaluate the urban hazard prediction assessment capability (HPAC) dispersion model. *J. Appl. Meteorol.* **2005**, *44*, 485–501. [CrossRef]
10. Milliez, M.; Carissimo, B. Numerical simulations of pollutant dispersion in an idealized urban area. *for different meteorological conditions. Bound. Layer Meteorol.* **2007**, *122*, 321–342. [CrossRef]
11. Gowardhan, A.A.; Pardyjak, E.R.; Senocak, I.; Brown, M.J. A CFD-based wind solver for urban response transport and dis-persion model. *Environ. Fluid Mech.* **2011**, *11*, 439–464. [CrossRef]
12. Yee, E.; Lien, F.S.; Ji, H. *A Building-Resolved Wind Field Library for Vancouver: Facilitating CBRN Emergency Response for the 2010 Winter Olympic Games*; Defense Research and Development: Suffield, AB, Canada, 2010.
13. Armand, P.; Christophe, D.; Luc, P. Is it now possible to use advanced dispersion modelling for emergency response? The example of a CBRN-E exercise in Paris. In *Air Pollution Modeling and its Application XXIV*; Springer: Cham, Switzerland, 2016; pp. 433–446.
14. Oldrini, O.; Armand, P.; Duchenne, C.; Perdriel, S.; Nibart, M. Accelerated Time and High-Resolution 3D Modeling of the Flow and Dispersion of Noxious Substances over a Gigantic Urban Area—The EMERGENCIES Project. *Atmosphere* **2021**, *12*, 640. [CrossRef]
15. Armand, P.; Duchenne, C.; Oldrini, O.; Perdriel, S. Emergencies Mediterranean—A Prospective High-Resolution Modelling and Decision-Support System in Case of Adverse Atmospheric Releases Applied to the French Mediterranean Coast. In Proceedings of the 18th International Conference on Harmonisation within Atmospheric Dispersion Modeling for Regulatory Purposes, Bologna, Italy, 9–12 October 2017; pp. 9–12.
16. OSGEO. Tile Map Service Specification. Available online: https://wiki.osgeo.org/wiki/Tile_Map_Service_Specification (accessed on 16 July 2021).
17. Oldrini, O.; Olry, C.; Moussafir, J.; Armand, P.; Duchenne, C. Development of PMSS, the Parallel Version of Micro SWIFT SPRAY. In Proceedings of the 14th International Conference on Harmonisation within Atmospheric Dispersion Modelling for Regulatory Purposes, Kos, Greece, 2–6 October 2011; pp. 443–447.
18. Oldrini, O.; Armand, P.; Duchenne, C.; Olry, C.; Tinarelli, G. Description and preliminary validation of the PMSS fast response parallel atmospheric flow and dispersion solver in complex built-up areas. *J. Environ. Fluid Mech.* **2017**, *17*, 1–18. [CrossRef]
19. Oldrini, O.; Armand, P.; Duchenne, C.; Perdriel, S. Parallelization Performances of PMSS Flow and Dispersion Modelling System over a Huge Urban Area. *Atmosphere* **2019**, *10*, 404. [CrossRef]
20. Oldrini, O.; Armand, P. Validation and sensitivity study of the PMSS modelling system for puff releases in the Joint Urban 2003 field experiment. *Bound. Layer Meteorol.* **2019**, *171*, 513–535. [CrossRef]

21. Moussafir, J.; Oldrini, O.; Tinarelli, G.; Sontowski, J.; Dougherty, C. A new operational approach to deal with dispersion around obstacles: The MSS (Micro-Swift-Spray) software suite. In Proceedings of the 9th International Conference on Harmonisation within Atmospheric Dispersion Modelling for Regulatory Purposes, Garmisch-Partenkirchen, Germany, 1–4 June 2004; Volume 2, pp. 114–118.
22. Tinarelli, G.; Brusasca, G.; Oldrini, O.; Anfossi, D.; Trini Castelli, S.; Moussafir, J. Micro-Swift-Spray (MSS) a new modelling system for the simulation of dispersion at microscale. General description and validation. In *Air Pollution Modelling and Its Applications XVII*; Borrego, C., Norman, A.N., Eds.; Springer: Boston, MA, USA, 2007; pp. 449–458.
23. Oldrini, O.; Nibart, M.; Armand, P.; Olry, C.; Moussafir, J.; Albergel, A. Introduction of Momentum Equations in Mi- cro-SWIFT. In Proceedings of the 15th International Conference on Harmonisation within Atmospheric Dispersion Modelling for Regulatory Purposes, Varna, Bulgaria, 8–11 September 2014.
24. Oldrini, O.; Nibart, M.; Duchenne, C.; Armand, P.; Moussafir, J. Development of the parallel version of a CFD–RANS flow model adapted to the fast response in built-up environments. In Proceedings of the 17th International Conference on Harmonisation within Atmospheric Dispersion Modelling for Regulatory Purposes, Budapest, Hungary, 9–12 May 2016.
25. Röckle, R. Bestimmung der Strömungsverhaltnisse im Bereich Komplexer Bebauungs-Strukturen. Doctoral Thesis, Technical University Darmstadt, Darmstadt, Germany, 1990.
26. Anfossi, D.; Desiato, F.; Tinarelli, G.; Brusasca, G.; Ferrero, E.; Sacchetti, D. TRANSALP 1989 experimental campaign—II. Simulation of a tracer experiment with Lagrangian particle models. *Atmos. Environ.* **1998**, *32*, 1157–1166. [CrossRef]
27. Tinarelli, G.; Mortarini, L.; Trini Castelli, S.; Carlino, G.; Moussafir, J.; Olry, C.; Armand, P.; Anfossi, D. Review and validation of Micro-Spray, a Lagrangian particle model of turbulent dispersion. In *Lagrangian Modelling of the Atmosphere, Geophysical Monograph*; American Geophysical Union (AGU): Washington, DC, USA, May 2013; Volume 200, pp. 311–327.
28. Rodean, H.C. *Stochastic Lagrangian Models of Turbulent Diffusion*; American Meteorological Society: Boston, MA, USA, 1996; Volume 45.
29. Thomson, D.J. Criteria for the selection of stochastic models of particle trajectories in turbulent flows. *J. Fluid Mech.* **1987**, *180*, 529–556. [CrossRef]
30. Archambeau, F.; Méchitoua, N.; Sakiz, M. Code Saturne: A finite volume code for the computation of turbulent incompressible flows-Industrial applications—Industrial Applications. *Int. J. Finite Vol.* **2004**, *1*, 1.
31. Skamarock, W.C.; Klemp, J.B.; Dudhia, J.; Gill, D.O.; Barker, D.M.; Duda, M.G.; Huang, X.-Y.; Wang, W.; Powers, J.G. *2008: A Description of the Advanced Research WRF Version 3*; NCAR Tech. Note NCAR/TN-475+STR; UCAR Communications: Boulder, CO, USA, 2008. [CrossRef]
32. Ahrens, J.; Geveci, B.; Law, C. *ParaView: An End-User Tool for Large Data Visualization*; Visualization Handbook; Elsevier: Amsterdam, The Netherlands, 2005; ISBN 978-0123875822.
33. Adobe Developers Association. *TIFF" Revision 6.0, Jun. 3, 1992*; Adobe Systems Incorporated: Mountain View, CA, USA, 1992.
34. Warmerdam, F. The geospatial data abstraction library. In *Open Source Approaches in Spatial Data Handling*; Springer: Berlin/Heidelberg, Germany, 2008; pp. 87–104.

atmosphere

Article

Optimization of HPC Use for 3D High Resolution Urban Air Quality Assessment and Downstream Services

Maxime Nibart [1,*], Bruno Ribstein [1], Lydia Ricolleau [1], Gianni Tinarelli [2], Daniela Barbero [2], Armand Albergel [1] and Jacques Moussafir [1]

1 ARIA Technologies, 92100 Boulogne-Billancourt, France; bribstein@aria.fr (B.R.); lricolleau@aria.fr (L.R.); aalbergel@aria.fr (A.A.); jmoussafir@aria.fr (J.M.)
2 ARIANET S.R.L., 20128 Milan, Italy; g.tinarelli@aria-net.it (G.T.); d.barbero@aria-net.it (D.B.)
* Correspondence: mnibart@aria.fr

Abstract: The number of cities, or parts of cities, where air quality has been computed using the PMSS 3D model now appears to be sufficient to allow assessment and understanding of performance. Two fields of application explain the growing number of sites: the first is the long-term air quality assessment required in urban areas for any building or road project. The geometric complexity found in such areas can justify the use of a 3D approach, as opposed to Gaussian ones. However, these studies have constraining rules that can make the modelling challenging: several scenarios are needed (current, future with project, future without project), the long-term impact implies a long physical time period to be computed, and the spatial extension of the domain can be large in order to cover the traffic impact zone of the project. The second type of application is dedicated to services and, essentially, to forecasting. As for impact assessments, the modelling can be challenging here because of the extension of the domain if the target area is a whole city. Forecast also adds the constraint of time, as results are requested early, and the constraint of robustness. The CPU amount needed to meet all these requirements is important. It is therefore crucial to optimize all possible parts of the modelling chain in order to limit cost and delay. The sites presented in the article have been modelled with PMSS for long periods. This allows feedback to be provided on different topics: (a) daily forecasts offer an opportunity to increase the robustness of the modelling chain; (b) quantitative validation at air quality measurement stations; (c) comparison of annual impact based on a whole year, and based on a sampling list of dates selected thanks to a classification process; (d) large calculation domains with widespread pollutant emissions offer a great opportunity to qualitatively check and improve model results on numerous geometrical configurations; (e) CPU time variations between different sites provide valuable information to select the best parametrizations, to predict the cost of the services, and to design the needed hardware for a new site.

Keywords: air quality impact study; 3D; PMSS model; high resolution grid

Citation: Nibart, M.; Ribstein, B.; Ricolleau, L.; Tinarelli, G.; Barbero, D.; Albergel, A.; Moussafir, J. Optimization of HPC Use for 3D High Resolution Urban Air Quality Assessment and Downstream Services. *Atmosphere* **2021**, *12*, 1410. https://doi.org/10.3390/atmos12111410

Academic Editors: Charles Chemel

Received: 22 September 2021
Accepted: 22 October 2021
Published: 26 October 2021

1. Introduction

1.1. General Context and Motivations

Air pollution is known to impact health strongly: the World Health Organization estimates today that it kills approximatively 7 million people per year. Exposure is especially alarming in dense urban areas, where both population and pollutant emissions are high.

Methodologies for estimating the exposure in cities is therefore fundamental to describe current and future situations, and define improvement strategies.

Air quality in cities results not only from local emissions, but is a complex system that can be described by the following simplifications [1]: the regional-scale contribution which have a uniform impact on the city; the city-scale contribution, including emissions from heating for example, that is uniformly considered within the city or within the different districts of the city according to the spatial resolution of the available emission inventory;

and the street-scale contribution, which are related mainly to local traffic emissions, and are highly non-uniform.

The present article is focused on the deterministic modelling of the street-scale contribution. As air quality observations provide concentration measurements of all the contributions, model validation is not performed with only the street-scale, but also with the larger scale contributions that can be named "background contribution".

Modeling provides additional and complementary information to air quality monitoring networks that are limited to a given number of measurement points: exposure maps detailing spatial gradients are one of the added values of modeling. Urban micrometeorology has been studied since the 1980s [2], and different types of approaches, ranging from simple analytical formulations (Gaussian models) to full fluid dynamics (CFD models), are used nowadays [3].

The approach is noticeably chosen depending on the objectives, the time, and budget constraints. The present article deals with applications where both 3D high resolution and limited CPU time are required, and for which the PMSS modeling system is an applied solution.

1.2. The PMSS Modeling System

Parallel-Micro-SWIFT-SPRAY (PMSS) is a flow and dispersion modelling system constituted by the SWIFT and SPRAY models. They became Micro-SWIFT and Micro-SPRAY when explicit consideration of the obstacles [4,5] was added based on the preliminary work of Rockle [6], with the principal objective of rapid modelling (faster than a full CFD model) of the accidental or malicious dispersion of pollutants in dense urban areas. PMSS has then been improved, and parallel versions of Micro-SWIFT and Micro-SPRAY (abbreviated as PSWIFT and PSPRAY) have been developed [7,8]. Recent validations in this scope of application are presented in [9,10].

The rapid nature of PMSS and the increase in power of the computational machines then enabled, in addition to the defense-oriented use, the study of air quality in dense urban areas, often highly impacted by road traffic. The model was first used on neighborhoods such as Bologna [5], and then on entire cities such as Paris [11]. In this type of application, the number of sources to be considered is much more important, and makes the calculation more CPU expensive.

PSWIFT [1] is a 3D diagnostic, mass-consistent, terrain-following model providing the wind, turbulence, and temperature following three main steps:

- A first guess computation based on the interpolation of heterogeneous meteorological input data, a mix of surface and profile measurements, and/or possibly meso-scale model outputs;
- The modification of the first guess using analytical zones defined around isolated buildings or within groups of buildings, based the approach originally proposed by [6];
- The mass-consistency (with impermeability condition at the ground and buildings) obtained by minimizing the difference of the wind field of the second step over the volume of the domain under the mass conservation constraint. The effect of atmospheric stability on the flow around obstacles is considered through a coefficient α applied to the vertical wind component terms during the minimizing process [12].

The method allows detailed flow fields in the urban environment to be generated within a matter of minutes [13]. To consider traffic-produced turbulence (TPT) in street canyons, a formulation based on the OSPM model [14] has been added, increasing the turbulent kinetic energy field predicted by PSWIFT in these zones according to vehicles fluxes, speed, and effective area.

PSPRAY is a 3D Lagrangian particle dispersion model (LPDM) [15], directly derived from the SPRAY code [16–22]. The dispersion of a pollutant (gas or fine aerosol) is simulated following the trajectories of a large number of numerical particles. The trajectories are obtained by integrating in time the particle velocity, which is the sum of a transport component defined by the local averaged wind provided by PSWIFT, and a stochastic component representing the dispersion due to the atmospheric turbulence. The stochastic component is obtained by solving a 3D form of the Langevin equation for the random velocity, and applying the stochastic scheme developed by Thomson [23].

To consider the transformation between NO and NO_2, a simple chemical scheme has been added in PSPRAY [24].

1.3. The Different Type of High Resolution Air Quality Modeling Applications

In France, the two main frameworks in which such modelling are used are regulatory impact studies for road or building projects, and forecasting (or now-casting) systems operated by territorial agencies. The number of neighborhoods or cities modelled in these contexts with PMSS now seems to be sufficiently large to gather different feedbacks and optimization methods of the computational resource in the present article, with the goal of satisfying temporal and budgetary constraints.

In the first section, different reasons making the use of HPC useful for long-term impact studies of air quality in cities are presented. In some examples, optimization methods of computing resources and modelling validations are also presented. It should be noted that for most impact studies, contractual context frequently does not allow for the publication of results. These studies, limited in terms of budget, also don't include modelling validation. Consequently, in this article, only the results obtained in the study of chronic pollution in the framework of collaborative research and innovation projects are presented.

In the second section, feedback is presented from forecasting systems incorporating the PMSS model. This type of system poses a significant performance constraint in terms of computational time in order to obtain results early enough.

Finally, the third section gathers information on the configuration parameters of the model, on the computational machines used, and on the performances obtained, in order to identify useful trends for specifying the resources required for future modelling of the same type.

2. Long Term Air Quality Assessment in Urban Areas

2.1. Why Is HPC Used?

In France, urban development projects can be subjected to environmental assessment or to a case-by-case analysis according to criteria and thresholds (decree No. 2020-844 of 3 July 2020 on the environmental authority and the authority responsible for the examination of the case). The content of the "air and health" section of impact studies is defined in the methodological guide provided with the technical note of "22 February 2019", related to consideration of the health effects of air pollution in road infrastructure impact studies. According to this methodological guide, air concentrations that are induced by the project must be estimated with the help of a dispersion model.

Several specificities of these development project impact studies on air quality in dense urban areas potentially lead to a model choice whose complexity and computational cost guide towards the use of HPC machines. The following section describes these different specificities, partially within the prism of French regulations.

2.1.1. Complex Geometry

The first element is the geometric context of this type of study: in a dense urban area, the volume occupied by buildings is important. Inside Paris, for example, 39% of land is covered with buildings (calculated from the BD TOPO database of the National Geographic Institute (IGN)). This significant percentage implies a calculation error of the pollutant concentration for modelling approaches that do not consider the volume occupied by buildings. This is the case, for example, for Gaussian models such as CALINE4 [25], for which concentration underestimations are recorded by [26] for a canyon street configuration.

The operational street pollution model, OSPM [27], takes into account the confinement effect in street canyons, while keeping an inexpensive approach in terms of computation time.

The high buildings density and their organization into streets, crossroads, and blocks of houses also involve channeling effects. The Gaussian approach to model the dispersion can then be upgraded, as in the SIRANE model [28], with pollutant mass exchange algorithms both at the top of canyon streets, and at crossroads between road segments. Modelling consequently remains inexpensive in terms of computing time.

These approaches are based on the geometric hypothesis of a classic or semi-open canyon street with a uniform building height along the street. To address any geometric configurations, it seems required to explicitly take into consideration the volumes of each building, for example, by constructing an unstructured mesh around it, or by projecting these volumes on a structured mesh. This spatial discretization is typical for CFD models, which are, however, more expensive in terms of computation time. In the atmospheric dispersion field, the family of models based on Rockle's work [6], such as PSWIFT, provides less expensive alternatives compared to full CFD models [29].

2.1.2. Unsteady Meteorology and Unsteady Emissions

A second specificity of impact studies is the temporal variability, both in terms of meteorological conditions, and pollutant emissions. This double variability raises questions concerning the use of steady or unsteady models, and increases the potential number of situations to be modelled.

Steady versus Unsteady Approaches

Straight-lined Gaussian models are stationary by construction, and the most complete models, such as the CFD models, can also be used in a stationary way within the atmospheric dispersion framework when the considered meteorological conditions and emissions are constants. To calculate an annual impact, a number (often limited for the more expensive models) of these steady-state situations is simulated, and an average, weighted by the frequency of occurrence of each situation, is used to estimate annual values. This approach was used in 2020, for example, on the impact study of the developing project around the Eiffel Tower (https://www.paris.fr/pages/grand-site-tour-eiffel-un-poumon-vert-au-c-ur-de-paris-6810/ (accessed on 10 September 2021)), and for the developing project at the "Porte de Montreuil" (https://www.paris.fr/pages/20-e-porte-de-montreuil-3329 (accessed on 10 September 2021)). It can be described as a frequency approach. The transitory aspect of the dispersion is neglected. To take it into account, it is necessary to use an unsteady dispersion model over sequences of several hours or several days for example, and ideally, over a full year. Today, calculating sequences over several days with a full CFD model seems to be too costly, in terms of computation time, to be compatible with deadlines and the total costs of impact studies, however, with intermediate approaches such as PMSS, it is reasonable, and already done for many studies. A complete one-year sequence is a possible approach (sequential approach), but in practice, the selection of a few dozen typical days (several 24 h sequences) is more affordable. The latter has been used, for example, in the impact study of the Porte Maillot project in 2019 (https://www.paris.fr/pages/projet-16e-17e-porte-maillot-4559 (accessed on 10 September 2021)). It is then both a sequential and frequential approach. This trade-off is detailed in a concrete study case in Section 2.2.

Input Data Cross-Variability

Time variability in meteorology and emissions implies, for a frequential approach, choosing the most representative combinations, a task which might not be trivial. For example, if low winds are rather nocturnal and, therefore, associated with low traffic periods, in winter, the night can last until the morning peak traffic. The situation then penalizes the air quality because of weak winds and large emissions, and might be considered in the selection. However, in a frequential approach, the limited number of cases often entails the impossibility of considering this type of specific combination because only average daily traffic values are used.

2.1.3. Purifying Systems

In order to bring a contribution to the air quality improvement in cities, some development projects include depollution systems. The use of a high-level model with a sufficient complexity is required in order to take into account the effects of the depollution systems, and quantify the extent of the affected zone in three dimensions.

2.1.4. Numerous Scenarios

In France, the methodological guide, coming with the technological note of "22 February 2019", related to the effects of air pollution on health in impact studies of road infrastructures, states the necessity of examining numerous scenarios: the current situation, and the future situation at different horizons (at least at the commissioning of the project, and up to 20 years after the commissioning, with and without the project)—at least five scenarios for each project should be examined. As deadlines to conduct impact studies are often short, it is necessary to use powerful computing machines to complete all of the calculations.

2.1.5. Domain Extents

For road development impact studies, French regulations require the computational domain to cover the zone for which road traffic is modified by 10%. Areas that must be modelled can then be far more extended than the development project itself. As an example, for the development of the Eiffel Tower neighborhood (Autorité Environnementale, 2020 Avis délibéré de l'Autorité environnementale sur l'aménagement du site de la Tour Eiffel (75)—N° Ae 2020-115), the project's extension is included in a 1.5 km × 1.5 km area, and the computation domain for the modelling of the air quality is 7 km × 5.3 km.

2.2. Annual Impact on a Coastal City with Very Complex Terrain—HPC Use Optimization with Classification

2.2.1. Presentation and Objectives

The computation times of PMSS can lead to costly annual impact studies of a site when many sources are being considered. As discussed in the previous section, for an urban site with emissions linked to road traffic, a classification of the input data (meteorological data for example), allows for the selection of a few entire days to model. Statistics, and, in particular, average concentrations, are therefore calculated from a few dozen days, but not from the whole year. The use of classified input data allows the computation time to be restricted, while keeping the benefits associated with the modelling of continuous sequences, taking into account diurnal cycle and emission modulations.

Self-organized maps (SOMs) is the classification method used in the study presented in this section. It is based on artificial neural networks that operate through unsupervised learning [30]. This method was introduced in the context of atmospheric studies in the late 1990s as a classification and shape recognition method. The article [31] reviews the applications of the SOMs method to meteorology and oceanography. This method is shown to be useful at very different spatial and temporal scales. The current study proposes verifying that the SOMs classification of the input data allows for the estimation of a yearly average and percentiles of concentrations close to those obtained without SOMs classification. Percentiles are important to assess the frequencies of exceedances for regulatory impact studies. Estimating these values with the classification is more challenging than annual averages because classifications tend to approximate statistical tails less accurately.

This study, in collaboration with Atmo Sud, deals with the air quality modelling of a port agglomeration in the south of France. The horizontal extent of the study domain is about 3.6 km by 4.6 km. Given the street types of the agglomeration, a 3 m resolution was chosen for the whole domain to efficiently model the dispersion of the pollutant in the urban fabric. The steep topography culminates at 770 m. The 3D buildings database (BD TOPO IGN) has 3637 polygons in the modelled domain. The road network is described in 1382 sections, on average 101 m long, but whose length can vary from a few meters up to 3.3 km. Hourly meteorological data from a station located in the computation domain is integrated as an input to the PMSS model. Pollutant emissions from road traffic were modelled with the COPERT 5 method, and were provided by Atmo Sud with time modulation profiles for business days, Saturdays, Sundays, and holidays. Atmo Sud also provided the emissions from an incinerator located in the modelled domain with the thermodynamic properties of the chimneys (height, temperature, expulsion speed of the discharges at the chimney output). Other sources of pollutants are taken into account by integrating background concentrations recorded at urban background stations located in the modelled area. These stations are supposed to be located outside of the direct influence of the sources that are explicitly modelled.

NO_2 concentrations are herewith presented. PM10 and PM2.5 concentrations have also been analyzed. As the conclusions from the SOMs classification assessment are similar, they are not presented. Background concentrations of NO_2 were recorded at two measurement stations, but only the minimum value is selected as the background concentration. O_3 background was also extracted from observations at background monitoring stations.

2.2.2. Model Performance without Classification

Days without observations were not modelled. Consequently, the year 2017 counted 359 days that could be modelled. In this first part of the study, all the available days were considered. Calculations were carried out at the CALMIP computation center (University of Toulouse). With PSWIFT, one day is modelled in 15 min by allocating 151 computation cores, whereas PSPRAY runs for 91 min with 240 computation cores. In comparison with a Eulerian model such as PSWIFT, the Lagrangian approach of the PSPRAY model implies a significant variability of the computation time, mainly according to meteorological situation. For example, the 15 March was modelled in 102 min, whereas the 17 March was modelled in 76 min.

In the context of the project, in 2018, Atmo Sud conducted a measurement campaign to complete the field diagnostic by bringing complementary data to the five permanent stations located in the modelled area.

Table 1 presents the scores obtained for hourly concentrations at the five stations. The high scores at station three can be explained by the fact that the values at this station were used as input data for the model.

Table 1. Statistical evaluation of the model performances for hourly NO_2 concentrations on five monitoring stations during 359 days of year 2017.

Station	Mean Observation (μg/m^3)	Mean Model (μg/m^3)	Bias (μg/m^3)	RMSE (μg/m^3)	r
Station 1 *	27.9	21.4	−6.8	15.46	0.78
Station 2	41.3	33.7	−7.6	24.44	0.56
Station 3 *	20.4	20.2	−0.4	5.33	0.95
Station 4	50.8	52.9	2.2	20.89	0.68
Station 5	35.3	32.8	−2.3	20.10	0.57

* background stations.

In addition to the five permanent stations, passive NO_2 measuring samplers were positioned on 37 locations in the modelled domain. The locations of the measurements aim to better characterize air quality at a fine scale on the modelled area. Particular attention has been paid to consider:

- Different neighborhoods, in order to evaluate the distribution of the concentrations, and to better consider the impact of the topography;
- Major road axes crossing the territory, and those with a canyon-type;

- The acquisition of measurements in the vicinity of atypical sources of emissions (proximity to heliports, gas stations, cruise ships quays, tunnel portals, etc.).

Sampling took place over two periods of the year. The first took place during the winter between the 22 January and the 26 February, whereas the second one took place between the 19 June and the 17 July in the summer season. Assessing two distinct periods allows for the inclusion of seasonal meteorological variations, as well as activity variations on the territory. Linear regression coefficients to estimate the 2017 annual average as a function of the values obtained during the campaign (average of summer and winter) were established for the permanent stations. These linear regression coefficients are then used to estimate average concentrations for 2017 of each passive sample.

Figure 1a shows the average observed concentrations for 2017 at the passive samplers for NO_2, and those computed with the PMSS model. The figure shows a significant heterogeneity in concentration levels, and this observation is generally valid for dense urban centers. For 86% of the points in Figure 1a, a vast majority of them, the PMSS model is less than 10 $\mu g{\cdot}m^{-3}$ from the observations. The regression line shows a slight underestimation of PMSS. The origin of this underestimation is difficult to isolate, and probably results from the different intermediate approximations. Figure 1b shows the annual average and 99.8th percentile estimated by PMSS at the five permanent stations. The calculation of the annual average also presents a slight underestimation. In the case of three stations out of five, an underestimation is observed by the model for the 99.8th percentile. The line «y = x/2» shows, however, that these values remain within a factor of 2 of the observations.

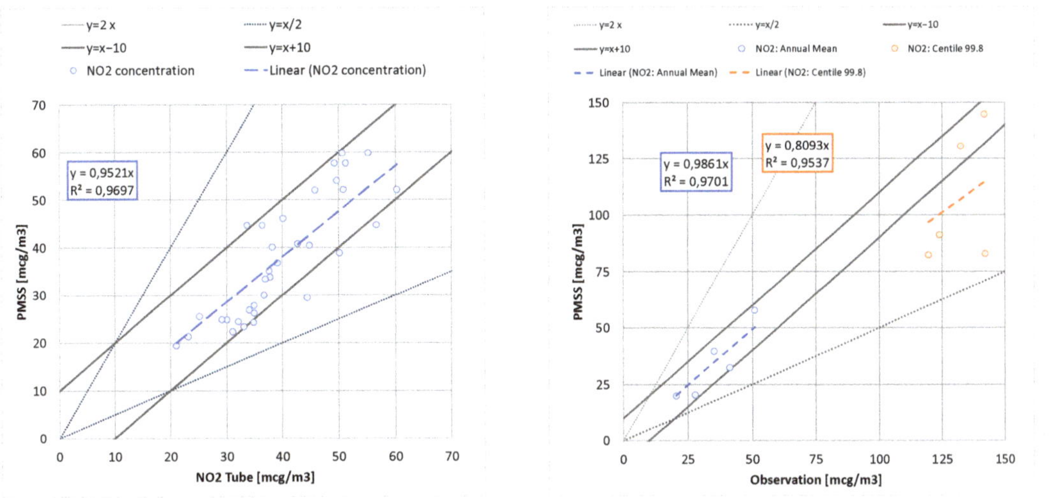

(a) (b)

Figure 1. Comparison of observed and computed NO_2 concentration for annual average at passive sensors (**a**), and annual average and 99.8th percentile at monitoring stations (**b**).

2.2.3. Results with SOMs Classification

All the elements that need to be classified (vectors) at the input of SOMs are of the same size: there are 359 days (vectors) described in hourly step, and each component of these vectors can be the 24 values of speed, and (or) of direction, and of temperature, etc. The SOMs classification is applied to two sets of surface data from permanent monitoring stations. In the c1 setup, the input data of SOMs are wind speed, its direction, humidity, and temperature, whereas setup c2 also includes background concentrations (PM10, PM25, NO, NO_2, O_3).

Input data processing is applied before the SOMs classification. The comparison between vectors is based on a Euclidian distance. In order to not privilege components with high values (temperatures in Kelvin for example) over those with low values (wind amplitudes in m·s^{-1} for example), the variables were standardized (zero mean and standard deviation of 1). The wind direction and its magnitude are preferred to wind components, in order to preserve the angle notion in the classification. This was not the case in this example, however, a shift of 90° might be considered if the north sector is a dominant direction of the wind (because of the proximity between 359° and 1°).

One of the parameters of the SOMs classification algorithm is the targeted class number Nc, defined as the product of two integers Nc = Xdim.Ydim. The SOMs method forms a grid with Xdim.Ydim nodes, where each node is associated with a vector. The classification is done through two successive learning phases, where the second one is finer than the first one. In this study, the number of neighbors of a node is fixed at 4 (rectangular grid), and all the nodes are modified by each input vector, with a respective weight that decreases with distance. Three-hundred fifty-nine vectors at the input are divided into Nc classes. A Euclidian distance between vectors and the barycenter of "their class" is calculated, and the vector with the smallest distance is acknowledged as the best representative of the class. The modelling of this day (=vector) with PMSS thus represents its class. It is then necessary to choose a method to rebuild a complete time history, in order to compute annual means and percentiles from the Nc best representatives modelled with PMSS. Two reconstruction methods are tested: an unmodelled day is represented by "its best" representative among those modelled (method m1); or by a weighted sum of the modelled days (method m2), with a weighting chosen from the Euclidean distance between a representative and the studied day (the weighting is described in [32]).

The symmetry of the SOMs classification has been checked. This means that a classification with Nc = Xdim.Ydim classes is identical to the classification with the Ydim.Xdim classes. It must, however, be noted that a number of classes can be obtained in several ways (40 = 5 × 8 = 4 × 10), and as many different classifications. For this study, and for the convergence analysis, the number of targeted classes is chosen as 5 × 5 = 25 and 10 × 10 = 100. Concentrations are extracted from the five air quality monitoring stations located in the calculation domain. Table 2 summarizes the presented elements.

Table 2. List of tested classifications parameters.

Setup	c1	Wind speed, direction, humidity, and temperature
	c2	Wind speed, direction, humidity, temperature, and background concentration (PM10, PM25, NO, NO_2, O_3)
Reconstruction	m1	A day is equal to the day associated with its representative among the modelled days
	m2	A day is equal to the weighted sum of the modelled days
Nc	5 × 5	
	10 × 10	

For the eight possible cross configurations (setup/reconstruction/Nc), Figure 2 shows the obtained values for the annual mean estimate (left) and the 95th percentile (right). Other statistics, such as the median and the 5th percentile, were also analyzed, and present similar conclusions. The regression lines allow an aggregation of the values at the five stations, and the correlation coefficients R^2 of the different lines are all superior to 0.995.

A greater gap with and without the classification is observed on the 95th percentile than on averages. A greater deviation also implies that the regression line moves away from the perfect correlation (y = x). The classification tends to underestimate the means and the 95th percentile values, and tends to overestimate the 5th percentiles. This is consistent with the complexity for the classifications to properly reproduce statistical distribution tails. However, errors remain in the order of 5 μg·m^{-3}, and are less than 2 μg·m^{-3} for the means. The consideration of the background concentrations (setup c2) substantially improves the means and the 95th percentile values estimation. It is consistent with the fact that air quality in urban areas is not only sensitive to local emissions, but also to background concentrations, with air quality being a multiscale issue. The 95th percentile estimate with the c1 setup does not seem to depend on the number of classes. In other cases, a decrease in the differences is observed overall with the increase of the number of classes. For the percentile's estimation in the c2 setup, this decrease is actually faster for the reconstruction method m1. This last finding is understandable because method 2 (m2) represents the unmodelled days by a weighted sum of all the representatives, and thus, smooths the complete time history more than the m1 method does. Annual means do not seem responsive to the reconstruction model, and this also remains true for each annual means at a specific hour (not shown here).

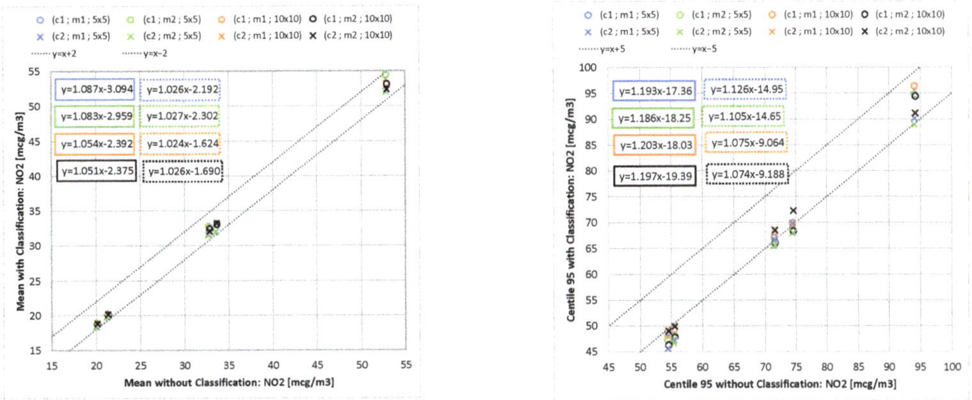

(**a**) (**b**)

Figure 2. Comparison of annual average (**a**) and 95th percentile (**b**) with and without SOMs classification, for different configurations (c1; c2), reconstruction method (m1; m2), and number of target classes (5 × 5; 10 × 10).

2.3. Grenoble Case—Validation with High Density Sensors Network

2.3.1. Context and Model Setup

A modelling at the local scale is operated on the Grenoble agglomeration by Atmo Auvergne Rhône Alpes (Atmo AURA), mainly based on the SIRANE model. In 2016–2018, the Mobicit'air project has also allowed the evaluation of different methodologies for the assimilation of concentration observations (https://www.atmo-auvergnerhonealpes.fr/sites/ra/files/atoms/files/rapport_final_mobicitair_lot3.pdf (accessed on 10 September 2021), published in 2017) with a relatively high density of sensors, thanks to the use of micro-sensors, which are less costly than reference stations with analyzers. In 2018–2019, as part of the FUI FAIRCITY project (https://www.axelera.org/fr/actualite/Projet-Faircity (accessed on 10 September 2021), published in 2019), this measurement campaign, including micro-sensors and reference stations, was used to assess the performance of the PMSS model, and develop a coupling methodology between SIRANE and PMSS. The coupling, which is not detailed in this article, aims to allow the application of PMSS on a sub-domain to the one of SIRANE. It is based on the creation of groups of emission sources in the SIRANE input in order to be able to access the effects of all sources, but also of all sources except the ones taken into account by PMSS. This specification in the larger scale modelling (here SIRANE) allows for the summing of the concentrations of PMSS and SIRANE in the PMSS subdomain without double counting. This summation assumes a linearity of the sources' contributions particularly, despite chemical reactions in the atmosphere, such as NO/NO_2 conversion.

The PMSS computation domain is 1770 m × 1671 m, located in the center of Grenoble (the SIRANE domain extent is, in comparison, 32 km × 44 km). The horizontal spatial resolution is 3 m. The vertical spatial resolution is 2 m between the ground and the first 10 m, then progressively decreases up to the computation ceiling, located at 2000 m. The mesh contains about 7.9 million cells. The domain was specifically chosen in order to include a large number of sensors: two reference stations and five micro-sensors. Concentration values used by the micro-sensors were post-processed by Atmo AURA, notably, thanks to the cross-comparisons before and after the measurement campaign (see https://www.atmo-auvergnerhonealpes.fr/sites/ra/files/atoms/files/rapport_final_mobicitair_lot3.pdf (accessed on 10 September 2021) published in 2017).

The emissions considered with PMSS are limited to the road traffic emissions. They are extracted from the emissions estimated by Atmo AURA and used in the SIRANE setup for the whole agglomeration. The potential contributions of other (and external) sources are considered with the help of the regional scale model (named CHIMERE [33]) and a kriging algorithm using concentration measurements from the entire Atmo AURA measurements network. Comparisons are made on NO_2 hourly concentrations over the continuous period from the 15 January 2017 to the 31 January 2017.

2.3.2. Results

The model-measurement deviations are inferior to 20%, except for the *MC_GRE_JPerrot* station, where the error exceeds 50%. A strong correlation is observed with an overestimation trend (see Figure 3). All the comparison stations are located close to traffic. Only the *Mob_Grenoble_caserne_Bonne* is an urban background type. For this station, where scores are the best (see Table 3), regional modelling quality might be preponderant. At the fixed station *Grenobles_Boulevard*, the hourly evolution of modelled concentrations reproduces well those of the observed concentrations (see Figure 4). If the absolute values of the concentrations are not properly reproduced on the peaks, the temporality is satisfactory.

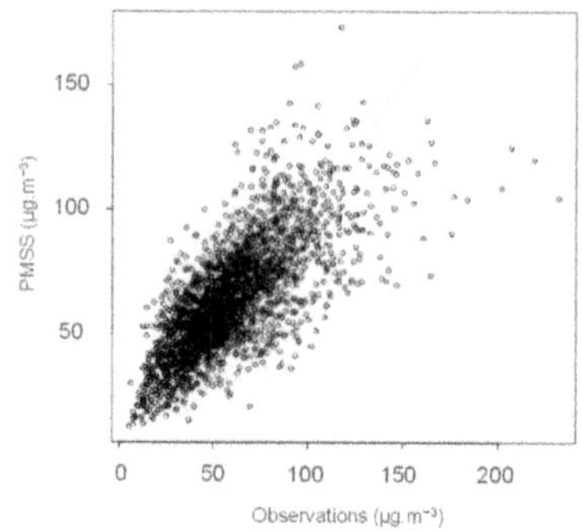

Figure 3. Scatter plot observation versus model for all the measurements points (two stations and five micro-sensors)—Hourly average concentration of NO_2 during the period from the 15 January 2017 to the 31 January 2017.

Table 3. Statistics of comparison between observed and computed hourly average concentration of NO_2 during the period from the 15 January 2017 to the 31 January 2017.

Station	Mean Observation (µg/m^3)	Mean Model (µg/m^3)	Bias (µg/m^3)	RMSE (µg/m^3)	r
Grenoble_Boulevards	64.9	76.1	11.2	19.9	0.74
MC_GRE_JJaures	53.4	61.0	7.6	13.5	0.84
MC_GRE_JPain	76.5	72.9	−3.6	24.7	0.73
MC_GRE_JPerrot	40.4	61.0	20.6	23.2	0.85
MC_GRE_Leclerc	52.9	45.6	−7.3	18.1	0.77
MC_GRE_VHugo	62.0	63.6	1.6	16.2	0.76
Mob_Grenoble_caserne_Bonne	53.8	53.7	−0.1	8.8	0.90

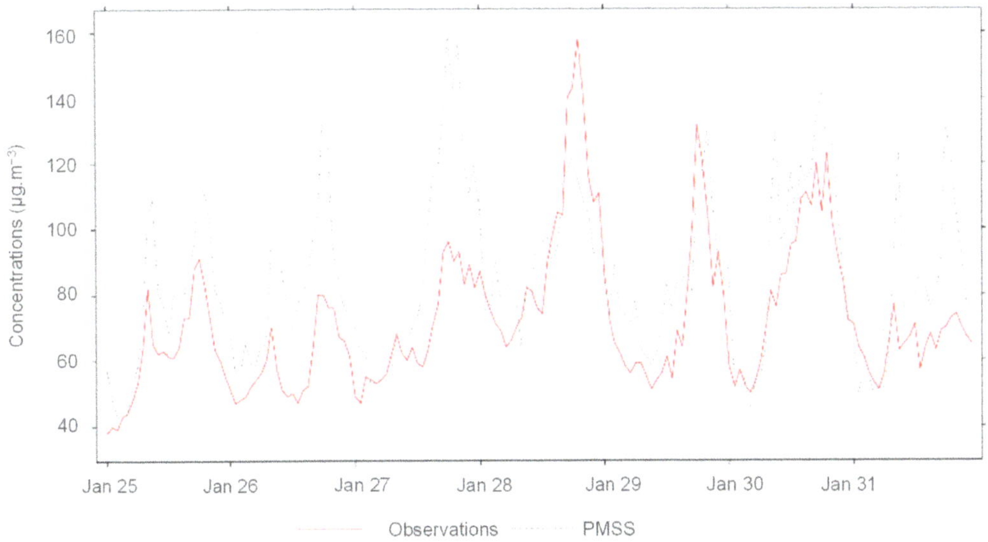

Figure 4. Time series of observed and computed concentration of NO_2 at *Grenoble_boulevards* station between the 25 January 2017 and the 31 January 2017.

PMSS NO_2 results are compared to SIRANE results obtained by Atmo AURA over the same computation period with strictly the same input data (emissions, meteorology, and background concentrations) (see Figure 5). Only the first vertical level of the PMSS model is compared to the concentrations of the SIRANE model, which correspond more by construction to spatial averages within each street.

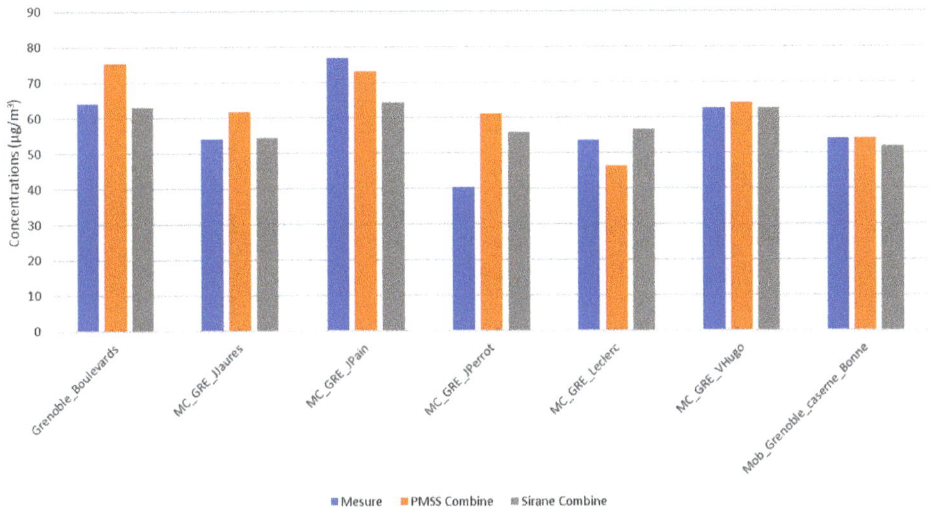

Figure 5. Average NO_2 concentrations during the period from the 15 January 2017 to the 31 January 2017 at measurements points—Observation in blue, PMSS model in orange, and SIRANE model in grey.

At the *MC_GRE_JPerrot* station, an overestimation with both models is observed. At the *MC_GRE_Leclerc* station, where PMSS presents an underestimation compared to SIRANE, PMSS input data was therefore reviewed, allowing for the notification of an underestimation of the pollutant mass rates in the neighborhood because of the position at the boundary of the calculation subdomain. The contribution of some near-road sections is missing.

At the *Grenoble_Boulevards* reference station, it is interesting to observe a good agreement for SIRANE, and an overestimation for PMSS. It is also interesting to analyze the geometric definition of the emission strands in this street, which is, in real life, organized into two traffic lanes separated by two tramway lanes at the center of the street. In the input data, the emissions from both directions are allocated to a single strand, which is itself off-centered and located near the measuring station on one of the sidewalks. In the database, most of the streets are taken into account by a single emission strand, as the SIRANE model considers a balance per street. This might explain the difference observed in the comparison between the two models.

Average concentration maps over the period (see Figure 6) have been calculated by only taking into account the results of the models, without adding the regional background. This enables a better comparison between the specificities of each model. The SIRANE map resolution is 10 m, whereas the PMSS one is 3 m. For the same color scale, concentrations are more contrasted with the PMSS model compared to the SIRANE model (higher concentrations on the road delineation, and zero in buildings). The decrease in concentrations obtained by moving away from a road is more abrupt with the PMSS model, where the SIRANE model further dilutes the concentrations within the streets. This difference not only originates from the resolution difference between the two models, but also, perhaps, from the different modelling principles.

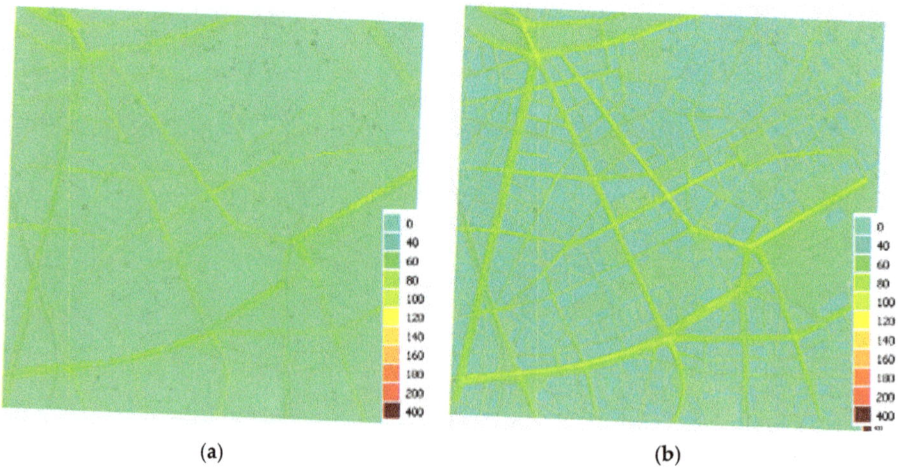

Figure 6. Average NO_2 concentrations maps during the period from the 15 January 2017 to the 31 January 2017 computed with SIRANE (**a**) and PMSS (**b**) models.

2.4. Rome Case

2.4.1. Context and Model Setup

In the framework of the BEEP project (Big Data in Environmental and Occupational Epidemiology, https://www.progettobeep.it/index.php (accessed on 10 September 2021), 2019), a long-term simulation at building-resolving scale over a large domain that covers most of the Rome conurbation has been run. The simulation has been conducted for the entire year 2015 over a 12 × 12 km urban domain with a high spatial resolution of 4 m grid step, and provides hourly ground concentration fields for different pollutants. The resulting fields account for the concentration values in cells between 0 and 3 m in height, but the domain extends in height up to 300 m, therefore, the calculation grid considers 1.8×10^7 cells.

The simulation has been carried out using a hybrid modelling approach to reproduce air quality of an urban area. The developed hybrid modelling system (HMS) couples PMSS with the Eulerian chemical transport model (CTM) and FARM (flexible air quality regional model) [34,35]. The latter reproduces the transport and the chemical interactions at regional scale of all the sources that are discretized only at the resolution of the CTM, such as the space heating. PMSS simulates, instead, traffic emissions within the city, which cause hotspots and strong concentration gradients typical of urban environments, and deals with the presence of buildings that lead to urban canyon effects. The two models have been run independently, and subsequently combined, allowing the application of each model over a different domain with appropriate grid resolution, and with proper emission inputs. The consistency between the models is ensured by using the same meteorological data, provided by the WRF meteorological model [36], as well as the same topography and land use data for both models. Consequently, FARM considers a 60 × 60 km domain centered over Rome with a horizontal resolution of 1 km, whereas PMSS is applied over the target domain described above.

The advantage of the models' independent execution is minimizing the computational effort, making it feasible to run long-term microscale simulations over large domains. In fact, a CTM, which requests a lower computational time, manages a greater number of sources, whereas a LPDM, which is more demanding, simulates only traffic emissions. Accordingly, PMSS computational time is representative of that of HMS. In this work, FARM computational time is equal to 11 min per simulated day on an HPC system with 1 node and 36 cores, whereas PMSS takes 3 h per simulated day on an HPC system with 5 nodes and 180 cores. Furthermore, PMSS computational demand is particularly low in respect to traditional implementations, thanks to the exploitation of the kernel method to calculate concentrations inside PMSS code [37]: for this simulation, the deployment of this method has allowed a reduction estimated at about 80% of the computational time compared to what would be obtained using the traditional PMSS code.

2.4.2. Results

The HMS achieved good performance when reconstructing the typical urban spatial variability, showing very diverse concentration levels in different neighborhoods across the city. The comparison of HMS outcomes with the results of a CTM run at urban scale (typically of 1 km) is particularly promising, mainly due to the fact that the HMS is capable of reproducing the presence of city hotspots, contrarily to a CTM, which is unable to capture these because of its coarser horizontal resolution. Figure 7 shows the comparison between observed and two predicted daily NO_2 concentrations during the entire year 2015 at Magna Grecia station, which belongs to the air quality monitoring system of this region, and which is located near a main road. The figure reports the concentrations calculated with FARM in green, run at 1 km of horizontal resolution, and taking into account all the emissions considered by the HMS, and the HMS concentrations are in blue. From this comparison, it is evident that the CTM itself underestimates NO_2 concentrations at this urban traffic station, whereas HMS, mainly thanks to PMSS, which better reproduces traffic contribution, has a good agreement with measurements.

Table 4 reports the statistical evaluation of HMS performance on the urban traffic station of Magna Grecia for daily NO_2 concentrations, carried out with the estimation of bias, RMSE, and correlation. It confirms the good agreement of HMS results with measurements.

Table 4. Statistical evaluation of HMS (FARM + PMSS) performance for daily NO_2 values on Magna Grecia station.

Station	Mean Observation (μg/m^3)	Mean Model (μg/m^3)	Bias (μg/m^3)	RMSE (μg/m^3)	r
Magna Grecia	65.2	61.6	−3.6	18.78	0.56

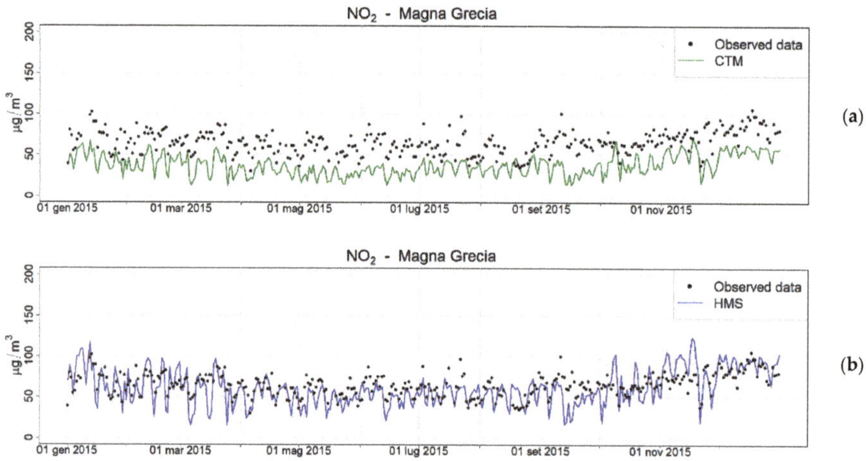

Figure 7. Comparison between observed (black dots) and modelled daily NO_2 concentrations by the CTM FARM run at 1 km (**a**) and by HMS (FARM + PMSS) (**b**) at monitoring station Magna Grecia in Rome.

3. REX from Different Forecast Systems

Using a full deterministic approach to forecast physical phenomena implies the constraint of a CPU time smaller than the real time: the whole modelling chain must be faster than real time.

Contrary to the context of the studies presented in the previous section, the computational time cannot be shortened by modelling only a number of days limited by a classification. Hybrid methods including a statistical part can be used [38]. In this paper, the focus is only on entirely deterministic approaches. Two operational systems on cities are presented in this section. The system operated during the Elise project in Turin in 2015–2016 gives an additional example, not detailed here, but described in [39].

3.1. Paris Forecast System

3.1.1. Context and Model Setup

A forecasting system was operated in Paris between September 2018 and January 2019 as part of the FUI FAIRCITY project. It partly takes the configuration setup previously used during the FEDER AIRCITY project [11]: the extent of the calculation domain of the PMSS model is 14,022 m × 11,499 m, including the whole city of Paris. The horizontal spatial resolution is constant, and equal to 3 m. The vertical spatial resolution is of 2 m between the ground, and the first 30 m then decreases until the calculation ceiling, located at 800 m. The mesh counts approximately 6.27×10^8 cells. The calculation chronology allows the calculation of the 24 h of the following day during the night. The system provides hourly concentrations.

Large-scale meteorological data and background concentrations of NO, NO_2, O_3, and PM10 are taken from the ESMERALDA forecast system of AIRPARIF (http://www.esmeralda-web.fr/accueil/index.php (accessed on 10 September 2021). They are available around 10 PM, local time. Emissions considered by the PMSS model are limited to those from road traffic. The pollutant emissions are estimated by AIRPARIF with the help of a classification by standard day (as opposed to quasi-real-time estimates using real-time-existing measurements of the vehicles fluxes that are not available for the forecasting system). All the input data are available at an hourly time resolution.

The period of several months was leveraged to make the computational workflow more robust. Malfunctions were analyzed and, if possible, led to calculation chain upgrades. Among the malfunctions, those that come from the hardware can be cited (under-sizing of the memory requested to the calculation server: a safety margin has been put into place;

crash of a computational node of the calculation server: one occurrence without possible patch; AIRPARIF ftp server does not provide access to the ESMERALDA forecast: two disfunctions), and those that come from the software can be cited (case of a very weak wind leading to PMSS error: three crashes before setting up a patch). The chronology of the malfunctions of the calculation system over the targeted period was: six in September, four in October, one in November, and none in December and January.

The results, in terms of NO_2 and PM10 concentrations, were quantitatively analyzed by comparison to observations at the location of AIRPARIF monitoring stations, qualitatively, through the observation of hourly average maps.

3.1.2. Results

Over the period between the 1 September 2018 and the 30 January 2019, the comparison of statistics at monitoring stations, from which we can distinguish those close to the traffic and those in the background, shows a significant over-estimation trend (see Table 5). The over-estimation is lower for the background stations, but present for NO_2 and PM10. The models' calculation cost has greatly limited the number of sensitivity tests that could be carried out afterward. The following paragraphs present some attempts to improve the modelling chain.

Background concentrations added hour by hour to the PMSS results could explain a part of the over-estimation. Modifying them does not require a recalculation, but only post-treatments. Scores at the measuring stations could be recalculated over the entire period with a different method of background estimation: instead of extracting the cell value of the ESMERALDA chain (horizontal resolution of 15 km in the available nesting level early enough for the PMSS forecast chain) located in downtown Paris, we have extracted the cell values upwind of Paris, according to the wind prediction from the ESMERALDA chain. This new calculation, therefore, remains compatible with a forecast mode. It allows for the avoidance of doubly counting emissions related to traffic in Paris, considered both with PMSS and ESMERALDA. Inversely, it does not allow for the inclusion of sources in the PMSS domain but not considered by PMSS, such as urban heating. This method permits significant improved RMSEs and biases, except for several background stations, particularly, with the emergence of negative biases for the NO_2 (*PA_12*, *PA_13*, and *PA_18*) and PM10 (*PA_18*) concentrations, which could be explained by the underestimate of sources other than the road traffic previously mentioned.

The significant size of the computation domain and the large size of the number of modelled days constitute a database with a large variability of both geometric and meteorological configurations. The analysis of concentration maps, more particularly, days with major deviations at stations, has enabled the identification of several issues. The main one is the emergence of a large number of concentration accumulation zones in front of buildings, yet not direct neighbors of an emission strand. The settings analysis of PMSS, and, more specifically, the comparison with other setups that do not present this artifact, highlighted the effect of the α stability coefficient in PSWIFT, used during the mass conservation step. In the Paris calculation chain, this coefficient was calculated during a pre-processing step to PSWIFT from the ESMERALDA temperatures profiles and a tabulation indexed on the vertical temperature gradient. This tabulation, derived from an in-house parametric study carried out with confidential data from a wind farm site with a complex terrain, is adapted for the consideration of the stability effect on obstacle avoidance with hill or mountain length scale, but not with a building length scale: for a Froude number lower than 1, an atmospheric flow bypasses an obstacle [40], but, even for a very stable meteorological case, such as a weak wind with a speed of $1\ m{\cdot}s^{-1}$ and a vertical gradient of potential temperature of $0.03^{\circ}{\cdot}m^{-1}$, the Froude number Fr is equal to 1.59 for an obstacle of height h = 20 m, such as Parisian buildings (Fr = 0.31 for h = 100 m). It is therefore convenient to use α coefficient close to 1 in order to calculate the flow around buildings in urban areas.

Table 5. Comparative statistics between observed and computed hourly average concentration of NO_2 and PM10 in $\mu g \cdot m^{-3}$—Config 1 corresponds to the initial parametrization of PMSS and background concentration extracted on Paris cell in the regional modelling; Config 1 Upwind background corresponds to the same parametrization of PMSS, but with background concentration extracted upwind Paris cell in the regional modelling; Config 2 corresponds to the parametrization of PMSS with a correction on the stability coefficient, and with background concentration extracted from the upwind Paris cell in the regional modelling; 5 months and 5 days corresponds, respectively, to the large period modelled initially, and to the sub list of 5 days used afterward.

Period	5 Months									5 Days							
	Obs	Model Config 1				Model Config 1 Upwind Background				Model Config 1				Model Config 2			
	Mean	Mean	RMSE	Bias	r	Mean	RMSE	Bias	r	Mean	RMSE	Bias	r	Mean	RMSE	Bias	r
NO2_AUT	81.6	124.4	77.2	43.7	0.34	103.4	69.2	31.4	0.35	127.3	93.6	53.2	−0.02	88.5	31.1	15.7	0.62
NO2_BONAP	48.5	67.7	22.2	12.1	0.65	44.8	19.9	−0.2	0.60	74.8	28.1	21.3	0.84	55.5	14.2	5.0	0.82
NO2_CELES	58.7	81.7	26.1	18.6	0.80	55.8	19.9	6.6	0.77	71.5	24.6	19.3	0.80	58.9	14.2	6.5	0.77
NO2_ELYS	47.5	73.1	36.5	25.9	0.54	57.6	30.1	13.7	0.48	70.1	33.0	24.9	0.52	66.2	29.4	21.2	0.72
NO2_HAUS	51.6	66.3	28.1	15.1	0.66	51.5	23.0	2.9	0.63	82.3	41.4	32.5	0.51	63.8	25.8	14.0	0.58
NO2_OPERA	64.9	99.9	47.5	35.5	0.66	81.9	39.5	23.4	0.63	104.7	54.5	43.3	0.56	94.6	45.1	33.5	0.68
NO2_PA04C *	39.3	46.9	17.9	7.8	0.61	33.8	16.7	−4.3	0.55	54.2	23.1	15.2	0.47	33.2	11.6	−6.3	0.78
NO2_PA07 *	34.8	51.1	29.4	16.4	0.35	37.6	25.8	4.2	0.28	57.9	31.5	25.7	0.67	36.3	18.1	3.9	0.61
NO2_PA12 *	38.2	40.8	17.1	3.0	0.58	28.2	19.1	−9.2	0.53	46.3	13.4	8.7	0.81	26.2	13.7	−11.4	0.87
NO2_PA13 *	35.8	38.4	16.2	3.0	0.58	26.1	18.7	−9.2	0.52	41.2	13.0	5.8	0.73	25.1	13.7	−10.5	0.82
NO2_PA18 *	41.5	34.9	14.6	−5.2	0.72	26.7	23.2	−17.4	0.64	35.3	9.2	−1.8	0.80	19.2	22.0	−17.7	0.68
PM10_AUT	37.2	51.0	29.9	14.3	0.44	44.7	27.1	10.1	0.43	53.1	35.5	18.4	0.31	38.8	19.2	4.4	0.45
PM10_ELYS	31.0	36.6	16.1	5.3	0.55	31.7	14.7	0.9	0.54	35.1	21.9	8.9	0.43	32.0	18.3	5.7	0.47
PM10_HAUS	30.6	34.4	15.1	4.0	0.56	29.7	13.9	−0.2	0.53	37.1	18.4	8.5	0.72	30.5	13.1	1.7	0.71
PM10_OPERA	29.7	41.0	19.0	11.7	0.63	35.7	16.1	7.5	0.61	41.8	24.3	14.2	0.64	37.0	20.2	9.4	0.60
PM10_PA04C *	21.3	28.9	16.3	7.9	0.48	24.7	14.5	3.7	0.42	28.7	19.2	10.4	0.69	21.6	13.2	2.9	0.67
PM10_PA18 *	21.1	17.9	8.4	−2.5	0.76	23.5	11.4	−6.7	0.67	15.9	6.4	−1.5	0.81	9.2	12.7	−8.3	0.51

* background stations.

Over the 5 months modeling period, 5 days were selected by including cases with good and bad scores, and recalculated with $\alpha = 1$. Accumulation zones have significantly decreased. Scores from the initial setup (called Config 1) limited to the 5 days are available in Table 5 next to the scores obtained with the α coefficient correction and the choice of the background concentration value in the upwind cell of Paris (named Config 2). Config 1 scores, limited to the 5 days, are generally worse than those for the 5-month period. The choice of 5 days is therefore rather penalizing. Over these 5 days, RMSE, bias, and correlation are improved between Config 1 and Config 2, except at the background stations *PA_12*, *PA_13* and *PA_18* for NO_2, and at *PA_18* for PM10. Low scores recorded at the Auteuil station are significantly improved. This station located on the edge of the ring road is not in the vicinity of a building. Nevertheless, it is close to a slope, the ring road there being located in a 10 m recessed zone. Modification of the α coefficient can therefore have an impact on the calculation of the flow in this area.

Another issue that was observed through the analysis of the hourly average concentration maps is the noisy appearance of the values with horizontal gradients from one cell to the other, even in open areas. Emissions might have been not well discretized because of the number of Lagrangian particles being too low. The calculation time constraint limits this number. Possible improvements include the use of the Kernel method [38] for the computation of concentrations from particles, as well as the optimization of the number of particles emitted by each road section. In the setup used for the Paris forecast system, all the strands emit the same number of particles, even if they emit different amounts of pollutants. On the setup presented in the next section, this distribution, according to mass rate, has been implemented and optimized. It allows the balance between computation time and noise in concentration fields to be improved.

Finally, another issue was observed on the hourly average concentration maps: concentration levels on some major roads appear to be potentially very high (exceeds 1000 $\mu g \cdot m^{-3}$ in NO_2 some days). These roads are associated with rather fast traffic and rather open areas, which are too wide to experience canyon effects (for example Grande Armée Avenue in Neuilly-sur-Seine or Paris Ring Road). The noise, due to the number of lagrangian particles being too low, could explain some of these high values, but it seems also that the impacted zones depend on the wind direction: particle accumulation along the road axis is maximized when the wind is aligned with the axis. One possibility for improvement could be to consider the turbulence induced by the traffic of these road axes, as already done in the canyon type streets (see Section 1), but with a formulation that would be adapted to less-confined areas.

3.2. *Antony Forecast System*

3.2.1. Context and Model Setup

A forecasting system for the city of Antony, located in the south suburbs of Paris, was operated between the 15 October 2019 and the 16 April 2020 in the framework of the "Numerical Challenge POC & Go on Air Quality". The extent of the computational domain of the PMSS model is 4300 m $\times$ 4800 m. The horizontal spatial resolution is constant, and is equal to 4 m. The vertical resolution is equal to 1.5 m between the ground, and the first 20 m then decreases gradually up to the calculation ceiling, located at 500 m. The mesh is composed of approximately 3.48×10^7 cells.

The computation chronology allows for calculation of the 24 h of the following day overnight. The chain provides hourly concentrations. Meteorological forcing of PMSS is driven by a forecast made with the WRF model, which, itself, is forced by the NCEP/GFS global forecast. Background concentration forecasts are extracted from the forecast of the COPERNICUS Atmosphere Monitoring Service (CAMS) implemented by ECMWF.

3.2.2. CPU Time Performance and Optimization

No reference station of air quality measurement is included in the domain. The validation of the pollutant concentration forecasts could not be done. Nonetheless, this case remains interesting for the optimization aspect of the computational time. The forecast chain has indeed been operated with a machine which has a limited computation capacity: one node of 12 cores. In order to obtain forecasts results on time (the targeted CPU time is 5 h for 24 modelled hours), the number of particles that discretize the dispersion in PSPRAY has been optimized. The particles emission time step, and the number of particles emitted per source (here, per section of road), were adjusted. By default, this number is identical for all sources, even if they have different pollutant mass rates. Some particles therefore have, with this configuration, a more important weight, which can lead to very noisy concentration fields. PSPRAY enables, though pre-processing or internally, to modulate the numbers of emitted particles according to the mass rate of the pollutant. A pre-processing adjustment was applied to the case of Antony, whose domain includes high traffic highways and low traffic residential lanes.

4. CPU Demand Analysis and Estimation

This section summarizes the computation times for the modelling described previously. Only the dispersion part (calculated by PSPRAY) is detailed because it is predominant in this type of application. For the meteorological part with PSWIFT, the optimization of the computation time could be done by the choice of the number/sizes of tiles (i.e., subdomains). However, this has not been carried out on the different cases. Computation times of PSPRAY can be found in Table 6, in hour and hour.core. They are given for the modelling of a day of physical time. Three of the cases were performed on the EOS server at the CALMIP computing center. Different servers were used for the two other cases. Part of the computing times variability can thus come from the variability of machine performances, but this has not been quantified.

Table 6. Modelling setup parameters, effective CPU demand for dispersion part with PSPRAY model, and estimation through a simple linear fit. The CALMIP calculation center server used is EOS. It is constituted of nodes with Intel IvyBridge @2.80 GHz (20 logical cores per node). The server1 and server2 are internal calculation servers, respectively constituted of two intel Xeon E5-2640 V4 @2.40 GHz (40 logical cores) and two Intel Xeon X5680 @3.33 GHz (12 physical cores without hyperthreading). The two last lines give estimations of CPU based on input parameters and a simple linear fit.

			Paris	Coastal City	Grenoble	Antony	Rome
Variable	**Short Name**	**Unit**					
Domain *X*-axis dimension	Lx	km	10	3.6	1.7	4.3	12
Domain *Y*-axis dimension	Ly	km	13	4.6	1.8	4.8	12
Horizontal resolution	dx	m	3	3	3	4	4
Number of tiles			120	9	1	9	36
Number of line sources per unit area	nblin	km^{-2}	327.7	1078.6	260.8	454.4	513.9
Number of emitted particles is function of mass rate			no	no	no	yes	yes
Kernel method			no	no	no	no	yes
Emission time step	dtmin	s	2	10	1	10	100
Synchronisation time step	dtsync	s	6	5	5	10	5
Number of cores			480	240	30	10	180
Machine			CALMIP	CALMIP	Server 1	Server 2	CALMIP
CPU time for 24 h (PSPRAY only)		hour	6.0	1.4	6.9	5.0	3.0
Number of hour·core for 24 h		hour·core	2880	333	208	50	540
Number of hour·core per day and per km^2	CPU_cost	hour·core. km^{-2}·day^{-1}	22.2	20.1	68.0	2.4	3.8
CPU_cost estimation = f(nblin/dtmin/dtsync/dx^2)	CPU_cost_estim1	hour·core·km^{-2}·day^{-1}	22.8	18.0	43.5	2.1	0.5
CPU_cost estimation = f(nblin/dtmin/dtsync/)	CPU_cost_estim2	hour·core·km^{-2}·day^{-1}	22.4	17.7	42.8	3.7	0.8

To compare the different cases, it is firstly assumed that the computation time increases linearly with the extent of studied area—the table thus provides CPU times per area unit. The resulting values range from 2.4 to 68 h·core·km^{-2}.

Other parameters that seem to significantly affect the computation time are provided: the number of emission sources, normalized by area, in order to compare sites of different sizes; emission time step and synchronization time step (the synchronization time step in PSPRAY allows recording of the position of particles at a regular frequency, which, apart from these moments, move with their own time steps); whether or not the Kernel method was used for the computation of concentrations. The latter takes more time for the concentration calculation, but can allow the number of particles that need to be transported to be reduced (see the larger emission time-step for the Rome study).

The table only provides a sample of five sites, and the number of parameters influencing the computation time is maybe too large to perform a multi-regression. However, a rough approximation of computation time by unit area (conjecturing the CPU time linearity with the calculation domain area) was done, assuming linearity with: (1) the emission sources number; (2) the inverse of the emission time step; (3) the inverse of synchronization time step; and possibly (4) the inverse of the square of the horizontal resolution. These rough approaches underestimate CPU times for the cases of Grenoble and Rome, but are satisfying for the three other cases. In the Rome study, the Kernel method was used, making this case special compared to others.

The goal of these estimates is to quantify, in principle, the CPU cost for a new site, in order to design the required size for the computing machines.

5. Discussion

The different sites presented in the paper provided both validation scores and valuable feedback to improve the modelling and its CPU cost. The scores' quality appears to be disparate between the sites, but also for the same site: modelling air quality at local scale means dealing with different configurations, even in the same city (narrow canyon streets, complex crossroads, half-buried highways, skyscraper zones, streets with complex emission distribution due to the presence of tramway lanes).

An overestimation tendency seems to be observed. Even if the possible origins are numerous, an underestimation of the turbulence due to road traffic could be one. This effect could be considered in the 3D turbulence fields, or, more simply, by enlarging the emission volumes.

The long-term results analysis has shown some patterns, and provides feedback that has led to model improvements, especially for making the Rockle approach more robust in more geometrical configurations.

The long-term calculations were also useful to analyze the performances of two methods used to improve the CPU cost:

- Classification with SOMs method to reduce the number of days to be considered. The study provides the quantification of the classification effect on annual average concentration and, more challengingly, on percentiles;
- Kernel method to compute concentration fields from particle clouds. The obtained CPU times during the BEEP project are attractive. The method, already implemented in other LPDM, such as FLEXPART [41] and LAPMOD [42], should be tested on other sites.

The CPU time of the different sites have been compared, and a tentative estimation has been made from input parameters and characteristics, such as the area, the number of roads, and particle emission time steps. The limited number of sites does not allow a true multi-regression to be calculated, so only a raw linear law is set. The CPU time database should be fed by all the impact studies performed with PMSS, but these values are not available because they are not stored by default.

The 3D approach presented in the paper allows access to detailed 3D concentration fields that open perspectives about health impact and exposure. Exposure, which is the integration of both concentration and population density, is usually performed based on a 2D raster method. The results provided with PMSS can be used differently: as the buildings are seen explicitly, the exposure can be computed by building. The concentration for each building could be the average and maximum concentration around it, for example. Moreover, the exposure can be computed level by level, quantifying the vertical gradient between the first and last floors. During the FAIRCITY project, a first attempt has been made on an area of Grenoble, with ATMO AURA showing some significant gradients (10% reduction between the ground and 18 m.a.g.l. for the NO_2 concentration), but no observation was available to validate the results. Observations in wind tunnel experiments are available [43], but might miss some of the processes present in a real street, such as traffic-induced turbulence, or convective flows due to the thermal and radiative effects of buildings. Having a field campaign with sensors at different heights on the façades of buildings, as is described in [44–46], would open a great validation perspective.

Author Contributions: Conceptualization: M.N.; Methodology: M.N.; Software: B.R. and D.B.; Validation: M.N., B.R., G.T. and D.B.; Formal Analysis: M.N., B.R., G.T. and D.B.; Writing—Original Draft Preparation: M.N., B.R., L.R., G.T. and D.B.; Writing—Review and Editing: M.N.; Visualization: B.R. and D.B.; Supervision: M.N.; Project administration: J.M. and G.T.; Funding Acquisition: J.M., G.T. and A.A. All authors have read and agreed to the published version of the manuscript.

Funding: The authors would like to thank the FUI, the Auvergne-Rhône-Alpes and Ile-de-France regions for the funding of the FAIRCITY project, and also the National Institute for Insurance against Accidents at Work for the project "BEEP" (project code B72F17000180005).

Institutional Review Board Statement: Not applicable.

Informed Consent Statement: Not applicable.

Data Availability Statement: Air quality observations on Paris are available on https://data-airparif-asso.opendata.arcgis.com/search?tags=mesure (accessed on 10 September 2021).

Acknowledgments: We truly appreciate the help of the three Air quality regional agencies who provided input data of their own modelling systems and participated in the analyses of the results obtained with the 3D model PMSS. The authors would also like to thank the CALMIP HPC teams for their support.

Conflicts of Interest: The authors declare no conflict of interest.

References

1. Harrison, R.M. Urban atmospheric chemistry: A very special case for study. *NPJ Clim. Atmos. Sci.* **2018**, *1*, 20175. [CrossRef]
2. Arya, S.P. *Introduction to Micrometeorology*; Academic Press: San Diego, CA, USA, 1987; p. 307.
3. Oke, T.R.; Mills, G.; Christensen, A.; Voogt, J.A. *Urban Climates*; Cambridge University Press: Cambridge, UK, 2017; p. 519.
4. Moussafir, J.; Oldrini, O.; Tinarelli, G.; Sontowski, J.; Dougherty, C. A new operational approach to deal with dispersion around obstacles: The MSS (Micro-Swift-Spray) software suite. In Proceedings of the 9th International Conference on Harmonisation within Atmospheric Dispersion Modelling for Regulatory Purposes, Garmish-Partenkirchen, Germany, 6–10 June 2004; Volume 2, pp. 114–118.
5. Tinarelli, G.; Brusasca, G.; Oldrini, O.; Anfossi, D.; Trini Castelli, S.; Moussafir, J. Micro-swift-spray (MSS) A new modeling system for the simulation of dispersion at microscale, general description and validation. In Proceedings of the 27th CCMS-NATO Meeting, Banff, AB, Canada, 24–29 October 2004.
6. Rockle, R. Bestimmung der Stromungsverhaltnisse im Bereich Komplexer Bebauungsstrukturen. Ph.D. Thesis, Vom Fachbereich Mechanik, der Technischen Hochschule Darmstadt, Darmstadt, Germany, 1990.
7. Oldrini, O.; Olry, C.; Moussafir, J.; Armand, P.; Duchenne, C. Development of PMSS, the parallel version of Micro SWIFT SPRAY. In Proceedings of the 14th International Conference on Harmonisation within Atmospheric Dispersion Modelling for Regulatory Purposes, Kos, Greece, 2–6 October 2011; pp. 443–447.
8. Oldrini, O.; Armand, P.; Duchenne, C.; Olry, C.; Tinarelli, G. Description and preliminary validation of the PMSS fast response parallel atmospheric flow and dispersion solver in complex built-up areas. *J. Environ. Fluid Mech.* **2017**, *17*, 997–1014. [CrossRef]
9. Trini Castelli, S.; Armand, P.; Tinarelli, G.; Duchenne, C.; Nibart, M. Validation of a Lagrangian particle dispersion model with wind tunnel and field experiments in urban environment. *Atmos. Environ.* **2018**, *193*, 273–289. [CrossRef]
10. Gomez, F.; Ribstein, B.; Makké, L.; Armand, P.; Moussafir, J.; Nibart, M. Simulation of a dense gas chlorine release with a lagrangian particle dispersion model (LPDM). *Atmos. Environ.* **2021**, *244*, 117791. [CrossRef]
11. Moussafir, J.; Olry, C.; Nibart, M.; Albergel, A.; Armand, P.; Duchenne, C.; Mahé, F.; Thobois, L.; Loaëc, S.; Oldrini, O. AIRCITY: A Very High Resolution Atmospheric Dispersion Modeling System for Paris. In *American Society of Mechanical Engineers, Fluids Engineering Division*; FEDSM: Chicago, IL, USA, 2014; Volume 1D. [CrossRef]
12. Geai, P. Méthode d'interpolation et de reconstitution tridimensionnelle d'un champ de vent: Le code d'analyse objective SWIFT. In *EDF/DER Internal Report HE34-87*; EDF Group: Chatou, France, 1987.

13. Hanna, S.; White, J.; Trolier, J.; Vernot, R.; Brown, M.; Gowardhan, A.; Kaplan, H.; Alexander, Y.; Moussafir, J.; Wang, Y.; et al. Comparisons of JU2003 observations with four diagnostic urban wind flow and Lagrangian particle dispersion models. *Atmos. Environ.* **2011**, *45*, 4073–4081. [CrossRef]
14. Berkowicz, R.; Hertel, O.; Larsen, S.; Sørensen, N. *Modelling Traffic Pollution in Streets*; Ministry of Environment and Energy, National Environmental Research Institute: Roskilde, Denmark, 1997.
15. Rodean, H.C. *Stochastic Lagrangian Models of Turbulent Diffusion*; American Meteorological Society: Boston, MA, USA, 1996; Volume 45.
16. Anfossi, D.; Desiato, F.; Tinarelli, G.; Brusasca, G.; Ferrero, E.; Sacchetti, D. TRANSALP 1989 experimental campaign Part II: Simulation of a tracer experiment with Lagrangian particle models. *Atmos. Environ.* **1998**, *32*, 1157–1166. [CrossRef]
17. Carvalho, J.; Anfossi, D.; Castelli, S.T.; Degrazia, G.A. Application of a model system for the study of transport and diffusion in complex terrain to the TRACT experiment. *Atmos. Environ.* **2002**, *36*, 1147–1161. [CrossRef]
18. Ferrero, E.; Anfossi, D. Comparison of PDFs, closures schemes and turbulence parameterizations in Lagrangian Stochastic Models. *Int. J. Environ. Pollut.* **1998**, *9*, 384–410.
19. Ferrero, E.; Anfossi, D.; Tinarelli, G. Simulations of Atmospheric Dispersion in an Urban Stable Boundary Layer. *Int. J. Environ. Pollut.* **2001**, *16*, 1–6. [CrossRef]
20. Trini Castelli, S.; Anfossi, D.; Ferrero, E. Evaluation of the environmental impact of two different heating scenarios in urban area. *Int. J. Environ. Pollut.* **2003**, *20*, 207–217. [CrossRef]
21. Tinarelli, G.; Anfossi, D.; Brusasca, G.; Ferrero, E.; Giostra, U.; Morselli, M.G.; Moussafir, J.; Tampieri, F.; Trombetti, F. Lagrangian particle simulation of tracer dispersion in the lee of a schematic two-dimensional hill. *J. Appl. Meteorol.* **1994**, *33*, 744–756. [CrossRef]
22. Tinarelli, G.; Anfossi, D.; Bider, M.; Ferrero, E.; Trini Castelli, S. A new high performance version of the Lagrangian particle dispersion model SPRAY, some case studies. In *Air Pollution Modelling and Its Applications XIII*; Gryning, S.E., Batchvarova, E., Eds.; Kluwer Academic/Plenum Press: New York, NY, USA, 2000; pp. 499–507.
23. Thomson, D.J. Criteria for the selection of stochastic models of particle trajectories in turbulent flows. *J. Fluid Mech.* **1987**, *180*, 529–556. [CrossRef]
24. Kaplan, H.; Olry, C.; Moussafir, J.; Oldrini, O.; Mahé, F.; Albergel, A. Chemical reactions at street scale using a lagrangian particle dispersion model (LPDM). In Proceedings of the 15th International Conference on Harmonization within Atmospheric Dispersion Modeling for Regulatory Purposes, Madrid, Spain, 6–9 May 2013.
25. Benson, P. CALINE4—A Dispersion Model for Predicting Air Pollutant Concentrations near Roadways. Internal Report No FHWA/CA/TL-84/15. 1989. Available online: https://trid.trb.org/view/215944 (accessed on 15 September 2021).
26. Malherbe, L.; Wroblewski, A.; Letinois, L.; Rouil, L. Evaluation of numerical models used to simulaite atmospheric pollution near roadways 13. In Proceedings of the International Conference on Harmonisation within Atmospheric Dispersion Modelling for Regulatory Purposes (HARMO 13), Paris, France, 1–4 June 2010; pp. 697–700.
27. Hertel, O.; Berkowicz, R. Operational Street Pollution Model (OSPM). In *Evaluation of the Model on Data from St. Olavs Street in Oslo, DMU Luft A-135*; National Environmental Research Institute: Roskilde, Denmark, 1989.
28. Soulhac, L.; Salizzoni, P.; Cierco, F.-X.; Perkins, R.J. The model SIRANE for atmospheric urban pollutant dispersion: PART I: Presentation of the model. *Atmos. Environ.* **2011**, *45*, 7379–7395. [CrossRef]
29. Armand, P.; Commanay, J.; Nibart, M.; Albergel, A.; Achim, P. 3D simulations of pollutants atmospheric dispersion around the buildings of an industrial site comparison of MERCURE CFD approach with Micro-SWIFT-SPRAY semi-empirical approach. In Proceedings of the 11th International Conference on Harmonisation within Atmospheric Dispersion Modelling for Regulatory Purposes (HARMO11), Cambridge, UK, 2–5 July 2007.
30. Kohonen, T. *Self-Organizing Maps*, 2nd ed.; Springer: Berlin/Heidelberg, Germany, 1997.
31. Yonggang, L.; Weisberg, R.H. A Review of Self-Organizing Map Applications in Meteorology and Oceanography. In *Self Organizing Maps: Applications and Novel Algorithm Design*; Mwasiagi, J.I., Ed.; IntechOpen: London, UK, 2011; pp. 253–272.
32. Wendum, D.; Moussafir, J. Méthodes d'interpolation spatiale utilisables pour le calcul d'écoulements atmosphériques à moyenne échelle. In *EDF Internal Report EDF/DER HE/32-85.22*; EDF Group: Chatou, France, 1985.
33. Menut, L.; Bessagnet, B.; Khvorostyanov, D.; Beekmann, M.; Blond, N.; Colette, A.; Coll, I.; Curci, G.; Foret, G.; Hodzic, A.; et al. CHIMERE (2013): A model for regional atmospheric composition modelling. *Geosci. Model Dev.* **2013**, *6*, 981–1028. [CrossRef]
34. Gariazzo, C.; Silibello, C.; Finardi, S.; Radice, P.; Piersanti, A.; Calori, G.; Cecinato, A.; Perrino, C.; Nussio, F.; Cagnoli, M.; et al. A gas/aerosol air pollutants study over the urban area of Rome using a comprehensive chemical transport model. *Atmos. Environ.* **2007**, *41*, 7286–7303. [CrossRef]
35. Silibello, C.; Calori, G.; Brusasca, G.; Giudici, A.; Angelino, E.; Fossati, G.; Peroni, E.; Buganza, E. Modelling of PM10 Concentrations Over Milano Urban Area Using Two Aerosol Modules. *Environ. Model. Softw.* **2008**, *23*, 333–343. [CrossRef]
36. Skamarock, W.C.; Klemp, J.B.; Dudhia, J.; Gill, D.O.; Barker, D.M.; Duda, M.G.; Huang, X.Y.; Wang, W.; Powers, J.G. *A Description of the Advanced Research WRF Version 3*; NCAR Technical Note NCAR/TN-475+STR; NCAR: Boulder, CO, USA, 2008. [CrossRef]
37. Barbero, D. Sviluppo e Applicazione del Metodo Kernel in Modelli Lagrangiani a Particelle a Scala Locale e a Microscala. Master's Thesis, Politecnico di Milano, Milan, Italy, 2019. Available online: http://hdl.handle.net/10589/146531 (accessed on 11 September 2021).

38. Bessagnet, B.; Couvidat, F.; Lemaire, V. A statistical physics approach to perform fast highly-resolved air quality simulations—A new step towards the meta-modelling of chemistry transport models. *Environ. Model. Softw.* **2019**, *116*, 100–109. [CrossRef]
39. Carlino, G.; Pallavidino, L.; Prandi, R.; Avidano, A.; Matteucci, G.; Ricchiuti, F.; Bajardi, P.; Bolognini, L.; Elise, P. Micro-scale modelling of urban Air quality to forecast NO2 critical level in traffic Hot-spots. In Proceedings of the Air Quality 2013, Milan, Italy, 14–18 March 2016.
40. Stull, R.B. *An Introduction to Boundary Layer*; Springer: Dordrecht, The Netherland, 1988.
41. Stohl, A.; Forster, C.; Frank, A.; Seibert, P.; Wotawa, G. Technical note: The Lagrangian particle dispersion model FLEXPART version 6.2. *Atmos. Chem. Phys.* **2005**, *5*, 2461–2474. [CrossRef]
42. Bellasio, R.; Bianconi, R. *LAPMOD—Manuale D'uso*; Technical report of Enviroware-Air Qual; Consult; Enviroware: Concorezzo, Italy, 2017.
43. Soulhac, L.; Méjean, P.; Perkins, R.J. Modelling the transport and dispersion of pollutants in street canyons. *Int. J. Environ. Pollut.* **2001**, *16*, 404–416. [CrossRef]
44. Wu, Y.; Hao, J.; Fu, L.; Wang, Z.; Tang, U. Vertical and horizontal profiles of airborne particulate matter near major roads in Macao, China. *Atmos. Environ.* **2002**, *36*, 4907–4918. [CrossRef]
45. Tuckett-Jones, B.T. Reade. In *City Quality at Height—Lessons for Developers & Planners*; Report from WSP; Parsons Brinkerhoff: London, UK, 2017.
46. Wong, P.P.Y.; Lai, P.-C.; Allen, R.; Cheng, W.; Lee, M.; Tsui, A.; Tang, R.; Thach, T.-Q.; Tian, L.; Brauer, M.; et al. Vertical monitoring of traffic-related air pollution (TRAP) in urban street canyons of Hong Kong. *Sci. Total Environ.* **2019**, *670*, 696–703. [CrossRef] [PubMed]

atmosphere

Article

A Study of Traffic Emissions Based on Floating Car Data for Urban Scale Air Quality Applications

Felicita Russo [1,*], Maria Gabriella Villani [1], Ilaria D'Elia [1], Massimo D'Isidoro [1], Carlo Liberto [2], Antonio Piersanti [1], Gianni Tinarelli [3], Gaetano Valenti [2] and Luisella Ciancarella [1]

1 ENEA, Italian National Agency for New Technologies, Energy and Sustainable Economic Development, SSPT-MET-INAT, 40129 Bologna, Italy; mariagabriella.villani@enea.it (M.G.V.); ilaria.delia@enea.it (I.D.); massimo.disidoro@enea.it (M.D.); antonio.piersanti@enea.it (A.P.); luisella.ciancarella@enea.it (L.C.)
2 ENEA, Italian National Agency for New Technologies, Energy and Sustainable Economic Development, TERIN-PSU-STMS, 00196 Roma, Italy; carlo.liberto@enea.it (C.L.); gaetano.valenti@enea.it (G.V.)
3 ARIANET, Via Gian Giacomo Gilino, 9, 20128 Milan, Italy; g.tinarelli@aria-net.it
* Correspondence: felicita.russo@enea.it

Abstract: Urban air quality in cities is strongly influenced by road traffic emissions. Micro-scale models have often been used to evaluate the pollutant concentrations at the scale of the order of meters for estimating citizen exposure. Nonetheless, retrieving emissions information with the required spatial and temporal details is still not an easy task. In this work, we use our modelling system PMSS (Parallel Micro Swift Spray) with an emission dataset based on Floating Car Data (FCD), containing hourly data for a large number of road links within a 1×1 km^2 domain in the city of Rome for the month of May 2013. The procedures to obtain both the emission database and the PMSS simulations are hosted on CRESCO (Computational Centre for Research on Complex Systems)/ENEAGRID HPC facilities managed by ENEA. The possibility of using such detailed emissions, coupled with HPC performance, represents a desirable goal for microscale modeling that can allow such modeling systems to be employed in quasi-real time and nowcasting applications. We compute NOx concentrations obtained by: (i) emissions coming from prescribed hourly modulations of three types of roads, based on vehicle flux data in the FCD dataset, and (ii) emissions from the FCD dataset integrated into our modelling chain. The results of the simulations are then compared to concentrations measured at an urban traffic station.

Keywords: air quality; urban scale; traffic emissions; micro-scale dispersion models; HPC

Citation: Russo, F.; Villani, M.G.; D'Elia, I.; D'Isidoro, M.; Liberto, C.; Piersanti, A.; Tinarelli, G.; Valenti, G.; Ciancarella, L. A Study of Traffic Emissions Based on Floating Car Data for Urban Scale Air Quality Applications. *Atmosphere* **2021**, *12*, 1064. https://doi.org/10.3390/atmos12081064

Academic Editor: Riccardo Buccolieri

Received: 12 July 2021
Accepted: 14 August 2021
Published: 19 August 2021

1. Introduction

Urban air quality is determined by complex atmospheric patterns, influenced by local emissions, and shaped by the three-dimensional structure of the built environment. The so called "street canyons" (which is a term frequently used for urban streets flanked by buildings on both sides) tend to entrap pollution near the ground, while in more open spaces (parks, squares, residential areas) the pollution levels take the form of an urban background, with increasing impact of more distant sources [1]. Among the sources of local emission, road traffic is usually the main contributor [2,3]. A study from The European Topic Centre on Air and Climate Change (ETC/ACC) [4] through a survey among European cities reported average percentage contributions from road traffic ranging from 40% (21%) to 49% (28%) of nitrogen dioxide (particulate matter) concentrations measured at background stations.

Microscale dispersion models have been used to evaluate pollutant concentrations at high spatial detail, featuring grids with metrical resolution in order to estimate citizen exposure more accurately than other air quality models at larger scales [5,6]. For species mainly driven by local emissions, such as nitrogen oxides (NOx), such a detailed model description of dispersion dynamics requires a coherent emission input, therefore reproducing the spatial variability at the single street/stack level and the temporal variability at hourly or even sub-hourly levels. For air quality applications, the calculation of street-level and hourly-level emissions covering large urban areas is not straightforward. Traffic flows and speeds on urban road networks at the street level are usually obtained as long-term averages by traffic assignment models calibrated from observations; as shown in the review by [7], these models are the typical input for Static Emission Models, commonly used for transportation planning purposes due to their relative simplicity but are mostly adequate when a high resolution is not required. Therefore, in order to obtain the hourly variations, modulation profiles are needed which generally come from dedicated measurements or the literature and are not street-specific [8,9]. Though it is able to obtain a comprehensive coverage of urban areas for long time periods, this approach has shown limitations in reproducing measured street level concentrations, namely in their hourly variability, and therefore it can lead to biased estimations of human exposure to pollutant concentrations. More recently, some studies have considered new types of measurements. Video-based systems allow both vehicle flows to be counted and the fleet composition to be retrieved [10], but at a fixed location and therefore with limited spatial coverage. Remote sensing of actual vehicle positions (FCD, Floating Car Data) can deliver very detailed vehicle passages on wider areas by tagging each single vehicle at very short time steps. e.g., 30 s [11,12]. This allows very detailed vehicle flows on individual streets and hourly intervals to be derived, improving emission calculations. Jing et al., reported a large part (60 km $\times$ 60 km) of the Beijing road network divided into segments according to the traffic speed, with an attribution of the traffic flow and speed for each segment [13]. On the other hand, depending on the area and the period considered, the storage and elaboration of FCD can require important resources, limiting their use on comprehensive urban road networks (e.g., Gately et al. used FCD only for vehicle speeds and in-road sensors for vehicle flows, while Jiang et al. studied the link between traffic speed and volume on a one ring expressway in Beijing) [14,15]. Therefore, there is at the moment no universal solution to evaluate microscale traffic emissions over an extended part of a city at a reasonable cost [8].

In this work, we describe a microscale simulation of NOx concentrations, conducted with the Parallel Micro Swift Spray (PMSS) model, and fed by an emission dataset based on FCD, containing hourly data for a large number of road links within the city of Rome for the month of May 2013. In this work, all the simulations were performed relying on the computational resources of CRESCO (Computational Centre for Research on Complex Systems) /ENEAGRID High Performance Computing infrastructure [16].

The main objective was to evaluate NOx concentrations simulated by PMSS using FCD-based emissions in comparison to NOx concentrations measured at the Magna Grecia urban air quality station. We computed NOx concentrations obtained by: (i) emissions coming from prescribed hourly modulations of three types of roads, based on vehicle flux data in the Floating Car dataset, and (ii) emissions from the Floating Car dataset integrated into our modelling chain. We also present an exploration of the FCD-emissions database, evaluating the feasibility of automatically integrating it into our modelling chain and possible alternative strategies for a more straightforward use as a model input.

Section 2 presents a detailed description of the emission database, the other input data, the modelling chain, and the simulations setup. Section 3 presents the results and discussions with an evaluation of the modelled NOx concentrations against the reference measurements (Section 3.1), and a detailed comparison of the traffic emissions processors used to retrieve the mass emitted starting from the vehicle flux information (Section 3.2). Section 4 presents our conclusions. Some useful additional graphics and definitions are included in Appendices A and B.

2. Methodology

2.1. Road Traffic SMTS Data

Nowadays, the use of massive FCD to extract traffic patterns and travel behaviors occurring in urban areas is extremely appealing [17–19]. It represents a reliable and cost-effective way to gather accurate traffic data over a wide-area road network and thus to improve many applications, such as location-based services [20], urban planning [21], and traffic management [22,23]. Despite this remarkable potential, FCD exploitation in urban transport is still at an early stage compared to other approaches [24,25], particularly due to the fragmented availability of transport/mobility data, institutional barriers, and data privacy/security issues. In this study we present ENEA's STMS (Systems and Technologies for Sustainable Mobility) laboratory FCD collected by Octo Telematics to obtain insights into the travel patterns of private cars and to estimate the traffic emissions in the case study. Octo Telematics is a company that offers data analytics for the auto insurance industry and provides other innovative connected user services including vehicle diagnostics, fleet management, road tolling, and real-time monitoring of traffic and environmental conditions [26]. The FCD used in this study represents about 5 percent of the circulating passenger cars in the study area. Similar datasets have been used in the past to calculate vehicle home locations to predict energy-oriented land use [27]. The residential population calculated with FCD was compared to a census population showing a R-squared value of 0.74, proving therefore that the initial FCD dataset was adequately representative of the mobility patterns of the entire road network, though covering a fraction of all vehicle movements. Cars were equipped with an OBU (On-Board Unit) that stores GPS measurements (position, heading, speed, quality) and, periodically, transmits them to the Data Processing Center. The OBU consists of a GPS receiver, a GPRS transmitter, a 3-axis accelerometer sensor, a battery pack, a mass memory, a processor, and a RAM. The OBU stores GPS measurements every 2 km travelled or, alternatively, every 30 s when the vehicle is running along a motorway or some main urban arterials. For each equipped vehicle, we then extracted the travel list performed and the most likely routes in the network by matching sequences of the positioning data to a street digital map. We reconstructed the route between each OD (origin–destination) pair by applying a map-matching algorithm that incorporates the street network topology, including prohibited maneuvers and turn restrictions information. The database at this point was used to evaluate the emitted mass per link per hour using the traffic emission processor ECOTRIP [28].

2.2. Emission Processors

In this work, we performed simulations using emission data obtained with two different emission processors: TREFIC [29], which is the native traffic emissions pre-processor for the model PMSS, and ECOTRIP. TREFIC is based on COPERT 4 [30] methodology for the calculation of the road vehicle emission factors. In order to calculate emissions, TREFIC takes into account vehicle type, fuel consumption, average travelling speed, and road type. TREFIC performs a reading and processing cycle for each road link. The input consists of 4 groups of files, related to the road network (geometry, speed, and volume of traffic flows, for each link of the network), vehicle fleet (split into COPERT 4 categories, for each of the road types or driving cycles), time modulations (tables of values which allow the time profiles of emissions to be quantified) and COPERT 4 methodology emission factors. Starting from the input information and for each road link, TREFIC calculates the emission factors (EFs) for each road type. These emission factors depend on fuel type, vehicle type, age and maintenance, road average speed, and driving cycle. If specific information is available, EFs can take into account the ambient average temperature (cold start and evaporative emissions), the average slope of the road, and the actual average load (for freight vehicles). Time modulation files, in text format, contain coefficients representing the time modulation factors of flow, speed, and temperature. These files allow a modulated input to be generated to the dispersion model and they support the user in generating time emission profiles for traffic flows, speed, and temperature. At each run, TREFIC generates at least three standard output files, containing aggregated emissions according to the temporal step in the input file, respectively, for traditional pollutants, particulate matter (PM) species, and evaporative losses.

ECOTRIP (Emission and Consumption Calculation Software Based on Trip Data Measured by Vehicle On-Board Unit) [28,31,32] is software developed by STMS to estimate atmospheric pollutant emissions (Carbon Monoxide, NOx, Non-Methane Hydrocarbons, and PM), greenhouse gas emissions (Carbon Dioxide), and fuel consumption produced by vehicle fleets. ECOTRIP is capable of carrying out a precise and georeferenced estimate of fuel consumption and polluting emissions produced by any type of vehicle in circulation equipped with on-board units. The innovative nature and originality of the ECOTRIP software derives from the ability to use data on actual routes, on the driving cycle, and on the characteristics of the vehicle, as well as the ability to operate on different levels of aggregation and detail and potentially in real time. ECOTRIP can be a valid tool to support vehicle traffic monitoring and mobility management activities.

ECOTRIP has been widely used and updated within the Electric System Research Programme supported by the Italian Ministry of Economic Development [28].

The evaluation procedures consider the speed-dependent hot emission factors described in the COPERT 4 guidebook [33] which were obtained from several experimental measurements collected in different European countries. These factors vary according to the fuel supply, the European Emission Standards, the engine size for passenger cars, and the weight for commercial vehicles and buses. ECOTRIP has been updated in order to include recent European Emission Standards (Euro 5 and Euro 6), as well as hybrid, electric, light-commercial and heavy-duty vehicles, buses, mopeds, and motorcycles. In addition to hot running emissions, ECOTRIP accounts for the "cold start" emissions, which occur when engines and catalysts are not (fully) warmed up and operate in a non-optimal condition. The estimation of the extra cold emissions refers to the methodology developed by INRETS (Institut national de recherche sur les transports et leur sécurité), which is based on several experimental tests performed in different European laboratories, as described in the ARTEMIS European Project [34]. In this study, ECOTRIP estimated the pollutant emissions from equipped cars. Estimates were carried out for each segment located between two consecutive GPS traces of a journey, combining the vehicle features together with the geographical information of the routes, then map-matched to the static road network.

2.3. Dispersion Model

Micro-SWIFT-SPRAY (PMSS) is a modelling system which reproduces primary pollutant transport and dispersion at the microscale (i.e., resolution of meters), and calculates the dry and wet deposition of airborne chemical species. PMSS is the parallelized version of the MSS model suite, which is fully described in several papers [35–37]. Here we provide a summary of its main characteristics, schematically represented in Figure 1.

The system has two pre-processing phases for the meteorological and the emission data, respectively. These modules prepare the input for the main processing models, the meteorological driver PSWIFT, an analytically modified mass consistent interpolator over complex terrain [38], and the three-dimensional Lagrangian particle dispersion model, PSPRAY. In the meteorological pre-processing phase, the meteorological data and the turbulence parameters provided at a local or regional scale are elaborated by using SURFPRO, the surface–atmosphere interface processor [39,40], to generate the input files required to run PSWIFT at a much higher spatial resolution (generally of the order of a few meters). In the emission pre-processing phase, the emission data and their spatial and temporal variations are used by the emission manager TREFIC to produce the emission input for PSPRAY. Different types of emission sources, such as a point, area, or line, can be simulated. Here we considered only line sources to study the primary pollutant dispersion at the urban scale caused by traffic. Obstacles, such as buildings, are directly considered in the model and are represented as filled cells in the meteorological field [35,36]. PSWIFT produces mass-consistent wind fields using data from a dispersed meteorological network or from simulated meteorological data at a lower resolution. PSPRAY calculates the pollutant concentration by means of "virtual" particles that carry a portion of the pollutant mass emitted by the sources. The velocity of the particles is calculated from a mean velocity component, defined by the local wind computed by PSWIFT, and a stochastic velocity component, representing atmospheric turbulence. PSPRAY can compute mean and instantaneous concentrations on a three-dimensional grid defined by the user, differentiating the calculation by both "chemical species" and "source". PSPRAY only simulates the dispersion of atmospheric compounds in the urban environment and cannot take into account the transformations due to chemical reactions. Recent developments have reported the implementation of several chemical models into PSPRAY to consider chemical reactions occurring at the urban scale [36]. However, in the present work, we considered only the dispersion characteristics of the PMSS system, neglecting chemical transformations. The modelling system PMSS is a commercially available software developed by ARIANET [41]. The codes PSWIFT and PSPRAY used here, in their versions PSWIFT-2.1.1 and PSPRAY-3.7.3, were compiled with an Intel16 compiler, using OPENMPI library.

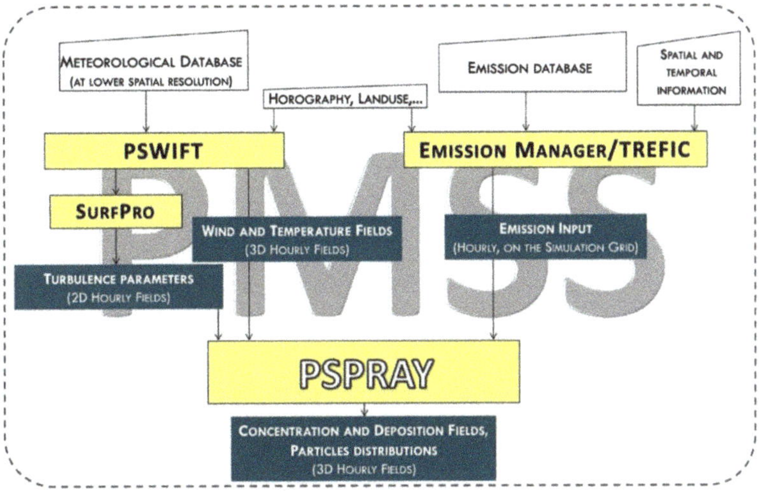

Figure 1. Scheme of the PMSS modelling system.

2.4. Simulation Setup

We performed simulations for a period of 29 days from 2 May to 30 May 2013. We used the hourly fluxes of passenger cars provided by the STMS database described in Section 2.1. We set a 2×2 km^2 horizontal domain that was centered on the urban traffic air quality (AQ) station Magna Grecia. The domain is indicated by the red square in Figure 2 and covered an adequate fraction of the emissions of the city while providing that the AQ station was far enough from the domain border, where the model uncertainty is generally higher. The spatial resolution of 3 m was chosen to ensure the highest resolution possible while keeping the run time within acceptable values, and the domain was composed of 667×667 grid points. For the vertical grid, we chose the following 25 levels above the ground: 0, 1.5 m, 3 m, 4 m, 6 m, 8 m, 10 m, 13 m, 16 m, 19 m, 22 m, 25 m, 30 m, 35 m, 42 m, 50 m, 60 m, 80 m, 120 m, 160 m, 200 m, 240 m, 300 m, 380 m, and 700 m.

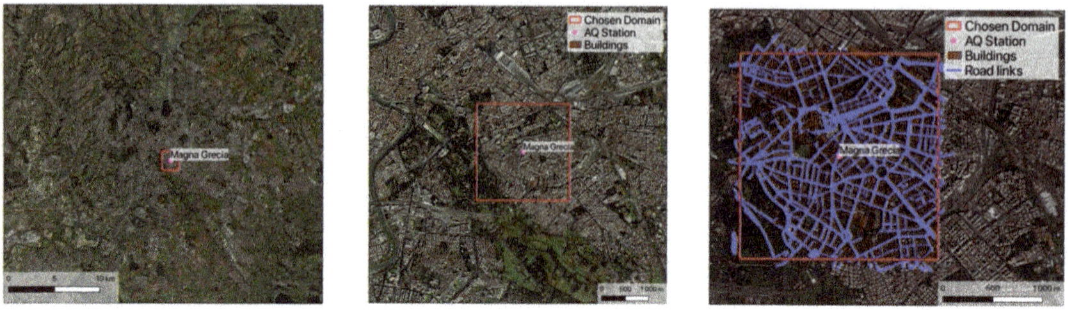

(**a**) (**b**) (**c**)

Figure 2. Simulation domain (red square, with a resolution of 3 m) centered in the AQ station of Magna Grecia (pink point). (**a**) shows the domain within the Rome urban area, (**b**) at increased zoom level, (**c**) includes (in blue) the street segments for which the FCD based emissions are available.

We conducted the simulations on the CRESCO/ENEAGRID High Performance Computing infrastructure funded by ENEA [25]. As the model system was structured, each 29-day-simulation consisted of 29 single model runs, each simulating 24 h. The restart option was applied, therefore, for each simulated day the values for the pollutant concentrations calculated for the last hour were saved and were used as initial conditions for the following run.

The simulation duration depended mainly on the number of emitted particles, and this in turn depended on the concentration resolution required, that is the concentration contribution given by a single particle in a concentration cell. In addition, the meteorological conditions can also play a key role in constraining a different number of particles to remain inside the computational domain depending on the mean flow and turbulence. In our simulations, using a concentration resolution of 0.5 $\mu g/m^3$ for the NOx species and using 528 cores for PSPRAY, which represents the most CPU demanding part of the system, the CPU time per core per simulated day was 8148 s.

2.4.1. Input Meteorological Data

Meteorological data used to feed the diagnostic model PSWIFT were provided by the Weather Research and Forecasting [42] mesoscale model using ERA5 reanalysis [43] as boundary conditions at a 28 km resolution and a 3-hourly time-step. WRF simulations, performed using the 3.9.1.1 model version, were based on two-way nesting over 3 grids, the coarser one covering the whole of Italy at a 9 km horizontal resolution (over Italy), then an intermediate domain at a 3 km resolution (over the Lazio region), and finally the target domain at a 1 km resolution over the city of Rome.

WRF parameterizations adopted for the simulation are summarized in Table 1.

Table 1. Parameterizations schemes used for the WRF meteorological simulation.

Parameterization	Reference
Cloud microphysics	Hong et al., 2004 [44]
Short wave radiation	Dudhia, 1989 [45]
Long wave radiation	Mlawer, 1997 [46]
Surface layer	Jimenez et al., 2012 [47]
Land surface model	Tewari et al., 2004 [48]
Planetary boundary layer	Hong et al., 2006 [49]
Cumulus convection (activated over coarse domain only)	Kain, 2004 [50]

Hourly data of the meteorological fields were used by PSWIFT to reconstruct the three-dimensional wind, temperature, and turbulent flow at a 3 m resolution.

2.4.2. Emissions

The Emission database described earlier includes the vehicle hourly fluxes determined by FCD. Therefore, the vehicle fluxes considered in the database were relative only to the circulating passenger cars. However, it was possible to estimate the emissions of the remaining part of the circulating fleet by considering the vehicle population in the city of Rome for the year 2013. Data relative to the circulating fleet needed for the emission estimate, such as vehicle type and fuel technology distribution, were retrieved from public registers of vehicle licenses [51]. By considering this information, we calculated a scaling factor for each of the remaining vehicle categories not included in the FCD database (motorcycles, light duty vehicles, heavy duty vehicles), in order to extrapolate their fluxes starting from the hourly fluxes of the passenger cars. These factors led to a vehicle fleet composition shown in Table 2.

Table 2. Fleet composition by vehicle type.

Vehicle Type	Fleet Composition
Passenger cars	77.3%
Motorcycles	17%
Light duty vehicles	5.4%
Heavy duty vehicles	0.3%

To calculate the emissions in terms of mass per hour, it was necessary to define at least one daily modulation profile for each vehicle category. TREFIC in fact requires as input the total flux per vehicle category and a separated time modulation profile. This allows for a variety of approaches regarding the time modulations (i.e., top-down emission definitions). Given the limitation on the number of input hourly profiles, a detailed inspection of the FCD was necessary to estimate the variability and the statistical significance of the modulation profiles computed from the emission database, with the aim to identify a small set of representative profiles.

The study of the most representative modulation profiles within the database was limited to the road links in close proximity to the monitoring station, shown in Figure 3, which were most likely to influence the hourly variability of the simulated NOx concentrations at the station.

The area examined covered about 300×300 m^2 and included 24 street links. The 2 May was chosen as a reference day and each hourly profile was scaled to its total daily fluxes for each day. From the database inspection, 3 average modulation profiles were determined, depending on the range of total flux, shown in Table 3. The profiles are shown in Figure 4.

Table 3. Average vehicle flux modulation profiles defined as input in the TREFIC model.

Profile ID	Vehicle Flux (Ftot) Range
1	Ftot < 3000
2	4000 < Ftot < 6000
3	6000 < Ftot < 11000

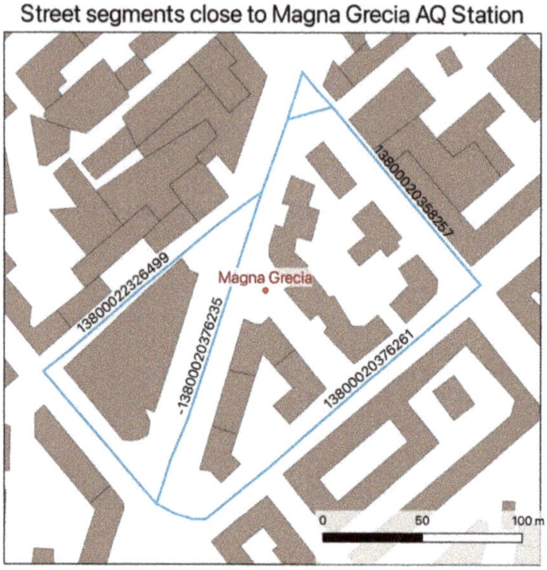

Figure 3. Street segments in proximity to the AQ Station Magna Grecia for which the time variability was studied. The labels indicate the ID number of the street segment within the database.

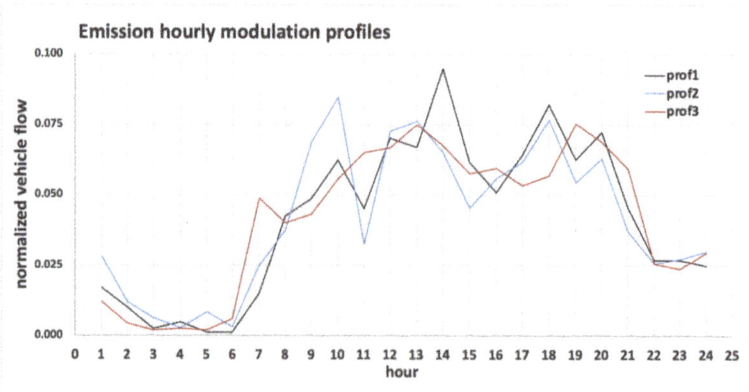

Figure 4. Vehicle flow modulation profiles used in TREFIC.

The profiles shown in the picture had similar behaviors, indicating a decrease in traffic during the night and an increase during the day, but they also had different behaviors for the hours at which the peaks were located:

- Profile 1 (prof1): shows a steady increase in traffic from the very early morning to midday when it started decreasing more slowly than during the morning increase;
- Profile 2 (prof2): shows 3 peaks, one in the morning at 10, one at 12–13, and the other at 18;
- Profile 3 (prof3): shows two relevant peaks: one at 13 and the other at 18.

These profiles are relative to 2 May. They were averaged over a few road links around the stations and are therefore representative of the traffic modulation in that particular location. Applying these profiles for the modulation of the dataset was certainly a significant approximation, the degree of which was tested through dedicated model runs as follows.

The validation of the observed concentrations was not the only goal of this study. A very interesting feature of the ENEA STMS emission database is its ability to potentially deliver an FCD-based emission database in near real-time. The possibility to incorporate this emission input in our PMSS modeling chain represents a foreseeable development of our modeling tools and needs to be investigated.

For these reasons, we performed different simulations with the different emission configurations that are listed in Table 4 and described hereinafter:

- Sim 1: we used the passenger cars traffic fluxes from the FCD database, the modulation profiles described in Figure 4, and the relative percentages listed in Table 2 to calculate with TREFIC the traffic emissions due to passenger cars, motorcycles, heavy duty vehicles, and light duty vehicles to be used as input to PMSS;
- Sim 2: this simulation was similar to Sim 1, and the calculation was performed only for passenger cars;
- Sim 3: we used as the input for PMSS the emissions already present in the FCD database in terms of emitted mass/unit time. These emissions were calculated using the emission processor ECOTRIP considering only passenger car traffic fluxes.

Table 4. Definition of the emission configurations used in the different simulations.

Simulation Name	Type of Vehicle	Emission Processor
Sim 1	Passenger cars Motorcycles Light duty Heavy duty	TREFIC
Sim 2	Passenger cars	TREFIC
Sim 3	Passenger cars	ECOTRIP

Among these simulations, only Sim 1 could be compared to the observations, having a complete coverage of the different types of emitting vehicles. However, Sim 2 and Sim 3 offered an interesting opportunity to compare the calculations of two different emission processors and to validate the time modulation assumption described earlier. In fact, Sim 2 and Sim 3 differed for two reasons:

- The operation of translating fluxes into emissions was completed using two different traffic processors, which despite being both based on COPERT 4 methodologies, still showed some differences (not shown here);
- Different hourly time modulations were used: in Sim 2 they followed Figure 4 profiles and were assigned following Table 3, while in Sim 3 they pertained to the variations of emissions in the database, and were therefore potentially different for each road link.

The comparison between Sim 2 and Sim 3 allowed the influence of the approximation of the time modulations to be understood.

2.4.3. NOx Background

Since our simulation only took into account primary NOx emitted in the study domain, to analyze how well our model reproduced the total NOx concentrations, we needed to estimate the background contribution, i.e., the NOx entering the domain from outside. One possible way to do this was to use measurements of the urban background NOx concentrations outside the domain, assuming that they were not influenced by sources internal to the domain (to avoid a double counting of emissions) [52,53]. In particular, in this study, the urban NOx background was calculated by taking into consideration the average of the NOx concentrations measured by the background air quality reference stations located around the simulation area, which are shown in Figure 5.

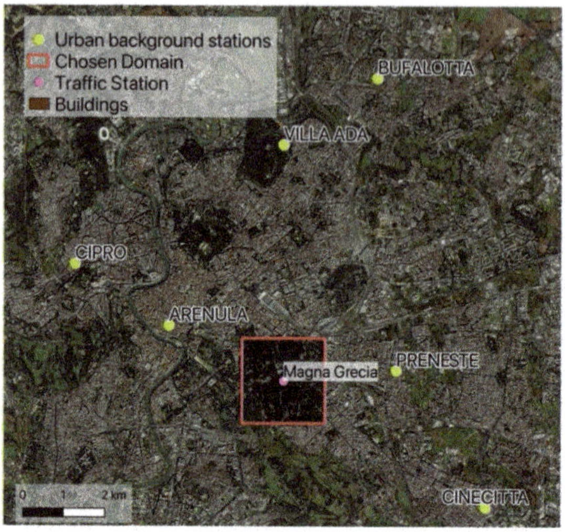

Figure 5. Maps reporting the location of urban background air quality stations. The yellow points were used for the calculation of the background NOx concentrations, while the pink point represents the urban traffic station of Magna Grecia inside the simulation domain (the area within the red square).

For these urban background AQ stations, we considered the data provided by the European Environmental Agency Air Quality portal [54] for May 2013 focusing on the pollutant "NOx as NO_2". By using R-cran base packages [55–58], we calculated the concentration time series for May 2013 of the main statistics parameters (mean and percentiles values of the distribution) from the concentrations of all the background stations. Figure 6 reports the average and median background concentrations (panel a) and the average, minimum, and maximum values of the background NOx concentrations (Panel b). We also included the NOx concentration measured at the urban traffic AQ station of Magna Grecia, as shown in Panel b in Figure 6.

The NOx concentration measured at Magna Grecia is always higher than the background with an average percentage difference of 39% and an absolute difference of 34 $\mu g/m^3$ between the monthly average values. Therefore, it is reasonable to assume that the measured concentrations observed at the Magna Grecia station are the sum of a background component and an additional component generated by local emissions in the immediate surroundings of the station. This background component was added, hour by hour, to the NOx concentration simulation to compare the results of Sim1 with the observations.

For Sim 2 and Sim 3, since the simulated NOx was not comparable with the NOx observed at the urban traffic station, the NOx background concentration was not added.

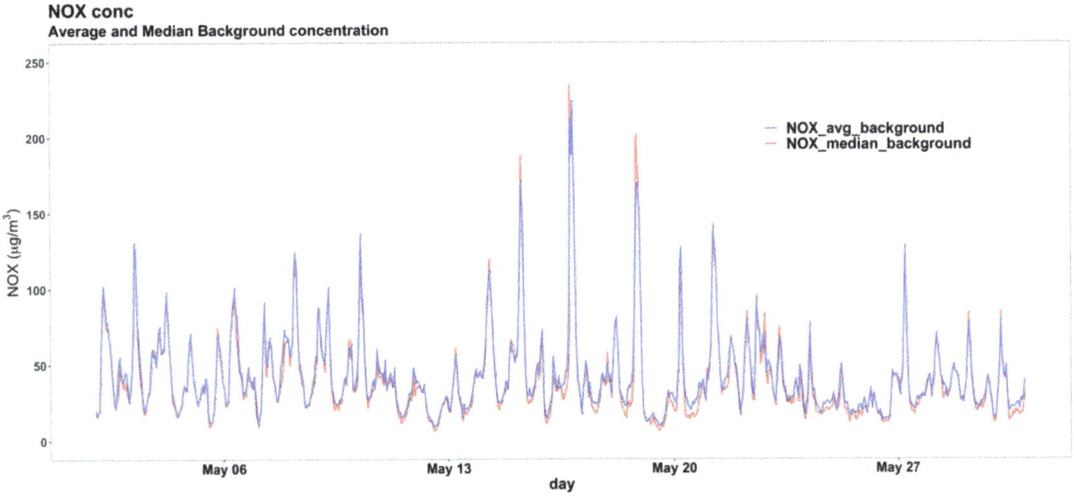

(a)

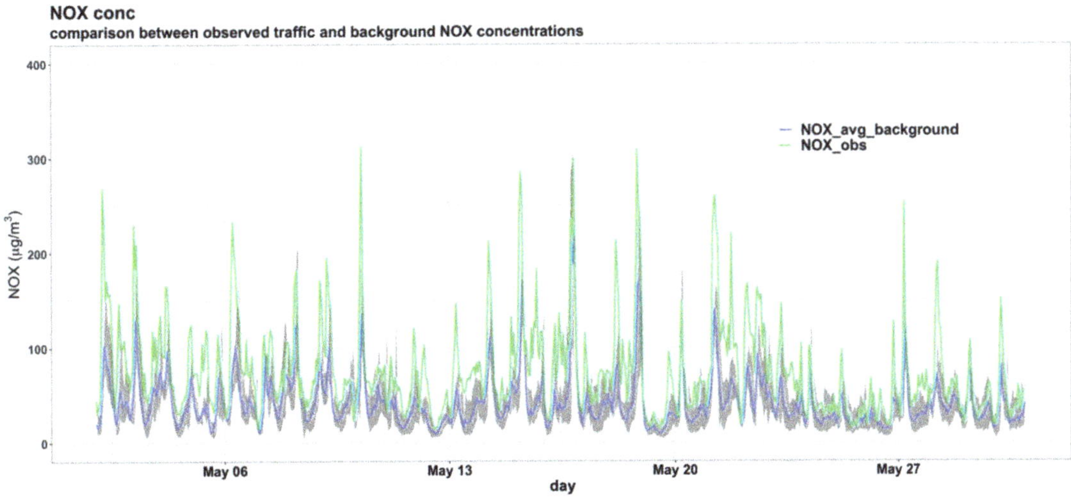

(b)

Figure 6. Time series of the main statistics parameters calculated for the NO_X concentrations of the urban background AQ stations surrounding the simulation domain. (**a**) Average background NO_X concentration compared to median concentration. (**b**) Average background in blue with the shaded grey area between the minimum and maximum values and the traffic station observed concentration in green.

3. Results and Discussion

In the following paragraphs, detailed comparisons between the observations and Sim 1 and between Sim 2 and Sim 3 are reported in Sections 3.1 and 3.2, respectively.

3.1. PMSS Traffic Emission Simulation (Sim 1)

Figure 7 shows the monthly average of the NOx concentration at ground level (1.5 m height). In this Figure we can see that the domain includes several busy roads, characterized by an average NOx concentration between 90 and 120 µg/m^3, and that the road next to the monitoring station is characterized by a significant traffic intensity with average NOx concentrations around 90 µg/m^3. Since the dispersion model has a very high spatial resolution (3 m), the high spatial variability of NOx concentrations near the road traffic emission sources can be appreciated. Concentrations are usually higher near the street central axes and lower towards the borders, with sharp spatial gradients in line with similar modelling studies in urban street canyons [5]. Concentrations vary in a wide range in different parts of the domain reflecting local traffic flows.

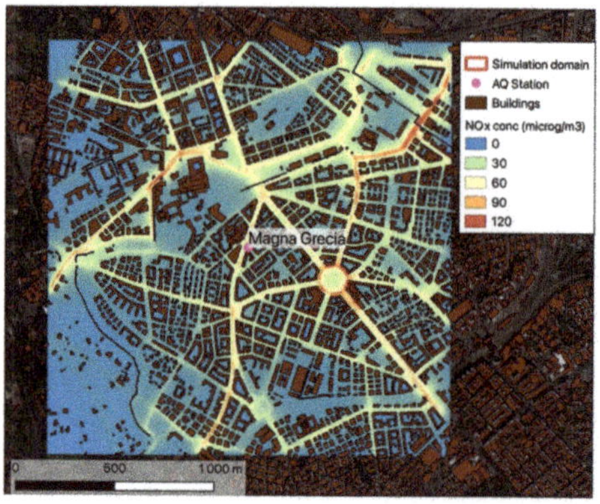

Figure 7. Average simulated NOx concentration in the month of May.

The comparison between the hourly observed NOx and the modelled concentrations of NOx at the Magna Grecia AQ station is shown in Figure 8, while in Figure 9 the absolute value of the daily fractional bias and the R-squared linear coefficient, whose definition is reported in the Appendix B, are shown. Here, we used the absolute value of the daily fractional bias as we were interested in discussing its magnitude rather that its sign. In general, Sim 1 showed a good agreement with the hourly NOx concentration variation (Figure 8) with an average fractional bias of 0.2 across the period. This was confirmed by the daily fractional bias that showed values below 20% for half of the days (Figure 9) and only for 7 days above 30%. Higher discrepancies are indicated by low values of R-squared. Generally, for almost half of the simulated days, R-squared values larger than 0.4 were found.

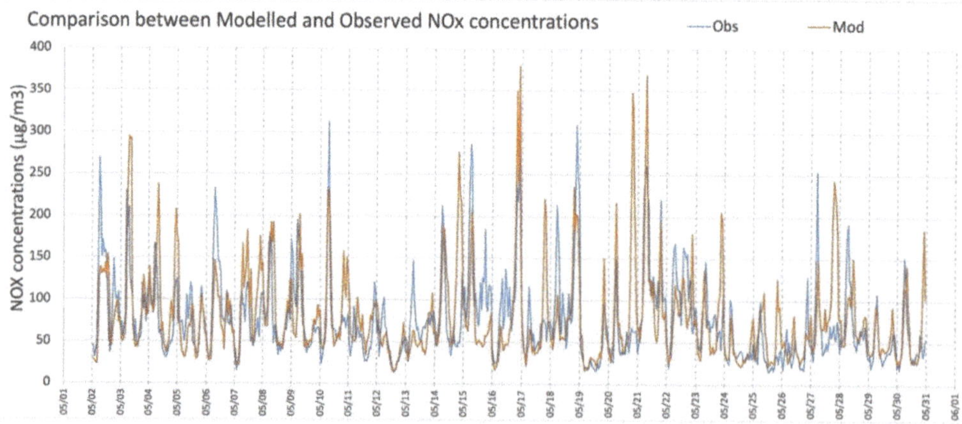

Figure 8. Hourly NOx concentration comparisons between observations (blue line) and simulations (orange line).

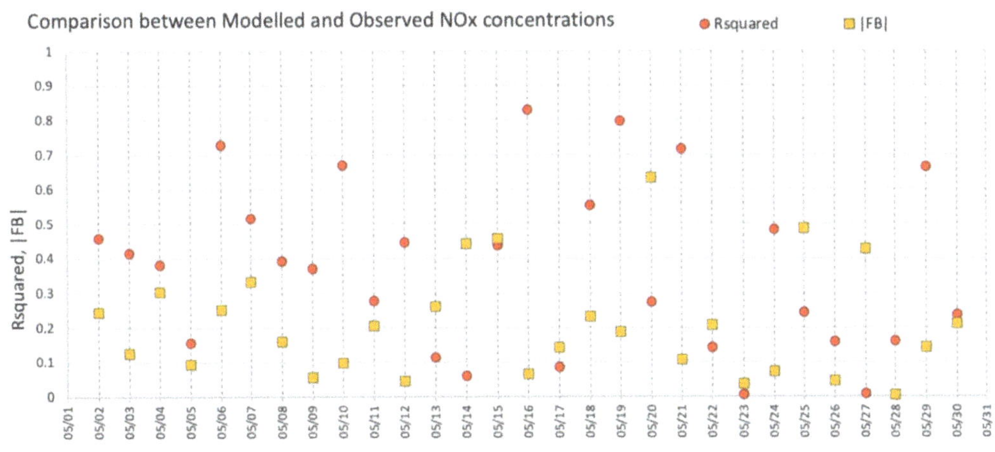

Figure 9. Daily linear correlation coefficients and the absolute value of the daily fractional bias relative to the time series shown in Figure 9.

The reasons behind the low correlations with a low fractional bias can be ascribed to the three chosen modulations that were not always the most representative of the vehicle flow variability in all the 29 days. These comparisons show nevertheless that the daily fractional bias was generally low, indicating that the background concentrations were fairly well transported by the model and that the reduced modulations used had little influence on most of the daily averages.

Figure 10a shows a regression plot of the modelled vs. observed NOx concentrations relative to all the simulated periods, and a R-squared coefficient of 0.35, indicating a significant influence of the few days with very low correlations. Daily plots on selected days, with a mean fractional bias lower than 20%, in the same Figure 10b–d, show that the daily-averaged correlation could reach significant values.

In general, there are no universally accepted rules to evaluate the performance of a model in the field of air quality. Usually, the metric to use and the acceptance levels to consider for such a task are matter of debate [58]. Recently, a growing number of studies have used metrics that are involved in the definition of general acceptance criteria for dispersion model evaluation, proposed by the authors of [59] for urban dispersion modelling. Among these studies, a recent paper [60], applied these criteria to evaluate the WRF-Chem NOx concentration simulations to use as background for a Micro Swift Spray simulation.

The acceptance criteria for dispersion model evaluation are based on the following metrics: Fractional bias (FB); Normalized Mean Squared Error (NMSE); Fraction of simulations within a factor of two of the measurements (FAQ2); and Normalized Absolute Difference (NAD). Their definitions, following the work in [59], are listed in Appendix B. A model will meet these acceptance criteria if the aforementioned indicators satisfy the values reported in Table A1 for urban and rural air quality stations.

The performances of our model are reported in Table 5.

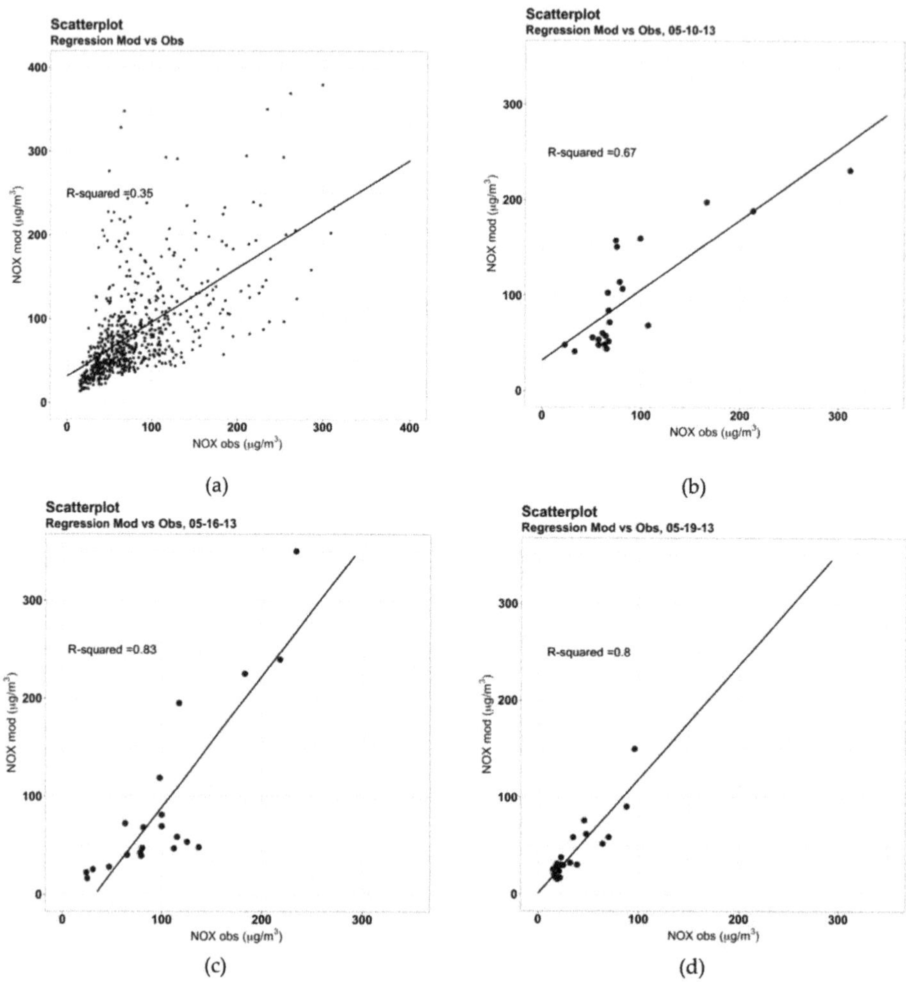

Figure 10. Regression plot of modelled vs. observed NOx hourly concentrations of the complete dataset (**a**) and of selected days 10 May (**b**), 16 May (**c**), 19 May (**d**) with a mean fractional bias lower than 20%.

Table 5. Model performances at Magna Grecia.

	| FB |	**NMSE**	**FAC2**	**NAD**
Magna Grecia (Urban)	0.049	0.37	0.87	0.20

The comparison between the performances of our model and the acceptance criteria indicated a good agreement between our simulation and the observations.

We also point out that Oldrini et al. 2017 [34] considered satisfying a simulation in which 68% of the predicted concentrations were within a factor of two, based on one of the same acceptance criteria cited previously, when we had 87%, as shown in Table 5.

Figure 11 shows a quantile–quantile plot of the modelled vs. observed NOx concentrations, which indicates that both the measured and modelled values follow the same statistical distribution except for values higher than 110 $\mu g/m^3$, when the model seemed to overpredict the NOx concentration. This feature was already noted in Villani et al., 2021 [61] who presented a comparison between the NOx concentrations simulated by PMSS and the concentrations observed in a street canyon in Modena during a field campaign. In that study, the Q–Q plot was very similar and the value at which the simulated and observed distributions started to significantly differ was above 50 μg/m^3. At the moment, no clear explanation for these differences has been identified.

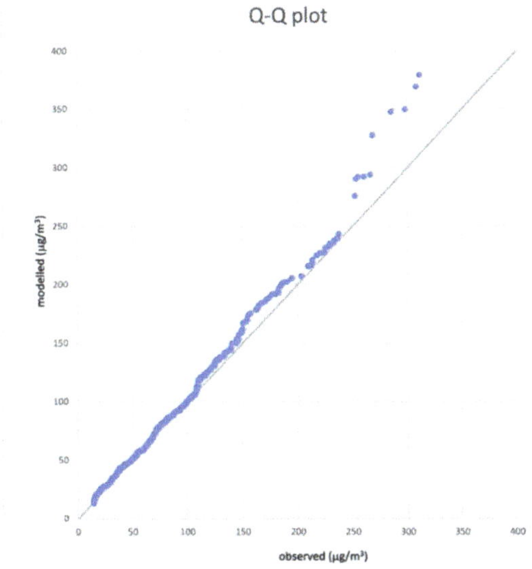

Figure 11. Q–Q plot of modelled vs. observed NOx.

The distribution of the differences between the observed and simulated NOx concentrations was generally normal, and slightly skewed towards positive values. Figure 12 shows the histogram of the hourly percentage difference combined with the cumulative probability of occurrence, indicating that more than 50% of the differences were between −10% and 20%. The R-squared values corresponding to some of these cases are shown in more detail in Figure 10. The results we show here are comparable to those obtained from the use of dedicated measured vehicle flow data during a field campaign [61].

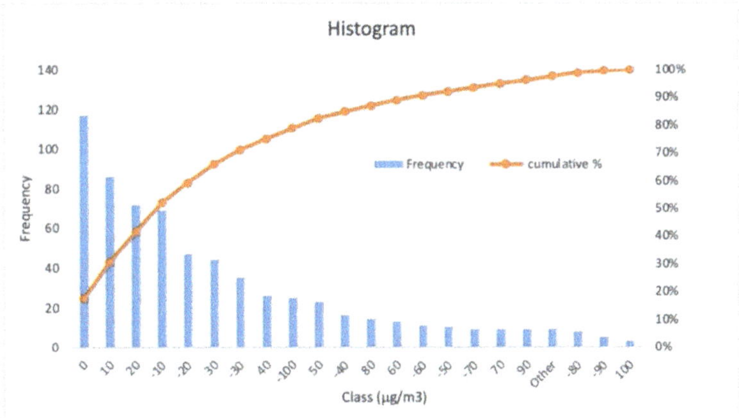

Figure 12. Histogram of the frequency of the differences of hourly modelled NOx concentrations vs. ob*Scheme 3*. 2. Passenger car simulations (Sim 2 and Sim 3).

3.2. Comparison between Emission Processors (Sim 2 and Sim 3)

In this section, we present the comparison between the two emission processors TREFIC and ECOTRIP. Figure 13 shows a general good agreement between the two simulations, except for a general overestimation of Sim 3 for high values of NOx concentrations. The percentage differences between Sim 2 and Sim 3 were mostly between −30% and 20% as shown in Figure A2 of Appendix A.

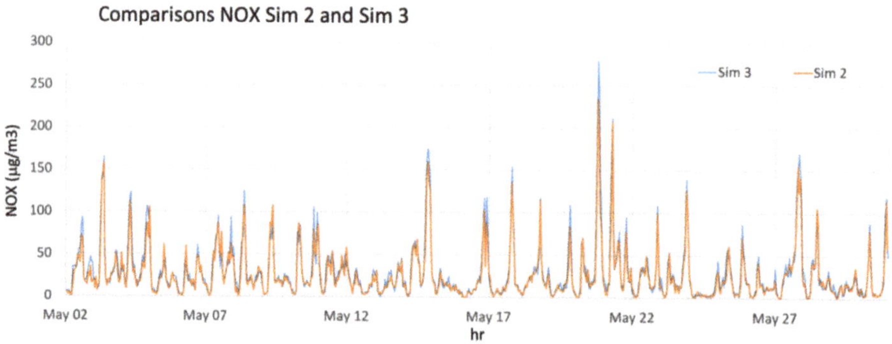

Figure 13. Comparison between Sim 2 and Sim 3 at Magna Grecia in terms of the concentration timeseries.

On average, the percentage difference between the NOx concentrations of Sim 2 and Sim 3 changed with the time of the day. In Figure 14 we mapped the period average of hourly concentrations at 19:00, which represented typical high traffic conditions.

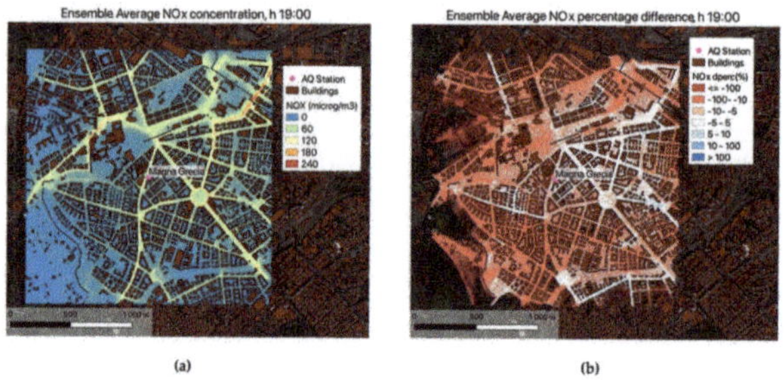

Figure 14. Maps of the ensemble average of the NOx concentration (Sim 3) at 19:00 (**a**) and the ensemble average of the NOx percentage difference at 19:00 (**b**).

We can observe that the percentage differences were often limited to values between −10% and 10% (Figure 14 panel b). On the other hand, and as expected, these differences could be significant at times where the NOx concentrations were lower. Therefore, depending on the aim of the study, the choice of the time modulation used as input in the data can have a significant impact on the simulated concentrations and the availability of an emission database with highly detailed hourly modulations possibly represents a significant advantage.

Once again, we find it necessary to point out that in these comparisons between Sim 2 and Sim 3, the NOx concentrations were generally lower than those shown in Section 3.1 for two reasons: fewer vehicle types were accounted for and no background evaluation was added to the final concentrations that were compared.

4. Conclusions

In this work, we applied FCD data for modelling NOx concentrations at a microscale level in Rome. We first used the hourly vehicle fluxes to generate our best estimate of the NOx concentration at the location of the Magna Grecia air quality station (Sim 1) from 2 May to 30 May of 2013. We made the assumption that three averaged time modulation profiles were sufficient to describe the variability within the database, with the purpose of comparing the simulated concentrations to the one measured in the air quality station. The comparison against the observation showed an acceptable agreement with the daily mean fractional bias below 20% for almost half of the days and above 30% for only 7 out of 29 days. This is an encouraging result that shows that we can successfully incorporate high resolution vehicle circulation data into our microscale modelling suite. We calculated the FB, NMSE, FAQ2, and NAD statistical indicators, the values of which satisfied the acceptance criteria [59] both for rural and urban air quality stations.

The innovative aspect of this work is the use of FCD computed emissions to feed our urban scale model. Although extremely appealing, the use of massive FCD to extract traffic patterns and travel behaviors in urban areas is still at an early stage compared to other sectors, particularly due to the fragmented availability of transport/mobility data, institutional barriers, and data privacy/security issues. As a consequence, the dataset used in this study was based on data collected by only 5% of the circulating passenger cars. Extrapolating this dataset to the entire fleet could represent an important approximation. For these reasons, 87% of the simulated values within a factor of two with respect to the measurements represents a remarkable success, even if the linear correlations were poor with only few days above 0.6.

Finally, the kind of agreement we found in this study was very similar to what we found in a recently published study using PMSS in the city of Modena, where traffic flows and NOx observations were available during a measurements campaign [61].

To have an indication of the effect of the simplified hourly profiles of vehicle flows on the simulated NOx concentrations, a test was made in which we used the FCD flows of passenger cars to compute the NOx emissions with TREFIC in the way described in Section 2.4.2 (Sim 2) and we compared the resulting NOx concentrations with the ones obtained using ECOTRIP emissions as input to PSPRAY (Sim 3). The added value of comparing Sim 2 (few modulation profiles for all the road links) vs. Sim3 (each road link with its own modulation) lies in the study of the influence of the modulation profiles used on the simulated NOx concentrations. These two simulations showed very similar concentrations at Magna Grecia and their differences were often between −30% and 20%. This further study allowed the direct implementation of this database into our microscale modelling suite, avoiding the assumption on the time modulations, and providing a valid tool to potentially enable the use of PMSS with emission inputs that can be provided in quasi-real time.

Future developments of this work could involve the use of more FCD data in our PMSS modelling chain. Moreover, if the availability of FCD data increases and is less fragmented, other pollutants and longer time scales could be explored.

Author Contributions: Conceptualization: F.R., M.G.V., A.P., I.D., C.L., G.V. and L.C.; Methodology: F.R., M.G.V., A.P., I.D., C.L., G.V. and L.C.; Software: F.R., M.G.V., M.D., G.T., C.L. and G.V.; Formal analysis: F.R., M.G.V. and A.P.; Investigation: F.R., M.G.V., A.P., I.D., C.L. and G.V.; Data curation: F.R., M.G.V., M.D., C.L. and G.V.; Writing-original draft preparation: F.R., M.G.V., A.P., I.D., C.L., G.V. and L.C.; Writing-review and editing: F.R., M.G.V., A.P., I.D., C.L., G.V., L.C., M.D. and G.T.; Visualization: F.R., M.G.V., A.P., I.D., C.L., G.V., L.C., M.D. and G.T. All authors have read and agreed to the published version of the manuscript.

Funding: This research received no external funding.

Data Availability Statement: Data are available upon request to the authors.

Acknowledgments: The computing resources and the related technical support used for this work have been provided by CRESCO/ENEAGRID High Performance Computing infrastructure and its staff [16]. CRESCO/ENEAGRID High Performance Computing infrastructure is funded by ENEA, the Italian National Agency for New Technologies, Energy and Sustainable Economic Development and by Italian and European research programs, see http://www.cresco.enea.it/english for information.

Conflicts of Interest: The authors declare no conflict of interest. The funders had no role in the design of the study; in the collection, analyses, or interpretation of data; in the writing of the manuscript, or in the decision to publish the results.

Appendix A

This section includes some material supplemental to the discussion in the main sections.

Appendix A.1. From ECOTRIP Data to PMSS Emission Input

To use the emissions calculated by ECOTRIP, it was necessary to adapt them to the PMSS input format. The information on the geographical domain and the position of each road segment included in the ENEA–SMTS database were coupled with the segments present in the PMSS input file usually used as PMSS emission input, containing the road segments with the associated hourly emissions. For each day of the simulations, a database containing details on the road segments and hourly emissions were compiled in the text format which was then processed to generate the emission input files for PSPRAY (see scheme in Figure A1). The procedure was defined using R-cran packages (base [55], lubridate [56], RGDAL [57]) and Fortran compiled libraries coded by ARIANET. The most significant difference in using the emission input generated with this procedure and the usual input generated by TREFIC, described in Section 2.4.2, was the possibility to use hourly emissions estimates for each road segment based on the traffic counts provided by the Octo Telematics without introducing any additional simplification on the modulation profiles.

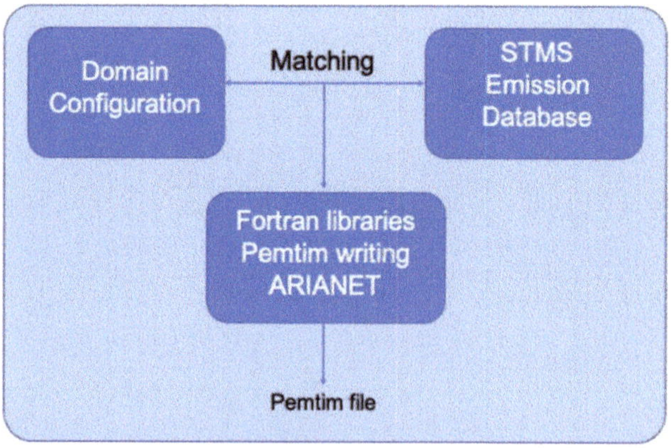

Figure A1. Scheme to illustrate the steps to create the files to use as input into PSPRAY.

Appendix A.2. Emission Processors Comparison

Figure A2 shows the percentage differences between Sim 3 and Sim 2 (i.e., (Sim 2–Sim 3) /Sim 2 × 100) with most of the values between −30% and 20% in agreement with what was found in Figure 14.

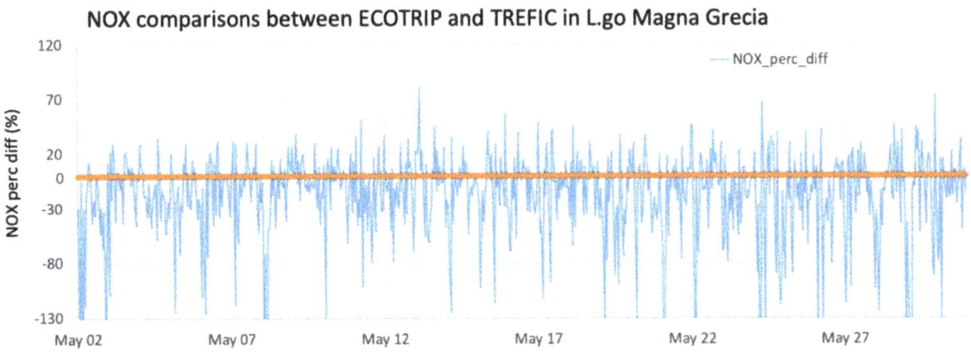

Figure A2. Comparison between Sim 2 and Sim 3 in terms of percentage difference. The orange line indicates the zero.

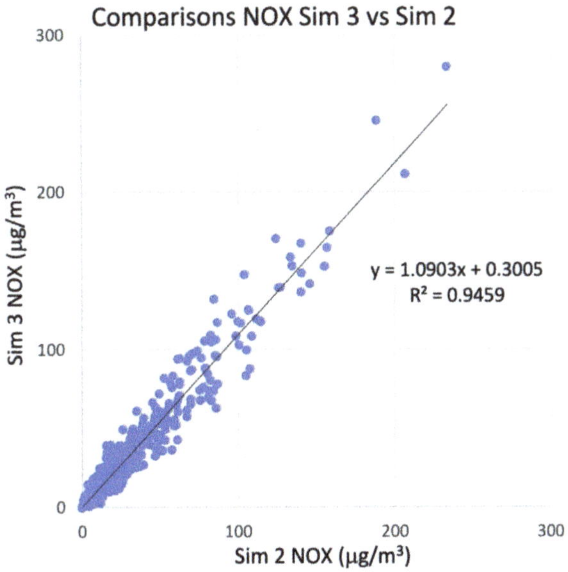

Figure A3. NOx concentrations scatter plot of Sim 2 vs. Sim 3.

Appendix B

In this work, to test the acceptance criteria of [59] we used the definition of average Fractional Bias (FB), the Normalized mean square error (NMSE), the Fraction of simulated values within a factor of two of the observed value (FAC2), and the Normalized Absolute Difference (NAD) as:

$$\mathrm{FB} = 2\frac{\overline{(C_o - C_s)}}{(\overline{C_o} + \overline{C_s})}, \tag{A1}$$

$$\mathrm{NMSE} = \frac{\overline{(C_o - C_s)^2}}{(\overline{C_o} * \overline{C_s})}, \tag{A2}$$

$$\mathrm{FAC2}\left(\text{fraction of data where } 0.5 < \frac{C_s}{C_o} < 2\right), \tag{A3}$$

$$\text{NAD} = \frac{\overline{|C_o - C_s|}}{(\overline{C_o} + \overline{C_s})} \quad \text{(A4)}$$

where *Co* and *Cs* are, respectively, the modelled and the observed concentrations. The acceptance criteria for dispersion model evaluation were defined for two categories of measurements stations based on their location. Hence, they were defined for rural and urban stations and are reported here in Table A1.

Table A1. Definition of the acceptance criteria for dispersion model evaluation.

Type of Measuring Station	∣FB∣	NMSE	FAC2	NAD
Rural	$\mid FB \mid \leq 0.3$	≤ 3	≥ 0.5	≤ 0.3
Urban	$\mid FB \mid \leq 0.67$	≤ 6	≥ 0.3	≤ 0.5

A model will meet the acceptance criteria if these statistical indicators satisfy the values reported in Table A1 for urban and rural types of monitoring stations.

Furthermore, as a metric for the linear regressions, we used the R-squared, that in a x-y linear regression is calculated as the square of the linear correlation coefficient (R) and indicates the amount of predicted variability explained by the observed variability. The value of R is calculated by the cor.test function in the package R-cran [55] which refers to [62,63].

References

1. Fenger, J. Urban air quality. *Atmos. Environ.* **1999**, *33*, 4877–4900. [CrossRef]
2. Harrison, R.M.; Vu, T.V.; Jafar, H.; Shi, Z. More mileage in reducing urban air pollution from road traffic. *Environ. Int.* **2021**, *149*, 106329. [CrossRef]
3. Marinello, S.; Lolli, F.; Gamberini, R. Roadway tunnels: A critical review of air pollutant concentrations and vehicular emissions. *Transp. Res. D Transp. Environ.* **2020**, *86*, 102478. [CrossRef]
4. Hak, C.; Larssen, S.; Randall, S.; Guerreiro, C.; Denby, B.; Horálek, J. Traffic and Air Quality—Contribution of Traffic to Urban Air Quality in European Cities. ETC/ACC Technical Paper 2009/12, 2010. Available online: https://www.eionet.europa.eu/etcs/etc-atni/products/etc-atni-reports/etcacc_tp_2009_12_traffic_and_urban_air_quality (accessed on 25 March 2021).
5. Vardoulakis, S.; Fisher, B.E.A.; Pericleous, K.; Gonzalez-Flesca, N. Modelling air quality in street canyons: A review. *Atmos. Environ.* **2003**, *37*, 155–182. [CrossRef]
6. Huang, Y.; Lei, C.; Liu, C.; Perez, P.; Forehead, H.; Kong, S.; Zhou, J.L. A review of strategies for mitigating roadside air pollution in urban street canyons. *Environ. Pollut.* **2021**, *280*, 116971. [CrossRef] [PubMed]
7. Wang, Y.; Szeto, W.Y.; Han, K.; Friesz, T.L. Dynamic traffic assignment: A review of the methodological advances for environmentally sustainable road transportation applications. *Transp. Res. Part B Methodol.* **2018**, *111*, 370–394. [CrossRef]
8. El-Fadel, M.; Sbayti, H.; Kaysi, I. Modeling of Traffic-Induced Emission Inventories in Urban Areas. Effect of Roadway Network Aggregation Levels Traffic Management and Technology (2002). In *Air Pollution Modelling and Simulation*; Sportisse, B., Ed.; Springer: Berlin/Heidelberg, Germany, 2020. [CrossRef]
9. Pinto, J.A.; Kumar, P.; Alonso, M.F.; Andreão, W.L.; Pedruzzi, R.; Dos Santos, F.; Moreira, D.M.; de Almeida Albuquerque, T.T. Traffic data in air quality modeling: A review of key variables, improvements in results, open problems and challenges in current research. *Atmos. Pollut. Res.* **2020**, *11*, 454–468. [CrossRef]
10. Quaassdorff, C.; Borge, R.; Pérez, J.; Lumbreras, J.; de la Paz, D.; de Andrés, J.M. Microscale traffic simulation and emission estimation in a heavily trafficked roundabout in Madrid (Spain). *Sci. Total Environ.* **2016**, *566–567*, 416–427. [CrossRef] [PubMed]
11. De Fabritiis, C.; Ragona, R.; Valenti, G. Traffic estimation and prediction based on real time floating car data. In Proceedings of the 11th International IEEE Conference on Intelligent Transportation Systems, Beijing, China, 12–15 October 2008; pp. 197–203.
12. Karagulian, F.; Messina, G.; Valenti, G.; Liberto, C.; Carapellucci, F. A Simplified Map-Matching Algorithm for Floating Car Data. In *Advanced Information Networking and Applications, AINA 2021, Lecture Notes in Networks and Systems*; Barolli, L., Woungang, I., Enokido, T., Eds.; Springer: Cham, Switzerland, 2021; Volume 227. [CrossRef]
13. Jing, B.; Wu, L.; Mao, H.; Gong, S.; He, J.; Zou, C.; Song, G.; Li, X.; Wu, Z. Development of a vehicle emission inventory with high temporal–spatial resolution based on NRT traffic data and its impact on air pollution in Beijing – Part 1: Development and evaluation of vehicle emission inventory. *Atmos. Chem. Phys.* **2016**, *16*, 3161–3170. [CrossRef]
14. Gately, C.K.; Hutyra, L.R.; Peterson, S.; Wing, I.S. Urban emissions hotspots: Quantifying vehicle congestion and air pollution using mobile phone GPS data. *Environ. Pollut.* **2017**, *229*, 496–504. [CrossRef]
15. Jiang, Y.; Song, G.; Zhang, Z.; Zhai, Z.; Yu, L. Estimation of Hourly Traffic Flows from Floating Car Data for Vehicle Emission Estimation. *J. Adv. Transp.* **2021**, *2021*, 6628335. [CrossRef]

16. Iannone, F.; Ambrosino, F.; Bracco, G.; De Rosa, M.; Funel, A.; Guarnieri, G.; Migliori, S.; Palombi, F.; Ponti, G.; Santomauro, G.; et al. CRESCO ENEA HPC clusters: A working example of a multifabric GPFS Spectrum Scale layout. In Proceedings of the 2019 International Conference on High Performance Computing Simulation (HPCS), Dublin, Ireland, 15–19 July 2019; pp. 1051–1052. [CrossRef]
17. Mastroianni, P.; Monechi, B.; Servedio, V.; Liberto, C.; Valenti, G.; Loreto, V. Individual Mobility Patterns in Urban Environment. In Proceedings of the 1st International Conference on Complex Information Systems, Rome, Italy, 22–25 April 2016; pp. 81–88. [CrossRef]
18. Altintasi, O.; Tuydes-Yaman, H.; Tuncay, K. Detection of urban traffic patterns from Floating Car Data (FCD). *Transp. Res. Procedia* **2017**, *22*, 382–391. [CrossRef]
19. Mastroianni, P.; Monechi, B.; Liberto, C.; Valenti, G.; Servedio, V.D.P.; Loreto, V. Local Optimization Strategies in Urban Vehicular Mobility. *PLoS ONE* **2015**, *10*, e0143799. [CrossRef]
20. Dölger, R.; Kleine, S.; Hoffmann, T.; Schreuder, M.; Ferrante, E.; Pressi, F.; Kaltwasser, J.; Ansorge, J.; Haspel, U.; Cheung, S.; et al. Floating Car Data, Report v. 01-00-00, FCD Workshop, Frankfurt 17th September 2019. Available online: https://www.its-platform.eu/filedepot_download/2218/6585 (accessed on 31 July 2021).
21. Cohn, N.; Bischoff, H. Floating Car Data for Transportation Planning: Explorative Study to Technique and Applications and Sample Properties of GPS Data, TomTom Presentation Slides, NATMEC, Dallas. 2012. Available online: http://onlinepubs.trb.org/onlinepubs/conferences/2012/NATMEC/Cohn.pdf (accessed on 31 July 2021).
22. Liberto, C.; Ragona, R.; Valenti, G. Traffic Prediction in Metropolitan Freeways. In Proceedings of the 7th International Conference on Traffic & Transportation Studies, ICTTS, Kunming, China, 3–5 August 2010; pp. 846–860, vol: Traffic and Transportation Studies, 2010. [CrossRef]
23. Ehmke, J.F.; Meisel, S.; Mattfeld, D.C. Floating Car Data Based Analysis of Urban Travel Times for the Provision of Traffic Quality. In *Traffic Data Collection and Its Standardization*; International Series in Operations Research & Management Science; Barceló, J., Kuwahara, M., Eds.; Springer: New York, NY, USA, 2010; Volume 144. [CrossRef]
24. Brockfeld, E.; Lorkowski, S.; Mieth, P.; Wagner, P. Benefits and Limits of Recent Floating Car Data Technology—An Evaluation Study. In Proceedings of the 11th WCTR Conference, Berkeley, CA, USA, 24–28 June 2007.
25. Leduc, G. Road Traffic Data: Collection Methods and Applications. *Work. Pap. Energy Transp. Clim. Chang.* **2008**, *1*, 1–55.
26. Octo Telematics Website. Available online: https://www.octotelematics.com/company/ (accessed on 9 May 2021).
27. Nigro, M.; Ferrara, M.; De Vincentis, R.; Liberto, C.; Valenti, G. Data Driven Approaches for Sustainable Development of E-Mobility in Urban Areas. *Energies* **2021**, *14*, 3949. [CrossRef]
28. Lelli, M.; Liberto, C.; Valenti, G. *Aggiornamento delle Librerie Software "ECOTRIP" per la Stima dei Consumi e delle Emissioni, Contratto di Servizio Tecnico-Scientifico tra ENEA e OctoTelematics SpA, Det. n.18/E/2016/DTE, C.A. CT4AAC*; ENEA e OctoTelematics SpA: Rome, Italy, 2017.
29. ARIANET. TREFIC—Traffic Emission Factors Improved Calculator. 2014. Available online: http://www.aria-net.it/front/ENG/codes/files/7.pdf (accessed on 19 April 2021).
30. Emisia SA. COPERT—COmputer Programme to Calculate Emissions from Road Transport. 2018. Available online: https://www.emisia.com/utilities/copert/documentation (accessed on 11 June 2020).
31. Liberto, C.; Valenti, G.; Orchi, S.; Lelli, M.; Nigro, M.; Ferrara, M. The Impact of Electric Mobility Scenarios in Large Urban Areas: The Rome Case Study. *IEEE Trans. Intell. Transp. Syst.* **2018**, *19*, 3540–3549. [CrossRef]
32. Valenti, G.; Lelli, M.; Liberto, C.; Orchi, S.; Messina, G.; Ortenzi, F.; Carapellucci, F. *Valutazione dei benefici ambientali della mobilità elettrica nell'area di Roma, Report Ricerca di Sistema Elettrico. RdS/PAR2015/213*; ENEA: Rome, Italy, 2016. (In Italian)
33. Ntziachristos, L.; Samaras, Z. EMEP/EEA Air Pollutant Emission Inventory Guidebook 2016, in 1.A.3.b.i-iv (Passenger Cars, Light Commercial Trucks, Heavy-Duty Vehicles Including Buses and Motor Cycles). June 2017. Available online: http://www.eea.europa.eu/publications/emep-eea-guidebook-2016/part-b-sectoral-guidancechapters/1-energy/1-a-combustion/1-a-3-b-i (accessed on 20 May 2017).
34. Joumard, R.; Andre, J.M.; Rapone, M.; Zallinger, M.; Kljun, N.; Andre, M.; Samaras, Z.; Roujol, S.; Laurikko, J.; Weilenmann, M. Emission Factor Modelling and Database for Light Vehicles—Artemis Deliverable 3, Document LTE 0523. 2007. Available online: http://hal.archives-ouvertes.fr/hal-00916945/document (accessed on 12 April 2017).
35. Oldrini, O.; Armand, P.; Duchenne, C.; Olry, C.; Moussafir, J.; Tinarelli, G. Description and Preliminary Validation of the PMSS Fast Response Parallel Atmospheric Flow and Dispersion Solver in Complex Built-Up Areas, Environ. *Fluid Mech.* **2017**, *17*, 997–1014. [CrossRef]
36. Tinarelli, G.; Mortarini, L.; Trini Castelli, S.; Carlino, G.; Moussafir, J.; Olry, C.; Armand, P.; Anfossi, D. Review and Validation of Microspray, a Lagrangian Particle Model of Turbulent Dispersion. *Geophys. Monogr. Ser.* **2012**, *200*, 311–327. [CrossRef]
37. Tinarelli, G.; Gomez, F. *PSPRAY General Description and User's Guide, 2017. Version Code 3.7.3.*; ARIANET/ARIA Technologies: Milan, Italy, 2017.
38. Nibart, M.; Armand, P.; Duchenne, C.; Olry, C.; Albergel, A.; Moussafir, J.; Oldrini, O. Flow and Dispersion Modelling in a Complex Urban District Taking account of the Underground Roads Connections. In Proceedings of the 17th International Conference on Harmonization within Atmospheric Dispersion Modelling for Regulatory Purposes, Budapest, Hungary, 9–12 May 2016.
39. Finardi, S.; Silibello, C. SURFPRO3 User's Guide (SURFace-Atmosphere Interface PROcessor, Version 3); Software Manual. Arianet R2011.31; Arianet: Milan, Italy. 2011. Available online: http://doc.aria-net.it/SURFPRO (accessed on 17 August 2021).

40. Finardi, S.; Morselli, M.G.; Brusasca, G.; Tinarelli, G. A 2-D meteorological pre-processor for real-time 3-D ATD models. *Int. J. Environ. Pollut.* **1997**, *8*, 478–488.
41. ARIANET. ARIA(NET). Available online: http://www.aria-net.it/ (accessed on 20 April 2020).
42. Skamarock, W.C.; Klemp, J.B.; Dudhia, J.; Gill, D.O.; Barker, D.M.; Duda, M.G.; Huang, X.-Y.; Wang, W.; Powers, J.G. *A Description of the Advanced Research WRF Version 3. NCAR Tech. Note 2008, NCAR/TN-475+STR*; University Corporation for Atmospheric Research: Boulder, CO, USA, 2008; 113p.
43. Hersbach, H.; Bell, B.; Berrisford, P.; Hirahara, S.; Horányi, A.; Muñoz-Sabater, J.; Nicolas, J.; Peubey, C.; Radu, R.; Schepers, D.; et al. The ERA5 global reanalisys. *Q. J. R. Met. Soc.* **2020**, *146*, 1999–2049. [CrossRef]
44. Hong, S.-Y.; Dudhia, J.; Chen, S.-H. A revised approach to ice microphysical processes for the bulk parameterization of clouds and precipitation. *Mon. Weather Rev.* **2004**, *132*, 103–120. [CrossRef]
45. Dudhia, J. Numerical study of convection observed during the Winter Monsoon Experiment using a mesoscale two–dimensional model. *J. Atmos. Sci.* **1989**, *46*, 3077–3107. [CrossRef]
46. Mlawer, E.J.; Taubman, S.J.; Brown, P.D.; Iacono, M.J.; Clough, S.A. Radiative transfer for inhomogeneous atmospheres: RRTM, a validated correlated–k model for the longwave. *J. Geophys. Res.* **1997**, *102*, 16663–16682. [CrossRef]
47. Jimenez, P.A.; Dudhia, J.; Gonzalez–Rouco, J.F.; Navarro, J.; Montavez, J.P.; Garcia–Bustamante, E. A revised scheme for the WRF surface layer formulation. *Mon. Weather Rev.* **2012**, *140*, 898–918. [CrossRef]
48. Tewari, M.; Chen, F.; Wang, W.; Dudhia, J.; LeMone, M.A.; Mitchell, K.; Ek, M.; Gayno, G.; Wegiel, J.; Cuenca, R.H. Implementation and verification of the unified NOAH land surface model in the WRF model. In Proceedings of the 20th Conference on Weather Analysis and Forecasting/16th Conference on Numerical Weather Prediction, Seattle, WA, USA, 12–16 January 2004; pp. 11–15.
49. Hong, S.-Y.; Noh, Y.; Dudhia, J. A new vertical diffusion package with an explicit treatment of entrainment processes. *Mon. Weather Rev.* **2006**, *134*, 2318–2341. [CrossRef]
50. Kain, J.S. The Kain–Fritsch convective parameterization: An update. *J. Appl. Meteor.* **2004**, *43*, 170–181. [CrossRef]
51. ACI. Available online: http://www.aci.it/laci/studi-e-ricerche/dati-e-statistiche.html (accessed on 29 March 2021).
52. Ghermandi, G.; Fabbi, S.; Baranzoni, G.; Veratti, G.; Bigi, A.; Teggi, S.; Barbieri, C.; Torreggiani, L. Vehicular exhaust impact simulated at microscale from traffic flow automatic surveys and emission factor evaluation. In Proceedings of the 18th International Conference on Harmonisation within Atmospheric Dispersion Modelling for Regulatory Purposes, HARMO 2017, Bologna, Italy, 9–12 October 2017; Volume 2017, pp. 475–479.
53. Sanchez, B.; Santiago, J.L.; Martilli, A.; Martin, F.; Borge, R.; Quaassdorff, C.; de la Paz, D. Modelling NOX concentrations through CFD-RANS in an urban hot-spot using high resolution traffic emissions and meteorology from a mesoscale model. *Atmos. Environ.* **2017**, *163*, 155–165. [CrossRef]
54. Air Quality e-Reporting (AQ e-Reporting), European Environmental Agency. Available online: httos://www.eea.europa.eu/data-and-maps/data/aqereporting-2 (accessed on 9 May 2021).
55. The R Base Package. Available online: https://www.rdocumentation.org/packages/base/versions/3.6.2 (accessed on 21 August 2020).
56. The R Lubridate Package: Make Dealing with Dates a Little Easier. Available online: https://cran.r-project.org/web/packages/lubridate/index.html (accessed on 28 August 2020).
57. The Rrgdal Package: Bindings for the 'Geospatial' Data Abstraction Library. Available online: https://cran.r-project.org/web/packages/rgdal/index.html (accessed on 28 August 2020).
58. Janssen, S.; Thunis, P. *FAIRMODE Guidance Document on Modelling Quality Objectives and Benchmarking*; EUR 30264 EN; Publications Office of the European Union: Luxembourg, 2020; JRC120649; ISBN 978-92-76-19746-1. [CrossRef]
59. Hanna, S.; Chang, J. Acceptance criteria for urban dispersion model evaluation. *Meteor. Atmos. Phys.* **2012**, *116*, 133–146. [CrossRef]
60. Ghermandi, G.; Fabbri, S.; Veratti, G.; Bigi, A.; Teggi, S. Estimate of Secondary NO2 Levels at Two Urban Traffic Sites Using Observations and Modelling. *Sustainability* **2020**, *12*, 7897. [CrossRef]
61. Villani, M.G.; Russo, F.; Adani, M.; Piersanti, A.; Vitali, L.; Tinarelli, G.; Ciancarella, L.; Zanini, G.; Donateo, A.; Rinaldi, M.; et al. Evaluating the Impact of a Wall-Type Green Infrastructure on PM10 and NOx Concentrations in an Urban Street Environment. *Atmosphere* **2021**, *12*, 839. [CrossRef]
62. Best, D.J.; Roberts, D.E. Algorithm AS 89: The Upper Tail Probabilities of Spearman's Rho. *Appl. Stat.* **1975**, *24*, 377–379. [CrossRef]
63. Hollander, M.; Wolfe, D.A. *Nonparametric Statistical Methods*; John Wiley & Sons: New York, NY, USA, 1973; pp. 185–194.

atmosphere

Article

Evaluating the Impact of a Wall-Type Green Infrastructure on PM_{10} and NOx Concentrations in an Urban Street Environment

Maria Gabriella Villani [1,2], Felicita Russo [1,*], Mario Adani [1], Antonio Piersanti [1,*], Lina Vitali [1], Gianni Tinarelli [3], Luisella Ciancarella [1], Gabriele Zanini [1], Antonio Donateo [4], Matteo Rinaldi [5], Claudio Carbone [6,7], Stefano Decesari [5] and Peter Sänger [8]

1 ENEA, Italian National Agency for New Technologies, Energy and Sustainable Economic Development, SSTP-MET-INAT, 40129 Bologna, Italy; mariagabriella.villani@enea.it (M.G.V.); mario.adani@enea.it (M.A.); lina.vitali@enea.it (L.V.); luisella.ciancarella@enea.it (L.C.); gabriele.zanini@enea.it (G.Z.)
2 ENEA, Italian National Agency for New Technologies, Energy and Sustainable Economic Development, SSTP-MET-INAT, 21017 Ispra, Italy
3 ARIANET S.r.l., Via Gilino n.9, 20128 Milano, Italy; g.tinarelli@aria-net.it
4 Istituto di Scienze dell'Atmosfera e del Clima, ISAC-CNR, 73100 Lecce, Italy; a.donateo@isac.cnr.it
5 Istituto di Scienze dell'Atmosfera e del Clima, ISAC-CNR, 40129 Bologna, Italy; m.rinaldi@isac.cnr.it (M.R.); s.decesari@isac.cnr.it (S.D.)
6 ENEA, Italian National Agency for New Technologies, Energy and Sustainable Economic Development, TERIN-PSU-ABI, 40129 Bologna, Italy; claudio.carbone@enea.it
7 PROAMBIENTE S.c.r.l., CNR Research Area, 40129 Bologna, Italy
8 Green City Solutions GmbH, 15741 Bestensee, Germany; p.saenger@mygcs.de
* Correspondence: felicita.russo@enea.it (F.R.); antonio.piersanti@enea.it (A.P.); Tel.: +39-051-6098406 (F.R.)

Citation: Villani, M.G.; Russo, F.; Adani, M.; Piersanti, A.; Vitali, L.; Tinarelli, G.; Ciancarella, L.; Zanini, G.; Donateo, A.; Rinaldi, M.; et al. Evaluating the Impact of a Wall-Type Green Infrastructure on PM_{10} and NOx Concentrations in an Urban Street Environment. *Atmosphere* **2021**, *12*, 839. https://doi.org/10.3390/atmos12070839

Academic Editor: Rafael Borge

Received: 29 April 2021
Accepted: 28 June 2021
Published: 29 June 2021

Abstract: Nature-based solutions can represent beneficial tools in the field of urban transformation for their contribution to important environmental services such as air quality improvement. To evaluate the impact on urban air pollution of a CityTree (CT), an innovative wall-type green infrastructure in passive (deposition) and active (filtration) modes of operation, a study was conducted in a real urban setting in Modena (Italy) during 2017 and 2018, combining experimental measurements with modelling system evaluations. In this work, relying on the computational resources of CRESCO (Computational Centre for Research on Complex Systems)/ENEAGRID High Performance Computing infrastructure, we used the air pollution microscale model PMSS (Parallel Micro-SWIFT-Micro SPRAY) to simulate air quality during the experimental campaigns. The spatial characteristics of the impact of the CT on local air pollutants concentrations, specifically nitrogen oxides (NOx) and particulate matter (PM_{10}), were assessed. In particular, we used prescribed bulk deposition velocities provided by the experimental campaigns, which tested the CT both in passive (deposition) and in active (filtration) mode of operation. Our results showed that the PM_{10} and NOx concentration reductions reach from more than 0.1% up to about 0.8% within an area of 10×20 m^2 around the infrastructure, when the green infrastructure operates in passive mode. In filtration mode the CT exhibited higher performances in the abatement of PM_{10} concentrations (between 1.5% and 15%), within approximately the same area. We conclude that CTs may find an application in air quality hotspots within specific urban settings (i.e., urban street canyons) where a very localized reduction of pollutants concentration during rush hours might be of interest to limit population exposure. The optimization of the spatial arrangement of CT modules to increment the "clean air zone" is a factor to be investigated in the ongoing development of the CT technology.

Keywords: urban air pollution; nature-based solutions; green infrastructure; PMSS Lagrangian model; NOx; PM_{10}

1. Introduction

With the proportion of the European population living in urban areas expected to rise to over 80% by 2050 [1], making European cities sustainable for human health has become a key challenge. This challenge includes efforts to improve air quality not only by reducing emissions, but also by modifying the urban morphology to reduce the exposure of the population to air pollution [2,3].

Beyond urban measures addressing air pollution mitigations, such as those aimed at reducing emissions from traffic, residential heating, industry and secondary particle formations [4], urban green infrastructures have proved to be promising in mitigating air pollution in urban areas of several cities around the world [5–8]. The work presented in [5] concerned modelling the annual potential of particulate matter and ozone removal from urban forests in Florence, amounting to 0.009–0.031 t/ha depending on the pollutant and the forest type. The results in [6] identified six specific policy interventions, underpinned by research, allowing green infrastructures to improve air quality with unambiguous benefits. The works of [7] analyzed a case study area in Melbourne, Australia, finding that trees provide the highest air pollution removal capability, but green roofs and green walls allow higher building energy savings. Finally, [8] provided in 2019 a scientific review of the link between air pollution, green infrastructure and human health, highlighting that the deployment of green infrastructure is critical to avoid unintended consequences such as trapping air pollution in street canyons or increasing the presence of bio allergens in the air. Therefore, it is crucial to develop design guidelines, vital for promoting and optimizing greening benefits.

Urban green infrastructures are networks of green spaces, water or other natural features within urban areas offering several important services to the urban population, such as reducing the risk of flooding [9], cooling high urban temperatures [10,11] and reducing human exposure to pollutants [8]. In particular, trees and vegetation can abate air pollution directly by trapping and removing fine particulate matter [12], and they may indirectly influence the urban climate by decreasing air temperatures [13,14]. The strength of the effect may depend on meteorological parameters, pollutants concentration, the type and state of vegetation, and the urban design [2,3,5–8,15]. As the actual impact of the green infrastructures on air quality is highly context-dependent, it is very important to assess such effects in realistic urban settings.

With the aim of addressing specific green infrastructure capability in relation to deposition and filtration processes, the Project CityTree Scaler [16] took place in Modena, Italy, during the years 2017 and 2018. A green infrastructure, the CityTree (hereafter indicated as "CT"), consisting of a vertical panel, sized $3 \times 0.6 \times 4$ m^3 [L × W × H], with vertical deposition and filtration mechanisms [17], was placed in viale Verdi-Modena (Italy), a road segment having the characteristics of a street canyon, close to a heavily polluted area of the city where traffic represents the main source of pollution. CTs are standalone units suitable for placement on curbsides and represent the ideal solution in architectural contexts in which the lack of free-soil space can prove problematic.

The Institute of Atmospheric Sciences and Climate of the Italian National Research Council (CNR-ISAC) and the Consorzio PROAMBIENTE, in collaboration with the CT manufacturer, Green City Solutions [17], examined the operational modes of the infrastructure during three field campaigns taking place in 2017 and 2018 and reported in [16]. In the cited work, the filtering and depositing capability of the urban green infrastructure were fully characterized. In particular, during these campaigns the bulk deposition coefficients (as in [18]) employed in our study were calculated. Hereafter, the field campaigns and observational datasets will be labelled "CNR".

As a companion paper of [16], the present study further examines the reduction of air pollutants, particularly PM_{10} and NOx concentrations, operated by the green infrastructure. We looked into the air pollution spatial distribution, by means of a Lagrangian dispersion and transportation numerical model. We chose this modelling approach as Lagrangian dispersion models have been successfully deployed to study the air flow within vegetation and forested environments [19–24], as well as within complex urban environments (see as an example: [25–28]).

The modelling tool we chose for this type of assessment is a fit-for-purpose model capable of reproducing passive deposition processes. We used the modelling system PMSS (Parallel Micro-SWIFT-SPRAY) [29–34], which has a demonstrated ability to reconstruct air transport in complex city environments, particularly in street canyons (please refer to: [35–43]).

The PMSS model operates at very high spatial resolution (meters), and it is able to simulate the effect of the presence of the infrastructure both on the pollutant concentration and on the reconstructed air flow within a street canyon. It is also able to consider the CT structure as an obstacle with measured deposition characteristics [44,45]. In fact, although our Lagrangian model cannot yet simulate filtration by obstacles directly, it can account for the porosity of the vegetation covering the vertical surfaces of the CT and the complex effects between air motion and the CT structure, already effectively included in the bulk deposition velocity values [18,44,45] used as input. In relation to the derivation of the deposition coefficients [16], these do represent a sort of "total deposition" as they include several effects such as particle interception, absorption and throughflow (please refer to [18]).

In the literature, a recent study [46] examined the impact of a similar CT studied in our work. This study aimed at investigating with a computational fluid dynamic model (CFD), the impact of a CT on the yearly pollutant abatement (particulate matter and nitrogen dioxide) in the city of Amsterdam. This study also considered different numbers of green structures and their configurations. However, it only involved CFD simulations based on statistics of pollutant concentrations and wind speed to guarantee realistic results, considering factory values of deposition efficiencies. The distinctive feature of our work, besides a different time period, is to rely on measurements available at the CT location, as well as on the adoption of a Lagrangian approach. The choice between an Eulerian and a Lagrangian approach depends highly on the objective of the study and on the characteristics of the problem under investigation [47]. The Lagrangian method, which describes the motion of the fluid by following individual fluid parcels, is typically used to predict the overall particle dispersion pattern [47,48] and the temporal variation of the mean concentration [47,49,50], with detailed particle spatial distributions [47]. On the other hand, the Eulerian approach is more suitable when the details of the air flux properties are relevant in a particular location, since it is based on the mass conservation equation and it can incorporate second and higher order chemical kinetic equations necessary to describe photochemical smog generation [51].

In this work we examine the abatement of air pollutants due to the green infrastructure focusing on two timeframes (12–31 May 2017 and 5–17 April 2018) when the CNR experimental campaigns provided a complete characterization of the deposition velocities and filtration efficiency of the CT.

Our study has three main objectives to assess: (i) the effective PM_{10} and NOx concentration reductions due to the CT; (ii) the characteristics of the area where the air pollutants abatement takes place; (iii) how well our model PMSS is able to reproduce the observed concentrations. In fact, the last objective (iii) represents a fundamental aspect to evaluate the model performances and is the basic step for objective (i) and (ii).

In this study, several simulations were performed and this required the implementation of the calculations on the High Performance Computing (HPC) infrastructure ENEA-GRID/CRESCO [52], whose clusters are distributed over six sites and have been running since 2008. CRESCO started with co-funding by the Italian Ministry of Education, University and Research (MIUR), in the framework of the 2001–2006 PON (European Regional Development Funds Program).

The methodology section describes the green infrastructure CT and the field campaigns performed by the CNR and Consorzio PROAMBIENTE. We explain the PMSS setup in details, including the urban settings (obstacles and buildings), the prescribed CT location and meteorological and emission features coherent with the experimental campaigns [16].

The results section focuses on the abatement of air pollutants produced by the CT and presented as percentage reductions in PM_{10} and NOx concentrations, with the description of their spatial characteristics. In Appendix A, we analyze the ability of the model to reproduce the measured concentrations. As will be explained in Section 3.1, the comparison is presented for only one of the pollutants (NOx), exploring the role of the background NO_2 concentrations and of the wind conditions.

2. Methodology

2.1. Green Infrastructure–CityTree Structure

The CT is a $3 \times 0.6 \times 4$ m^3 ($L \times W \times H$) panel with the two largest vertical sides covered with a combination of hydroponic cultures of mosses and non-vascular plants that act as depositing surfaces for particulate matter and pollutant gases (see [17] for details). The mosses were predominantly Amblystagium varium "Plattenmoos" and Lucobryum Glaucum "Polstermoos" types. The first type was placed in the inner side of the panel since it benefits from reduced sunlight. The second type, better suited for withstanding direct sunlight, was placed on the external part of the panel. Irrigation in CTs is provided by a fully automated system that relies on temperature and relative humidity measurements to ensure the highest efficiency for moss cultures. The newest types of CT host hydroponic cultures of mosses only. Being permeable to air and with their mesh-like texture, mosses can capture atmospheric particulate matter by impaction and deposition when flowed with ambient air. Air flow can be forced in CTs by the internal venting system. As discussed in the companion paper, the "filtration mode" results in enhanced removal of PM but not of NOx with respect to when the venting system is switched off ("passive mode"), most likely due to an increased probability of impact and deposition of solid particles induced by the higher flow rate, not involving the gaseous species [16]. Since filtration requires an energy supply, CTs were operating most of the time in passive mode during the field campaigns in Modena.

2.2. Experimental Activities

Since 2017, three field campaigns have been performed in order to obtain a complete characterization of the filtration efficiency and deposition velocities of the CT in a real environment, consisting of a urban street canyon, and different meteorological conditions. We considered the western side of the road named viale Verdi (44° 38′ 34.94″ N and 10° 56′ 07.81″ E) in Modena (Italy), (see Figure 1).

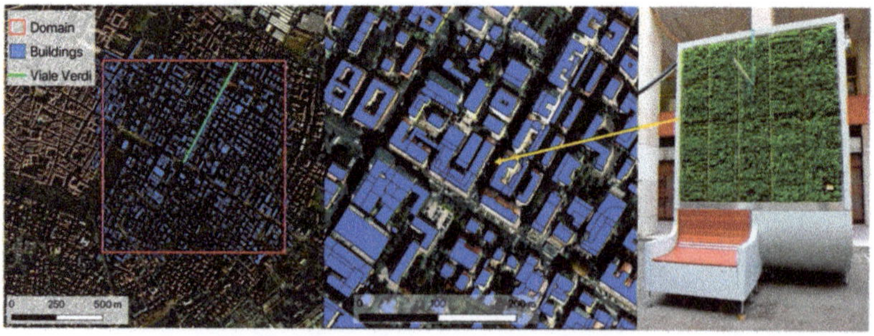

Figure 1. Simulation domain and location of the CT unit within the urban area of Modena, which includes viale Verdi, indicated by a green line.

In the companion paper [16], data and results from the experimental activities are fully described. For the reader's convenience, here we report a brief description of the experimental activities. The first measurement campaign was performed between 12 May and 27 June 2017, for a total duration of 37 days. This campaign focused on the determination of the CT deposition velocities for selected air pollutants in spring-summer meteorological conditions. The second measurement campaign was performed from 9 November to 25 November 2017. This campaign aimed at the determination of the CT filtration efficiency when the ventilation system was active.

The last campaign was performed from 27 March to 17 April 2018. The measurements in deposition mode were performed from 6 April to the end of the period for a total duration of 12 days. This campaign focused on studying the removal rates of particulate and gaseous pollutants in winter-spring conditions both in deposition and filtration modes.

The mobile laboratory was equipped with state-of-the-art instrumentation for air quality and meteorological observations. Particle concentrations were measured using a condensation particle counter (CPC, Grimm Aerosol Model 5.403, 1 Hz) for total particle number concentration (PNC) and an optical particle counter (OPC, Grimm 1.109, 31 channels) for particle number size distribution and particle mass size distribution. Both CPC and OPC collected samples with 1 min resolution. The system was able to measure PNC in the range between 0.009 and 1 μm. The OPC was used to measure the particle number size distribution in the size range between 0.25 and 32 μm. The OPC provides estimates of the particulate mass in the PM_{10}, $PM_{2.5}$ and PM1 size ranges. Gaseous concentrations of NO, NO_2, and NOx were measured with a temporal resolution of 1 min. NO, NO_2, and NOx were measured using a Teledyne-API (200E analyzer). Black Carbon (BC) was measured by a Thermo Fisher Scientific Multi-Angle Absorption Photometer (MAAP). The instrument provides the atmospheric concentration of the equivalent black carbon (BC) with 1 min time resolution. Furthermore, the mobile laboratory was equipped with a three-dimensional ultrasonic anemometer (R3, Gill Instruments), installed at about 3 m above the ground on a telescopic mast, and a slow response thermo-hygrometer Rotronic MP100A (Campbell Scientific).

Statistically significant passive deposition velocities (first quartile, median, and third quartile of the distribution were provided) for PM_{10} and NOx, measured during the first and the third field campaigns were used in the PMSS model to evaluate potential reductions of air pollutants concentration in the area nearby the CT. In addition, traffic observations (consisting of vehicle counting) were used to estimate the traffic emission input for the simulation.

2.3. *Micro-SWIFT-SPRAY (PMSS)*

The modelling system Micro-SWIFT-SPRAY (PMSS) is a suite for simulations of primary pollutant transport and dispersion. PMSS is the parallelized version of the MSS model suite, which is fully described in [29–34]. The system is composed of two main model units: PSWIFT, an analytically modified mass consistent interpolator over complex terrain [30], and PSPRAY, a three-dimensional Lagrangian particle dispersion model. The modelling suite can be used for both local scale and microscale simulations, with complex terrain or obstacles such as buildings, which are represented as filled cells in the meteorological field [32,33]. PSWIFT produces mass-consistent wind fields using data from a dispersed meteorological network or from simulated meteorological data at lower resolution.

In PSPRAY the pollutant concentration is simulated by generating a certain number of "virtual" particles, each of them carrying a portion of the pollutant mass. The velocity of the particles is composed by a mean velocity component defined by the local wind computed by PSWIFT, and a stochastic velocity component, characteristic of the atmospheric turbulence. PSPRAY can compute mean and instantaneous concentrations on a three-dimensional grid defined by the user, differentiating the calculation both by "chemical species" and by "source".

PSPRAY can simulate the dispersion of pollutants from point, area or line sources. The model reproduces the transport, the dispersion and the dry and wet deposition of the airborne chemical species emitted. In particular, PSPRAY is also able to compute dry deposition due to interactions with roofs, walls and ceilings, by inserting species dependent deposition velocities into the model [33].

In its current version and differently from a CFD model [15,53], our model cannot compute the air flow through a porous medium, such as a tree, since the porosity of obstacles has not yet been implemented. However, as already mentioned in the introduction, the bulk deposition coefficients measured in [16] represent a sort of "total deposition" as they include the main effects such as particle interception, absorption and throughflow (see [18]). Moreover, they already take into account the porosity of the vegetation and the complex effects between air motion and vegetation as found in the literature [18,44,45].

In our model simulations, the CT structure was represented as an obstacle with deposition characteristics, which we can parametrize in PMSS. When the filtration mode was activated in the CT, the calculated bulk "effective filtration" deposition velocity was assumed to be the deposition velocity capable of producing the impact of the CT measured in filtration mode. In this way, we could use PMSS to study the spatial characteristics of the filtration efficiency.

As a Lagrangian dispersion model, PSPRAY performs the dispersion of chemical compounds (NOx) in the urban environment. However, chemical reactions occur at this scale, affecting the concentrations of the chemical species [33]. In order to represent this, recent developments have addressed the implementation of several chemical models into PSPRAY (see [33] for more details). However, in the present work, we decided to consider only the dispersion characteristic of the PMSS system, and the chemical models were not implemented.

The PMSS modelling system is a commercial software developed by ARIANET [54]: PSWIFT and PSPRAY codes, in the versions PSWIFT-2.1.1 and PSPRAY-3.7.3, compiled with Intel16 compiler, using OPENMPI library.

2.4. Simulation Set-Up

As mentioned in the introduction, we focused on two-time frames: 12–31 May 2017; and 5–17 April 2018. In these time frames the CNR experimental campaigns provided a complete characterization of the bulk deposition velocities for NOx and PM_{10}. Since these measurements did not show a significant variation during the campaigns, their average values were used for the simulations. Only for the PM_{10} concentrations was the filtration efficiency retrieved of the CT operating in active mode. The reason for selecting PM_{10} lies in the fact that [16] reported neglectable abatement for the NOx concentrations when the CT operated in active mode.

We performed two sets of simulations:

1. For the year 2017, we ran PMSS to simulate vertical deposition due to the CT operating in passive mode (only deposition).
2. For the year 2018 we reproduced the air pollution abatement linked to the deposition vertical velocities when the CT operated in passive mode (as in 2017). In addition, we studied the PM_{10} concentration abatement produced when the CT filtration mode was activated. Here we used the passive deposition velocity, calculated from measurements analysis, able to produce the same pollutant deposition measured with the CT in filtration mode (see [16]).

2.4.1. Input Meteorological Data

Meteorological data used to feed the diagnostic model PSWIFT for 2017 and 2018 periods were provided by RAMS (Regional Atmospheric Modelling System, [55]), the meteorological model driving the national Air Quality modelling system MINNI [56–58] and the air quality forecast system FORAIR-IT [59,60].

Hourly data of meteorological fields were used by PSWIFT to reconstruct the three-dimensional wind, temperature and turbulent flow at 2 m resolution.

The meteorological parameters that are used in the model (such as relative humidity and precipitation) affect only the state of the atmosphere and cannot change the deposition characteristics of the CT, which are described by the bulk deposition velocity that is constant for each pollutant.

2.4.2. Simulation Domain, Area and Obstacles

We set a 1×1 km^2 horizontal domain that was centered in viale Verdi, where the CT was located (Figure 1). The domain is shown by the red square in Figure 1 and covered an adequate fraction of the emissions of the city while providing that the CT unit was far enough from the domain border, where the model uncertainty is generally higher. The spatial resolution of 2 m was chosen to ensure that the model represented the CT as closely as possible while keeping computational time within acceptable values. For the vertical grid, we chose the following 25 levels above the ground: 0, 1.5 m, 3 m, 4 m, 6 m, 8 m, 10 m, 13 m, 16 m, 19 m, 22 m, 25 m, 30 m, 35 m, 42 m, 50 m, 60 m, 80 m, 120 m, 160 m, 200 m, 240 m, 300 m, 380 m, and 700 m. Since buildings are represented by the model as filled cells in the computational grid, vertical levels were chosen in order to properly describe the range of the heights of the buildings within the domain.

As mentioned in the introduction, in our simulations the CT was represented as an obstacle with deposition characteristics. This is a sound assumption, as [44] showed that, rather than representing every leaf, air flow and particle-leaf interactions are represented statistically by using a continuum approach. As for the case of the CT that contains a sufficiently large number of single vegetated elements, the effects of individual vegetated items are irrelevant in the average transport through the volume. Moreover, each vegetation item of the CT surfaces has dimensions much smaller than our model grid size. On this basis, we represented the CT as a barrier with proper measurable parameters describing vegetation (see for example [44,45]), which are fully provided in our modelling study by the measurement campaigns of [16].

2.4.3. Bulk Deposition Velocities for PM_{10} and NOx

The bulk deposition velocities used in our simulations are of two kinds:

- Bulk Deposition Velocities with the CT in passive mode (filtration switched off)

We considered the main statistical parameters of deposition velocities resulting from the measurements of both the 2017 and 2018 campaigns. Upon observing that the values did not change significantly, the results from both campaigns were included for a more robust statistic.

Measurements of air pollutants were performed, alternatively, through a computer-controlled valve switching system, at two different horizontal distances on both sides of the CT (at the front and at the rear of the wall). Figure 2 shows the sampling system that was composed of four identical conductive silicon sampling tubes connected to a common inlet through four electronic valves for automatically switching the inlet (P1, P2, P3 and P4) every 7 min, resulting in a complete cycle of 28 min. Inlet P1 and P2 were positioned in proximity (at 5 and 30 cm respectively) of the surface of moss and leaves on the roadside of the CT panel. Inlet P3 and P4 were in a specular position on the curb side of the CT panel.

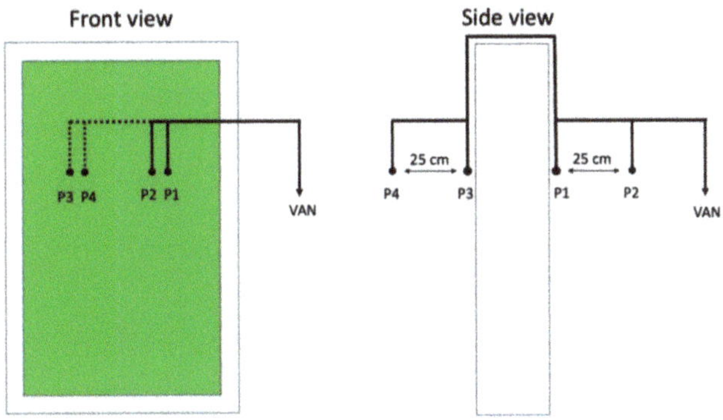

Figure 2. Diagram showing the sampling points used to measure Bulk Deposition Velocities.

The statistical parameters considered for the estimate of the deposition velocity were the first quartile, the median and the third quartile, and are presented in Table 1. These values are consistent with the deposition velocity values found in the literature [61].

Table 1. Summary of PMSS set-up.

Parameters	2017	2018
Days Simulated	12–31 May 2017	5–17 April 2018
Pollutant Simulated	PM_{10}, NOx	PM_{10}, NOx
Horizontal Domain	1 km × 1 km	1 km × 1 km
Horizontal Resolution	2 m	2 m
Emissions: Vehicle Input	Flow model for 2017 + vehicle Measurements 2017 in viale Verdi	Flow model for 2017 + vehicle Measurements 2018 in viale Verdi
Obstacle Position	Viale Verdi 21	Viale Verdi 21
Meteo Input	RAMS [55] 4 km × 4 km, May 2017	RAMS [55] 4 km × 4 km, April 2018
Bulk Velocity Deposition (PM_{10}) CT passive mode: 1st quartile, median, 3rd quartile	vd -> 0.0005 m/s, 0.0012 m/s, 0.003 m/s	vd -> 0.0005 m/s, 0.0012 m/s, 0.003 m/s
Bulk Velocity Deposition (PM_{10}) CT active mode	Not simulated	vd filtration ~0.024 m/s
Bulk Velocity Deposition (NOx) CT passive mode: 1st quartile, median, 3rd quartile	vd -> 0.0004 m/s, 0.0011 m/s, 0.0025 m/s	vd -> 0.0004 m/s, 0.0011 m/s, 0.0025 m/s
Model Versions	PSWIFT-2.1.1, PSPRAY-3.7.3, Intel16 compiler, OPENMPI library	PSWIFT-2.1.1, PSPRAY-3.7.3, Intel16 compiler, OPENMPI library

- Bulk Deposition Velocities for PM_{10} with the CT in filtration mode

The filtration mechanism occurs in the CT by means of fans operating inside the structure, and it is a more efficient mechanism than the passive deposition (e.g., [16,62,63]). In particular, [16] showed that the aerosol removal efficiency of the CT was from ~3 to almost 20 times higher in filtration than in deposition mode.

We provided an estimate of the effect of the CT in filtration mode on particulate matter concentrations to study its influence at increasing distance from the structure. Considering PM_{10} measured concentrations, CNR and Consorzio PROAMBIENTE calculated an "equivalent deposition velocity" as the bulk deposition velocity that the CT should have in order to produce the observed pollutant reduction in the active filtration mode. The average value of about 0.024 m/s was obtained and is consistent with the largest values of deposition velocity intervals found in the literature.

2.4.4. Emissions

We computed the traffic emissions in viale Verdi from the hourly vehicle flows measured during the campaigns from 12–31 May 2017 and from 5–17 April 2018. Hourly modulations of buses passages were calculated according to the public transportation timetable at the bus stop. In all the other roads of the simulation domain, total daily vehicle flows generated by a flow assignment model run for 2017 by the Municipality, combined with hourly traffic modulation measured in viale Verdi, were used (e.g., Figure 3).

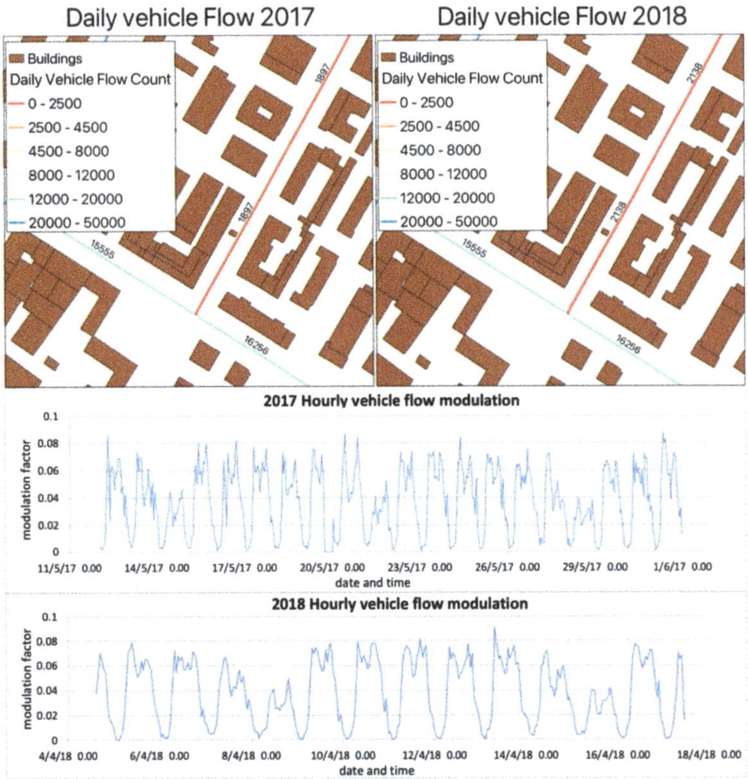

Figure 3. Hourly vehicle flow modulation measured during the 2017 and the 2018 campaign.

The composition of the vehicle fleet, in terms of age, emission standard, displacement, and fuel, was retrieved from the public registry of motor vehicles for the year 2017 (which was the latest published online at the time of the study) and for the city of Modena. The emission input for PSPRAY has been calculated by TREFIC (Traffic Emission Factors Improved Calculation), a software designed by ARIANET [64] for the calculation of road emissions. TREFIC calculates emission factors (EFs) using the COPERT 4 methodology [65], based on vehicle type, fuel consumption, average travelling speed and road type.

2.4.5. Parallel Run Details CRESCO

PMSS allows for parallelization in time for PSWIFT and for PSPRAY, both with respect to the domain, referring to a multi tiles domain, and to sources/particles decomposition [31]. Our domain is an area of 1×1 km^2 at the resolution of 2×2 m^2, which is suited to be represented by a single tile [32]. Therefore, our PSPRAY simulations were handled by a single tile of 501×501 cells, and the computation was parallelized with respect to particles.

We conducted the simulations on the CRESCO/ENEAGRID High Performance Computing infrastructure funded by ENEA [52]. As the model system was structured, each 12-day-simulation consisted of 12 single model runs, each simulating 24 h. The restart option was applied, i.e., for each simulated day the values of the pollutant concentrations calculated for the last hour were saved and used as initial condition for the following run.

The simulation duration depends also on the number of emitted particles and on the concentration resolution required. In our simulations, the total number of emitted particles reached peaks of about 34.3 million particles per day, with a typical particle number per time step of 4000. Here we used 264 cores and for PSPRAY, which represents the most

CPU demanding part of the system, the CPU time per core per simulated day was 470 s. A summary of the whole simulation set-up is provided in Table 1.

2.4.6. NOx and PM_{10} Reduction Operated by the CT

In order to quantify the abatement due to the presence of the CT on PM_{10} and NOx concentrations, we set up two simulations for each mode (deposition for 2017 and 2018, and filtration-like mode for 2018) that were equivalent to considering the cases with and without the CT. In particular, we considered the measured deposition properties of the CT as an on and off parameter in the model runs, similarly to what was done in the study [46]. Then, we calculated the percentage difference of the PM_{10} and NOx concentrations with and without the deposition on the CT as metrics of its impact on the air quality.

2.5. Reference Air Quality and Meteorological Data

The model validation represents a key factor to evaluate the model performances, and thus to assess the reliability of the results. Since our simulation took into account only primary emissions, to analyze how well our model reproduces NOx concentrations, we needed to estimate the secondary contribution due to atmospheric chemicals processes. Our best indicator available for secondary contribution to NOx concentrations were the measurements of urban background NO_2 concentrations, consisting of NO_2 hourly concentrations measured in the air quality station of Parco Ferrari (10° 54′ 22.8″ Lon, 44° 39′ 2.2″ Lat, 34 m a.s.l (see [66]).

Concerning the meteorological data, we used wind speed and direction data measured by ARPAE station located in (10° 55′ lon, 44° 39′ lat, 35 m altitude a.s.l.). Both stations are presented in Figure 4.

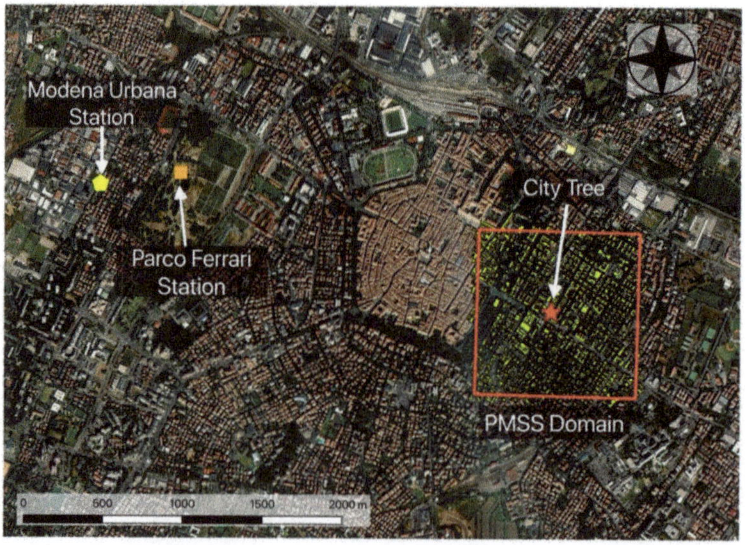

Figure 4. Map reporting PMSS domain, the CT position (red star symbol), the reference meteorological station Modena Urbana (yellow pentagon symbol) and the air quality station at Parco Ferrari (orange square symbol).

2.6. Tools for the Analyses

For the analyses performed, we used the FAIRMODE IDL-based DELTA software tool version 5.4 [67,68] provided by the Forum for Air Quality Modelling in Europe [67], and R-cran based scripts [69,70].

3. Results

3.1. NOx Modelled and Measured Data Evaluation

To evaluate the PMSS performances, we studied the NOx concentration time series for the year 2017. We focused on this time period since the measurement campaign from 12 May to 27 June 2017 was the longest (37 days). We compared hourly NOx concentrations measured during the CNR campaign with the simulated NOx after adding the background concentrations. The reason for choosing the pollutant NOx relies on the fact that we used our PMSS modelling system studying the dispersion of pollutants, neglecting chemical reactions (see Section 2.3). Therefore, it was possible to properly represent air pollutions from primary sources only. In our case, the NOx concentration is produced mostly by traffic, thus the approximation that is made neglecting the chemistry has a smaller influence on the simulated concentration that can be corrected by adding a background concentration, which represents most of the secondary contribution.

Hourly measurements of NO_2, provided by the urban background air quality station located in Parco Ferrari (see Section 2.5), at a distance of about 2 km from viale Verdi, were used as background concentration values, since data for the concentrations of NO and NOx were not available at this station.

In Appendix A we present detailed analyses to assess our model performances.

3.2. Green Infrastructure Abatement

The results of the simulations are presented here as averages over the two simulation periods (19 days for 2017 and 12 days for 2018), in order to show the pattern of the area interested by the effect of the presence of the CT.

Here the differences between the simulations with and without the CT are presented as mean PM_{10} and NOx percentage concentration differences. All the graphics reported show results using the median of the measured deposition velocities. Similar maps corresponding to the values obtained using the first and third quartile of the deposition velocity distributions for each case simulated did not differ significantly, therefore they are not shown.

3.2.1. CT in Passive Mode-Deposition

Median values of the deposition velocities were 0.0012 m/s and 0.0011 m/s for PM_{10} and NOx, respectively (see Table 1).

From Figure 5, the effect of the CT in passive mode (deposition) translates into a maximum concentration reduction of about 0.8% for both PM_{10} and NOx concentrations. Though our simulations are not directly comparable to the CFD simulations in the TNO study, nevertheless our results for PM_{10} are in agreement with those found in Amsterdam [46]. For a given deposition velocity, we can define a "region of influence" of the CT as the area where the concentration reductions are larger than 10% of the maximum value reached. For the CT operating in passive mode, the region of influence has concentration reductions larger than 0.1%, and it is found in proximity to the CT, with an extension of about 10×20 m^2 more or less centered on the CT position, as shown in Figure 6.

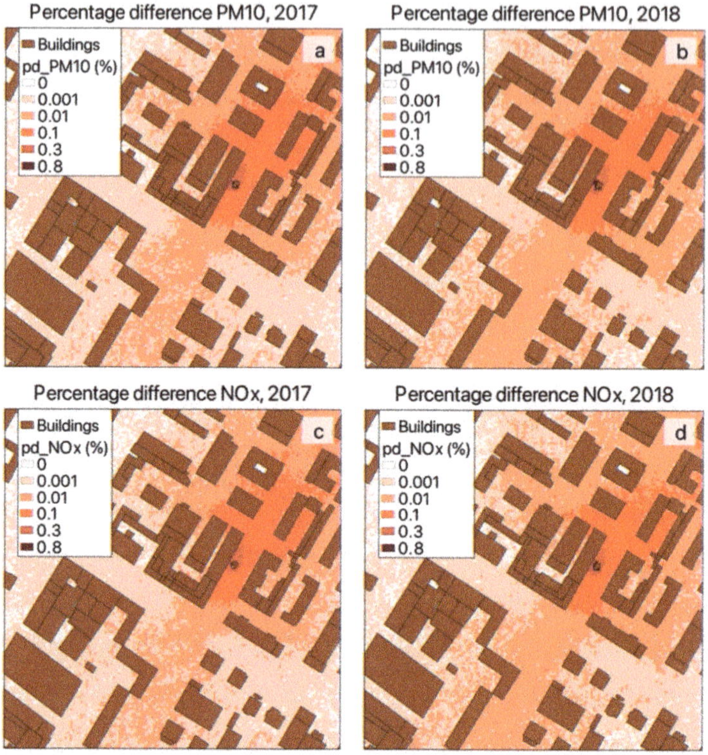

Figure 5. Simulation results for the case of PM_{10} (top panels) and NOx (bottom panels) showing the percentage difference between the concentration simulated without and with the CT for the two simulation periods 2017 (**a**,**c**) and 2018 (**b**,**d**). The values in the legend indicate the upper limit of the corresponding interval.

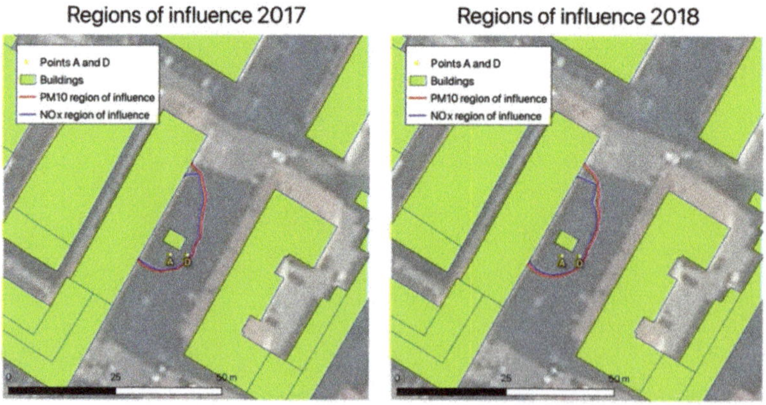

Figure 6. Maps for 2017 (**left**) and for 2018 (**right**) representing the areas where the percentage difference is larger than 0.1%.

Figure 6 also indicates the location of the two points (Figure A1), A and D, in which we studied the NOx concentration comparisons with measurements (presented in detail in Appendix A) and in which we extracted the vertical profiles shown in Figures 7 and 8.

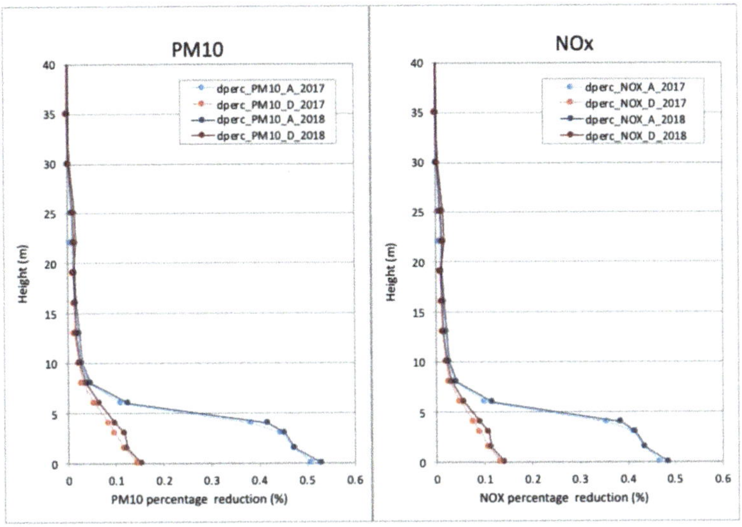

Figure 7. Vertical profiles for percentage difference reduction for PM_{10} (**left panel**) and NOx (**right panel**) in passive mode.

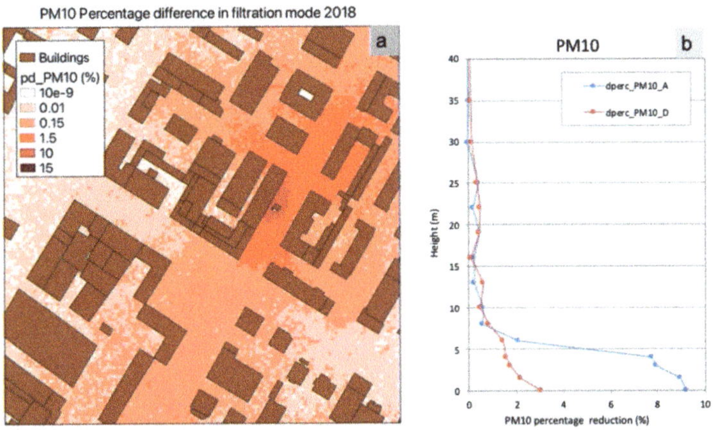

Figure 8. Impact of the CT operating in filtration mode on the PM_{10} concentration (**a**) and an example of vertical dependence of such impact in the sampled points A and D (**b**). The map corresponds to the deposition velocity vd = 0.024 m/s. The numbers in the legend indicate the upper limit of the corresponding interval.

3.2.2. Vertical Profiles of Pollution Reduction in Passive Mode

The mean vertical profiles for the pollutant reduction are reported in Figure 7. In particular, we took into consideration two points, A and D, which were close (point A) and 4 m away (point D) from the structure (see Figure A1).

We observe:

- Reduction profiles for PM_{10} and NOx have the same characteristics. They have the maximum reduction values in the first few meters above the ground. The reduction decreases significantly at heights above the green infrastructure (above 6 m).
- For both NOx and PM_{10}, the maximum pollutant reduction is obtained very close to the green infrastructure, at point A, with values of about 0.5%. At point D, the maximum pollutant reduction decreases to 0.1–0.2%.

- Air pollutant reduction decreases rapidly, moving away from the green infrastructure, both in the vertical and in the horizontal directions.

3.2.3. CT in Active Mode (Filtration) for PM_{10} Concentrations in 2018

The results for the CT in active mode of operation (filtration) for PM_{10} concentrations in 2018 are shown in Figure 8.

The space distribution of the impact did not significantly change with respect to the previous simulations with lower deposition velocities (here the value is 0.024 m/s, instead of about 0.001 m/s). However, the values of the maximum PM_{10} concentration reduction increased to 15% very close to the CT. In analogy with CT in passive mode (deposition) cases, we observe a very similar region of influence with different values and largest concentration reductions at the southwest CT corner. Here, in the areas where the percentage difference is higher than 1.5%, we observe a much larger extension covering most of viale Verdi and some of the lateral roads (approximately 10×30 m^2). The vertical distribution of the impact of the CT (Figure 8b) has not significantly changed with respect to the passive deposition case (Figure 7).

4. Discussion

When we take into consideration the pollutant abatement caused by the CT in passive mode (deposition), we observe that reductions of NOx and PM_{10} concentrations are smaller than 1% in an area of about of 10×20 m^2 around the CT and at the same height as that of the green wall. Comparing our results to those obtained in other studies is not an easy task. The only other modelling study published on a similar CT as that deployed in Modena, was performed by TNO [46]. The simulations in this study were performed by considering a realistic situation using factory specifications for the CT, therefore not considering input measurements for the model, as done in the study presented here. Although our modelling results are not directly comparable to those of the TNO study, nevertheless our results for PM_{10} in the case of passive mode of operation are in agreement with those found by TNO [46].

When the CT operated in active mode (filtration), our modelling estimate of the impact suggests pollutant reductions of a larger order of magnitude, reaching values of about 15%. As for the previous result, in this case finding similar results in the literature for comparison was not an easy task. A recent study [71] involving a similar type of green wall with filtering capabilities placed at the edge of a highway showed measurements that can be compared to the filtration efficiency measured during the CNR campaigns. These measurements are in some agreement, in the experimental estimate of the filtration efficiency for PM2.5, with the results of the CT experimental campaigns [16].

We observed that the region of influence has about the same shape and dimensions for both 2017 and 2018, with the CT in passive mode, as well as for the case with the CT in active mode of operation. Given the differences in the setups for the 2017 and 2018 simulations, especially regarding emissions and meteorological conditions, we expect that this area is more significantly influenced by the building pattern and distribution than by CT position in relation to the buildings. Indeed, in preliminary studies conducted for the year 2017, with the same setup with the exception of the CT position, the area of influence showed a shape which did not significantly differ from the one here presented in terms of contours, size and pollutants abatement.

We can conclude that a single CT will not have a significant impact in the reduction of particulate matter and nitrogen oxide concentrations in the whole urban extension. This finding is consistent with the work of [46] where several hypothetical configurations were analysed, including set ups with many CT structures. When more units of CT were represented, [46] obtained pollutant removal in the area considered was less than 20–30%. However, when looking at air pollution in hotspots and/or considering specific building arrangements, these abatements could be achieved.

In the present work, the reliability of the simulated reduction operated by the CT depends on the quality of our model simulations, e.g., we assumed that our model could represent absolute concentrations well, given our specific traffic primary emissions as input. The analyses of the measured and observed NOx concentrations in Appendix A aim to investigate the soundness of our results. The validation of a pollutant such as particulate matter is more complex, not having considered as input all the precursors of the PM secondary components, nor the chemical reactions in the atmosphere that generate these components.

In Appendix A.2, we observed that the values of the NO_2 background concentrations were generally close to 20 $\mu g/m^3$, and they were associated with prevailing easterly winds (up to 6 m/s). These concentration values were possibly generated by air being transported from Modena city center that, due to strict traffic regulations, is less affected by traffic than peripheral areas of the city. Differently, larger NO_2 background concentrations (above 50 $\mu g/m^3$) were measured with SW-SE winds of less than 3 m/s, and they were possibly brought from major traffic roads located southward, some kilometers away from Parco Ferrari.

The NOx observed and modelled concentration analyses reported an overall good agreement (see Figures A4–A7). The modelled values are generally close to the observations (see Figure A4). However, in specific cases the simulated concentrations are higher than the measurements (Figures A5 and A6). In particular, the measured and modelled hourly data show a dispersion that changes with the concentration values: for concentrations of less than 50 $\mu g/m^3$, the two distributions are very similarly normally distributed (e.g., Figure A6), while for higher values (specifically for more than 100 $\mu g/m^3$) the model tends to overpredict the NOx concentrations.

To further investigate the comparisons of the NOx concentrations, we took into account groups of days (Figure A9) and local and urban wind directions classified according to four sectors (Figures A10 and A11) (Appendix A.1). We found that:

- Wind measurements taken at the urban scale (at Modena Urbana) and close to the CT (CNR measurements inside the urban canyon) are significantly "decoupled" as already well documented in the literature (e.g., [72]). It was very difficult to establish a relation between the urban and local wind datasets, likely to be due to the air circulation specific to the street.
- The agreement between modelled and measured data is significantly more evident during the days 25–31 May, characterized by urban south-easterly wind and NOx concentrations between 20–100 $\mu g/m^3$. During the days 26–31 May, the Taylor diagram shows the best agreement with correlation values of about 0.85 and normalized standard deviations close to one.

In the previous settings, to assess if the NOx concentration distributions would spatially differ close to the CT, we analysed the concentrations at the points A and D (see Appendix A.1). However, we concluded that the distributions at the two locations did not differ significantly (e.g., we obtained similar statistical scores for both A and D, from the Taylor diagrams).

We did not identify the reasons for the good agreement obtained during 26–31 May. In addition, we need to further investigate the reasons why our model tends to overpredict NOx concentrations higher than 100 $\mu g/m^3$.

Note that the NO_2 background concentration measurements did show a very good correlation with the CNR local measurements. This represented a key factor for the overall good agreements between measured and simulated NOx concentrations at the local scale.

5. Conclusions

We presented an application of the modelling suite Micro-SWIFT-SPRAY (PMSS) as a reasonable tool for an estimate of the impact of a single CityTree (CT) on the abatement of air pollution concentrations, such as PM_{10} and NOx, in viale Verdi, Modena (Italy).

We applied the Lagrangian model Spray considering:

- The specific urban setting centered in viale Verdi, Modena (obstacles and buildings);

- A prescribed location of the CT consistent with the experimental field CT installation;
- Meteorological data and emission features coherent with the first and the third research campaigns;
- Values of bulk deposition velocities deriving from the experimental campaigns (and in agreement with results obtained in the literature).

PMSS simulations showed that the passive operational mode of the CT may produce air pollutant concentrations abatement larger than 0.1% and up to about 0.8% in an area of 10×20 m^2 around the CT and along the vertical extension of the green infrastructure (4 m). The area may depend on the urban infrastructures design.

When the CT operated in filtration mode, by using a value of bulk deposition that produced the same PM_{10} concentration abatement, in terms of concentration difference between the values measured in front and at the rear of the panel, we obtained a PM_{10} concentration reduction of about 15% close to the CT. This is the best estimate of the active CT mode operation on PM_{10} concentration abatement achievable with the PMSS suite. Furthermore, this finding suggests that active filtration mode can abate particulate matter concentrations at least one order of magnitude more efficiently with respect to the only deposition removal and that the abatement is present along the entire vertical extension of the CT.

The novelty of our results lies in the use of measured passive deposition velocities during three intensive field campaigns in the city of Modena (Italy) [16]. In addition, an assessment of model performance was developed with reference to NOx modelled and measured concentrations. For the year 2017, good agreement resulted between observations and modelled data reporting overall correlation indexes of about 0.6 and normalized standard deviations of about 1.25.

The results of the present study indicate that a single CT will not have a significant impact in improving the air quality of a street along its full length. These findings, tested with different meteorological conditions and different emissions input for the experimental time periods in 2017 and 2018, show that, according to the value of the deposition velocity, the major reduction effect occurs in a specific area in the neighborhood of the infrastructure (the "region of influence") in both deposition and filtration modes. Therefore, an effect in reducing particulate matter and nitrogen oxides concentrations, and thus population exposure, can still be achieved for specific environments (e.g., sidewalk benches, small gathering areas, bus stops, school or hospital entrances).

On the basis of these results, we expect that a cluster of CT modules, properly located in an urban context, could design restricted densely populated areas with consistent air pollution reduction, the so-called "clean air zones", where citizens are expected to spend time during rush hours. For practical applications, parameters such as energy and water consumption per operating hour are important. Such technical parameters of CTs were not optimized at the time of our field campaigns, while later generation CTs have been accurately engineered to account for power and water supply. A comparison of the technical features of the new CT model with respect to the one deployed in Modena (a CT 1.0 version) is available in Appendix B (Figure A12). In new CT 2.0 models, energy consumption is around 120–150 W per operating hour and water consumption is between 2 and 8 L per operating hour depending on local climate conditions. In recent experimental evaluations the filtration capacity for CT 2.0 has been estimated as 31 mg of fine particulate matter per hour. These results support the potential of CT technologies for protecting air quality in local urban contexts.

Author Contributions: Conceptualization, M.G.V., F.R., M.A., A.P., L.V., G.T., L.C. and S.D.; methodology, M.G.V., F.R., M.A., A.P., L.V., G.T. and L.C.; software, M.G.V., F.R. and G.T.; validation, M.G.V. and F.R.; formal analysis, M.G.V. and F.R.; investigation, M.G.V., F.R., L.C. and S.D.; resources, L.C., G.Z., S.D. and P.S.; data curation, M.G.V., F.R., M.A., A.P., L.V. and G.T.; writing—original draft preparation, M.G.V., F.R., L.C., A.D., M.R., C.C. and S.D.; writing—review and editing, M.G.V., F.R., L.C., A.D., M.R., C.C. and S.D.; visualization, M.A., A.P., L.V. and G.T.; project administration, L.C.,

G.Z., C.C., S.D. and P.S.; funding acquisition, G.Z., C.C., S.D. and P.S. All authors have read and agreed to the published version of the manuscript.

Funding: This research was funded by EIT (European Institute of Innovation and Technology) within the Climate-KIC project "CityTree Scaler".

Institutional Review Board Statement: Not applicable.

Informed Consent Statement: Not applicable.

Data Availability Statement: Data set available on request to corresponding authors.

Acknowledgments: The computing resources and the related technical support used for this work have been pro-vided by CRESCO/ENEAGRID High Performance Computing infrastructure and its staff [52]. CRESCO/ENEAGRID High Performance Computing infrastructure is funded by ENEA, the Ital-ian National Agency for New Technologies, Energy and Sustainable Economic Development and by Italian and European research programs, see https://www.eneagrid.enea.it/CRESCOportal/ (accessed on 29 June 2021) for information.

Conflicts of Interest: The authors declare no conflict of interest. The funders had no role in the design of the study; in the collection, analyses, or interpretation of data; in the writing of the manuscript, or in the decision to publish the results.

Appendix A

In this section we assess the ability of the model to reproduce the measured NOx concentrations considering the urban background NO_2 concentrations and the role of the wind conditions.

Appendix A.1. Viale Verdi Orientation and Geographical Sectors

To assess the model performances and to assess the pollutant reductions, we compared the simulated concentrations with measurements. In particular, we took into consideration features like the orientation of viale Verdi with respect to the northerly direction (about 32° northward) to define four sectors of wind direction, which are parallel to viale Verdi:

- NE (347–77°),
- SW (167–257°);

and perpendicular:

- NW (257–347°),
- SE (77–167°).

To examine the air pollutant reduction caused by the CT, we decided to study two cells of the simulation domain: A and D as indicated in Figure A1. The points were both close to the cell that represents the CT in the model grid: point A is adjacent to the CT and can be considered as representative of the air volume directly in contact with the CT, while point D is more related to the air volume within the street canyon. Point D is only four meters away from point A.

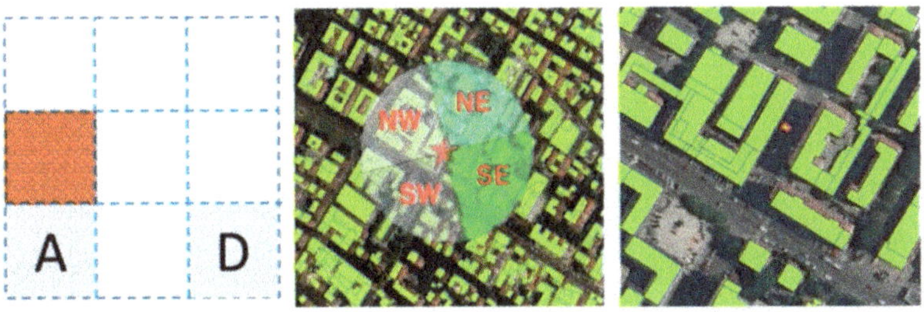

Figure A1. Left: positions of the A and D points with respect to the CT indicated here with the brown square. Centre: sectors considered according to the viale Verdi orientation. Right: position of the CT, in red, and the buildings and streets nearby.

Appendix A.2. NO_2 Background Concentration

We analysed the background NO_2 concentration provided by the reference station in Parco Ferrari considering the wind observed at the background station Modena Urbana. The results are shown in Figure A2. According to the diagram we outline:

- The prevailing wind directions were easterly. However, significant contributions were also from north-west and south-west directions.
- Mean NO_2 background concentrations were generally close to 20 μg/m^3. These values were detected particularly with easterly winds, which were stronger (up to 6 m/s) than in the other directions where intensities were generally lower than 4 m/s (Figure A3).

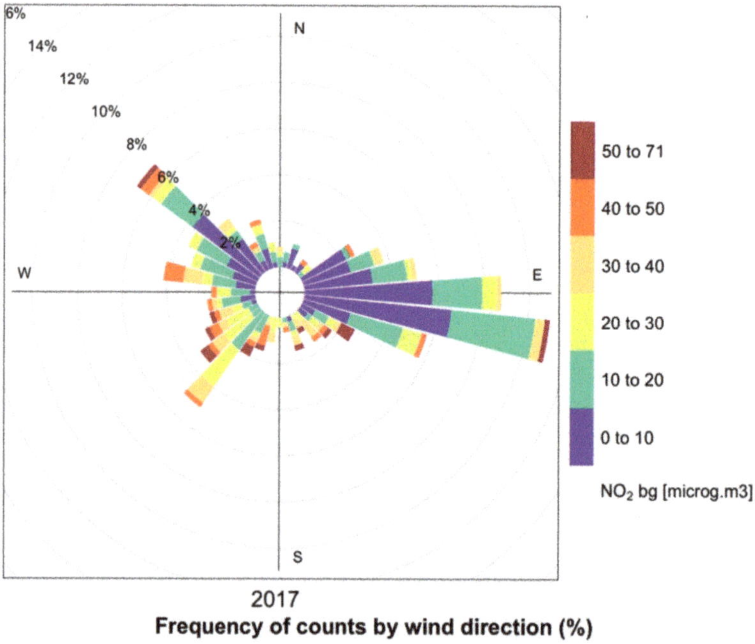

Figure A2. Concentration wind rose diagrams expressing the dependence of Parco Ferrari background NO_2 concentration on wind direction from Modena Urbana station (2017).

The Modena city center is located in an eastward direction with respect to the station and, because of traffic restrictions, is usually affected by less traffic than the peripheral areas. Larger NO_2 background concentrations, reaching hourly concentrations above 50 μg/m^3, were measured with SW-SE winds of less than 3 m/s. Winds blowing from these directions carry pollutants from major traffic roads such as the motorway A1/E45, which is located about three km away from Parco Ferrari.

Appendix A.3. Wind Distributions

As a first step, we studied the wind distributions of the data measured at the stations of Modena Urbana (background station) and from the CNR and Consorzio ProAmbiente, in viale Verdi. The resulting wind rose diagrams are presented in Figure A3. When urban and local wind distributions are compared, many differences can be outlined relating to both wind directions and speed. The local wind directions were mainly between NE and SW, which was probably due to the microscale circulation driven by the presence of the buildings (see Figure A1). Wind speeds were generally lower than 2 m/s with the majority of the values below 1 m/s. Stronger wind speeds up to 2 m/s were measured with westerly and north-easterly winds, the latter corresponding to the north-related orientation of viale Verdi (see Figure A1).

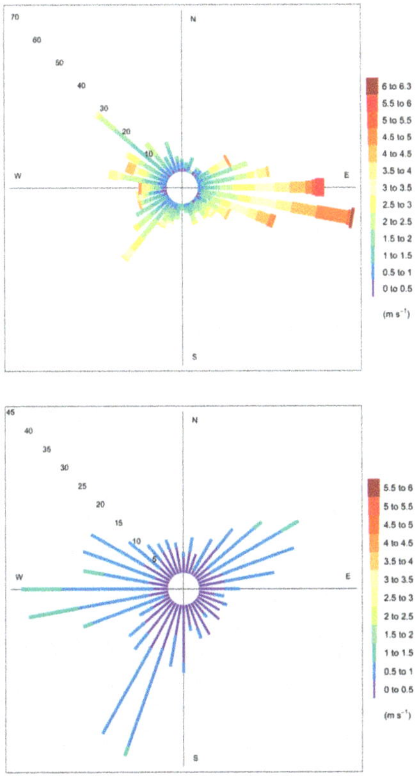

Figure A3. Wind rose distribution for wind measured at Modena Urbana (**top**) and in viale Verdi (**bottom**).

Appendix A.4. NOx Concentrations—Model Evaluation

We conducted the analyses in two steps:

- Dataset exploration:

We applied the FAIRMODE IDL-based DELTA software tool version 5.4 [67,68] provided by the Forum for Air Quality Modelling in Europe (FAIRMODE) [67,73,74]. The tool was developed under the FAIRMODE framework specific evaluation protocols to support the use of air quality models when they are applied for officially reporting on national air pollution levels and for examining compliance with regulations (assessment, forecasting, planning) [73].

The FAIRMODE IDL-based DELTA software tool is based on an approach relying on paired modelled and monitored data to offer diagnostics of model performances using various statistical indicators and diagrams [73]. Given the nature and the time length of our case study, we applied the FAIRMODE IDL-based DELTA software tool for exploratory purposes [68] since the user can analyze different statistical metrics and diagrams.

Our dataset was composed of the measured data from the CNR group, and modelled data extracted in two points close to the green infrastructure, A and D, as seen in Figure A1.

We used the software to obtain: (i) timeseries plots, in order to analyze the dataset with respect to time; (ii) scatterplots, for the correlation between measured and modelled values, including their mean values; (iii) quantile plots, to compare the probability distributions by plotting their quantiles against each other; (iv) and Taylor diagrams, to quantify the degree of correspondence between the modelled and observed behaviour in terms of Pearson correlation coefficient, and the normalized standard deviation.

- Further dataset Analysis:

We took into account hourly and daily meteorological data for a more detailed analysis by applying R-Cran library-based scripts [75], and the Taylor diagram package [70]. The meteorological measurements were provided by the station Modena Urbana (see Section 2.5), and the observed data from the CNR-Consorzio ProAmbiente measurement campaigns.

- Exploration with the FAIRMODE IDL-based DELTA software tool

Regarding the exploration with the FAIRMODE IDL-based DELTA software tool, Figures A4–A7 show the diagrams resulting from the application of the software tool when applied in the exploratory mode to our datasets [68].

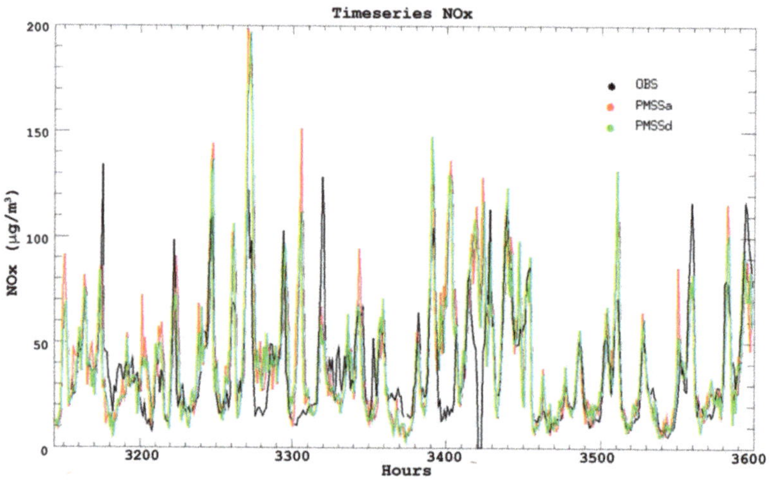

Figure A4. Timeseries diagram from FAIRMODE IDL-based DELTA software tool with hourly CNR observations (black), modelled data in A (red), and in D (green).

The timeseries plot (Figure A4) reports on the x-axis the time in hours. The figure shows an overall good agreement in trend between observed and modelled values (e.g., see the values between the hours 3480–3570, 26 May up to 29 May 18:00). With the exception of few cases, such as the hours 3170–3180, (13 May 2017, 02:00–12:00 h), 3330 (19 May 2017, 18:00), 3560 (29 May 2017, 08:00), the modelled values are in line with the observations and in some cases they are higher than the observations (e.g., hours before 3170, 13 May 2017 02:00, and between 3390–3490, 22 May 2017 06:00 up to 26 May 2017 10:00).

From the scatterplot diagram in Figure A5 and the quantile distribution in Figure A6, measured and modelled hourly data show a dispersion that changes with the concentration values. For NOx values of less than 50 $\mu g/m^3$ the two distributions are very similar and normally distributed. For higher values of NOx concentrations, the model tends to over predict, particularly when NOx concentrations are higher than 100 $\mu g/m^3$. The overall R^2 is quite low (0.4) in both the points A and D.

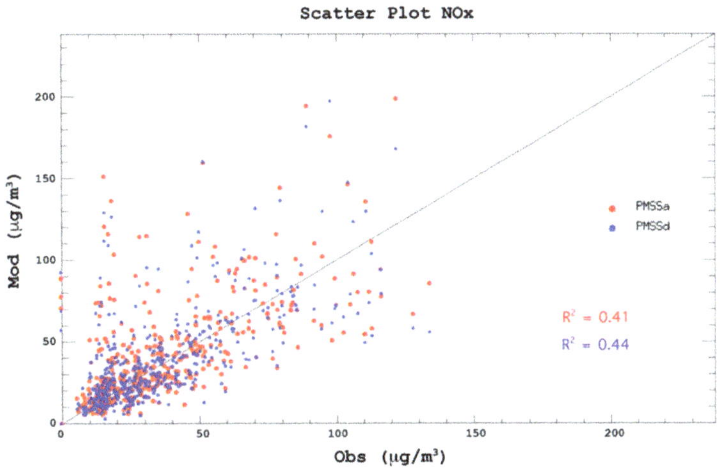

Figure A5. Scatterplot diagram from FAIRMODE IDL-based DELTA software tool with hourly modelled data in A (red), and in D (blue).

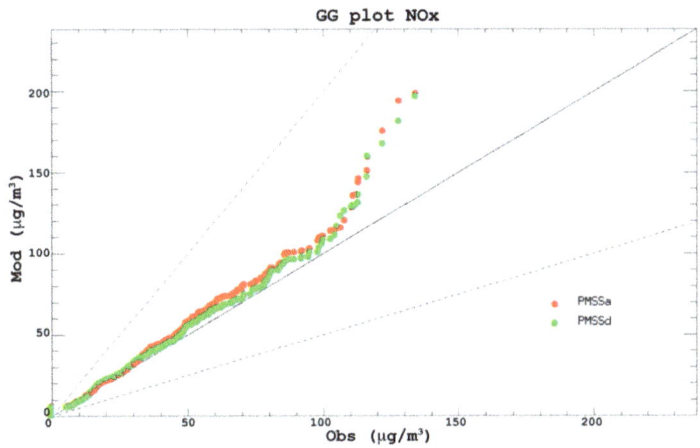

Figure A6. Quantile plot from FAIRMODE IDL-based DELTA software tool with hourly modelled data in A (red), and in D (green).

The Taylor diagram in Figure A7 summarizes the features of the modelled and observed distributions found in the previous plots. Here, we obtain correlation coefficients larger than 0.6 for both A and D points, the latter having R slightly larger. The ratio sigma modelled over sigma observed, e.g., normalized standard deviation is close to 1.25, indicating that the modelled values are more dispersed than the observations.

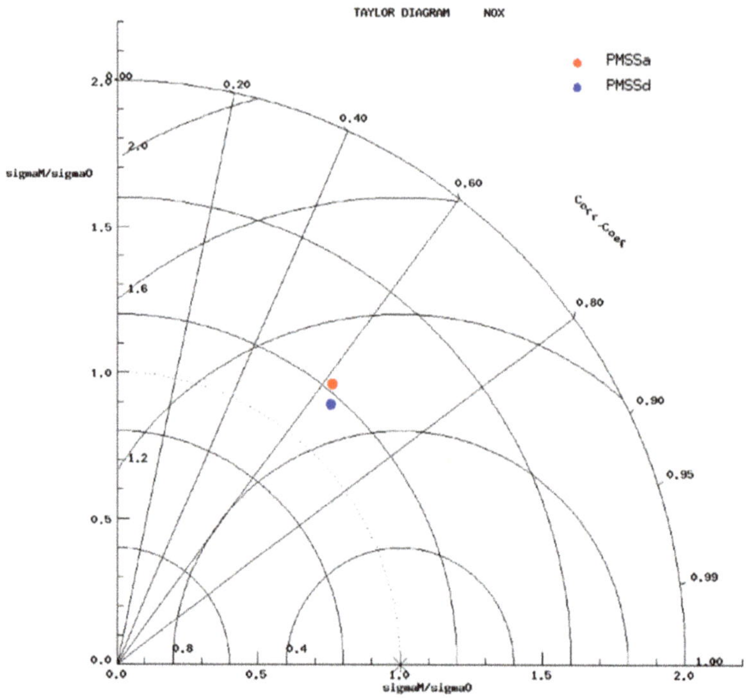

Figure A7. Taylor diagram from FAIRMODE IDL-based DELTA software tool with hourly modelled data in A (red), and in D (blue).

- NOx Concentration Comparisons considering Wind Directions

Figure A8 shows the comparisons between timeseries of modelled and measured CNR data with the urban wind directions in the background.

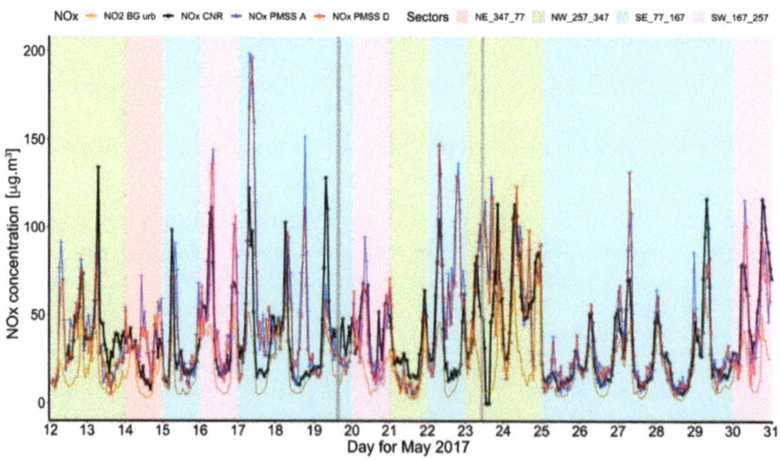

Figure A8. NOx concentrations (orange: NO_2 concentration measured by the urban background station; black: observed NOx concentration; blue: NOx concentration simulated at point A, and red: NOx concentration simulated at point D) with daily prevailing wind directions from Modena Urbana according to the sectors specified in Appendix A.1. Grey vertical bands indicate rainy periods.

Figures A9–A11 show the Taylor diagrams obtained for the datasets (modelled points A and D in red and blue, respectively, and the measured urban NO_2 background, in green) taking the CNR data as reference measurements. Data were classified by:

- day range within the month of May 2017 (12–16; 16–21; 21–26; and 26–31);
- urban wind direction within four sectors (NE 347–77; NW 257–347; SE 77–167; SW 167–257);
- local wind direction in the same four sectors as defined in Figure A1.

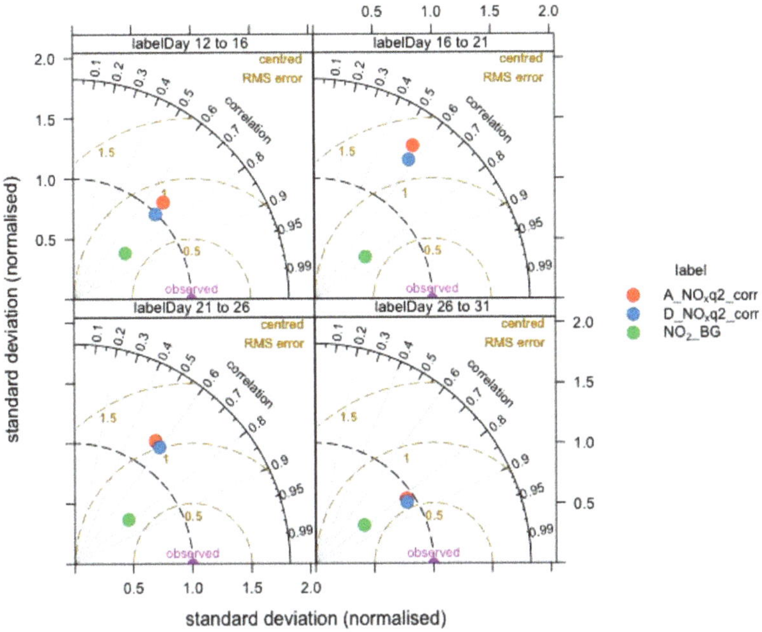

Figure A9. Taylor diagram with hourly modelled data in A (red), and modelled data in D (blue), and the measured background NO_2 concentration (green), according to four different groups of days.

The agreement between modelled and measured data (Figure A8) is much more evident after 25 May, where the days are generally characterized by south-easterly wind, and NOx concentrations are between 20–100 μg/m^3, which are generally lower than on other days. During 26–31 May, the Taylor diagram shows the best agreement with correlation values of about 0.85 and normalized standard deviations close to one, indicating a very good agreement in the values distribution for modelled and measured data.

All the Taylor diagrams show a very good correlation with the urban background measured NO_2 concentrations (green dot) which, the distribution being lower in value than the CNR observations, has a normalized standard deviation about half that of the measured one.

Figure A9 indicates that the best model performance is found during the period 12–16 May, with correlation of about 0.7 and normalized standard deviation about 1; and the period 26–31 with correlation R above 0.8 and normalized standard deviation close to 1. In the period between 16–21 and 21–26, the correlation R is between 0.5–0.6 and the normalized standard deviation is about 1.25, indicating a general over-prediction of the CNR standard deviations.

In Figure A10, NO_2 concentration measured by the urban background station (green dot) is generally always well correlated with the CNR NOx measurements (correlation between 0.6 and 0.85), and it has normalized standard deviations between 0.3 (SW_167_257 sector) and 0.6 (NE 347-77 sector). With the modelled data (A in red, D in blue), the best agreement is obtained in the SW 167-257 sector with a correlation close to 0.8, and the normalized standard deviation of about 1.2. The other sectors are characterized by correlation values generally close to 0.6, and normalized standard deviation between 1.2 and 1.5.

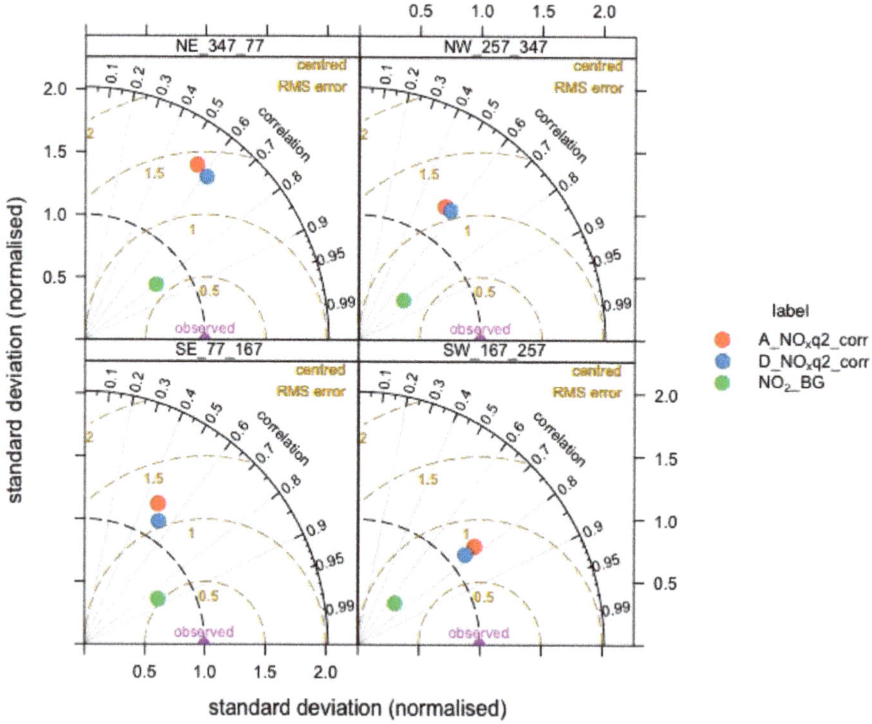

Figure A10. Taylor diagram with hourly modelled data in A (red), and modelled data in D (blue), and the measured background NO_2 concentration (green), according to the urban wind sectors defined as in Figure A1.

Figure A11 shows background NO_2 values for all the sectors (green dot) with correlations close to 0.8 and normalized standard deviation values in a limited range (0.4–0.5). Differently, the modelled data (red and blue dots) show a larger variability according to the wind sector. The best agreement is found in the SE 77–167 sector where the correlation is close to 0.8 and the normalized standard deviation is 1, followed by the SW 167–257 sector, with R close to 0.6 and normalized standard deviation slightly larger than one. Sector NE 347–77 has a correlation above 0.6, but the normalized standard deviation is larger than 1.2. The sector which shows the worst model performance is NW 257–347 where R is smaller than 0.4 and the normalized standard deviation is close to 1.8–2.

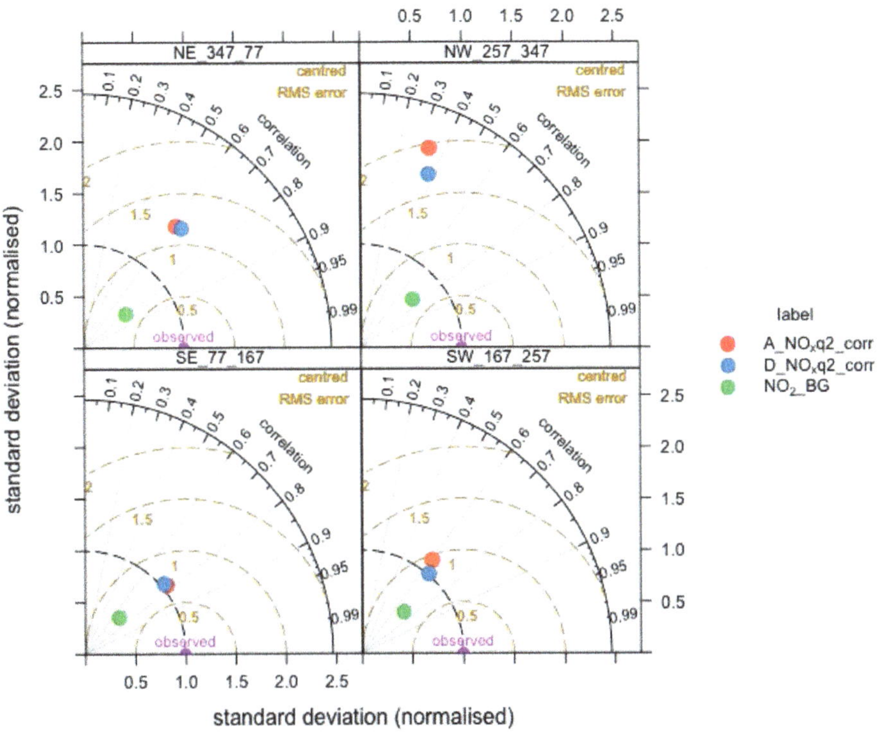

Figure A11. Taylor diagram with hourly modelled data in A (red), and modelled data in D (blue), and the measured background NO_2 concentration (green), according to the local wind sectors defined as in Figure A1.

Appendix B

Here we show a diagram (Figure A12) representing the evolution of the CT design from its first version (CT 1.0) that is the one studied in the work presented here, and the newest generations of CT (CT 2.0) available at the moment in which this paper was submitted. Different design features, such as for example the presence of a wooden cover to provide shade for the moss, have been improved and the development has been oriented at lowering the costs of operation, such as energy and water consumption. In the new CT 2.0 models, energy consumption is around 120–150 W per operating hour and water consumption is between 2 and 8 L per operating hour depending on local climate conditions. These results support the potential of CT technologies for protecting air quality in local urban contexts.

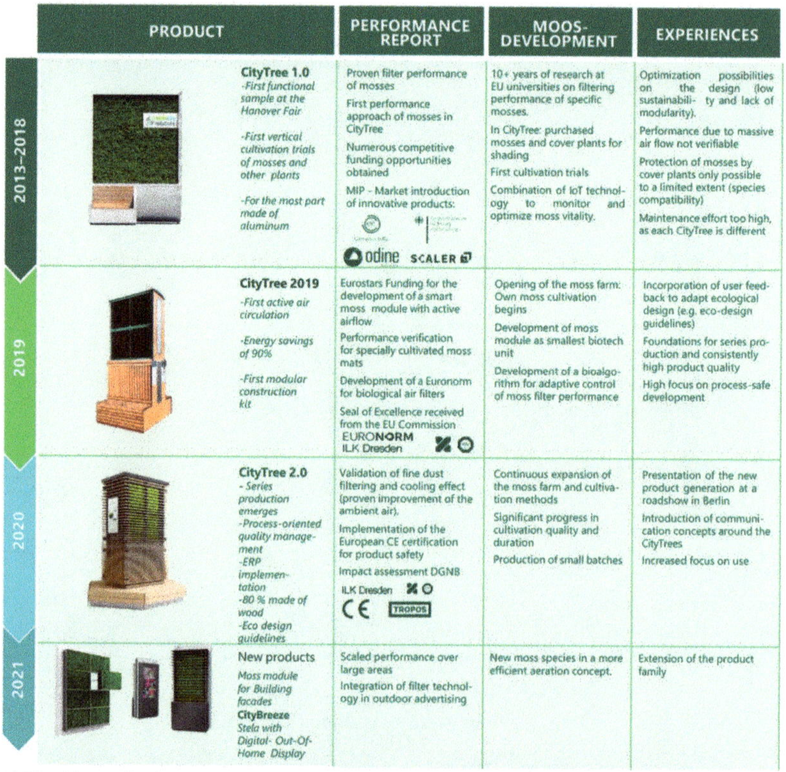

Figure A12. Timeline of the CT improvements from 2013 to present. Many design features have improved, included the use of wood in substitution of the cover plants.

References

1. United Nations, Department of Economic and Social Affairs, Population Division. Available online: https://population.un.org/wup/Publications/Files/WUP2014-PressRelease.pdf (accessed on 9 June 2020).
2. Oke, T.R. Street Design and Urban Canopy Layer Climate. *Energy Build.* **1988**, *11*, 103–113. [CrossRef]
3. Oke, T.R. Towards Better Scientific Communication in Urban Climate. *Theor. Appl. Climatol.* **2006**, *84*, 179–190. [CrossRef]
4. AIRUSE. Air Quality Mitigation Measures in Urban Areas from Southern Europe, Report for Life Project: Airuse, Testing and Development of Air Quality Measures in South Europe, LIFE11/ENV/ES/584. Available online: https://ec.europa.eu/environment/life/project/Projects/index.cfm?fuseaction=home.showFile&rep=file&fil=LIFE11_ENV_ES_000584_LAYMAN.pdf (accessed on 11 June 2020).
5. Bottalico, F.; Chiricia, G.; Giannetti, F.; De Marco, A.; Nocentini, S.; Paoletti, E.; Salbitano, F.; Sanesi, G.; Serenelli, C.; Travaglini, D. Air Pollution Removal by Green Infrastructures and Urban Forests in the City of Florence. *Agric. Agric. Sci. Procedia* **2016**, *8*, 243–251. [CrossRef]
6. Hewitt, C.N.; Ashworth, K.; MacKenzie, A.R. Using green infrastructure to improve urban air quality (GI4AQ). *Ambio* **2020**, *49*, 62–73. [CrossRef] [PubMed]
7. Jayasooriya, V.M.; Ng, A.W.M.; Muthukumaran, S.; Perera, B.J.C. Green Infrastructure Practices for Improvement of Urban Air Quality. *Urban For. Urban Green.* **2017**, *21*, 34–47. [CrossRef]
8. Kumar, P.; Druckman, A.; Gallagher, J.; Gatersleben, B.; Allison, S.; Eisenman, T.; Hoang, U.; Hama, S.; Tiwari, A.; Sharma, A.; et al. The nexus between air pollution, green infrastructure and human health. *Environ. Int.* **2019**, *133*, 105181–105186. [CrossRef] [PubMed]
9. Dige, G.; Eichler, L.; Vermeulen, J.; Ferreira, A.; Rademaekers, K.; Adriaenssens, V.; Kolaszewska, D. *Green Infrastructure and Flood Management: Promoting Cost-Efficient Flood Risk Reduction via Green Infrastructure Solutions*; European Environment Agency—European Union: Luxembourg, 2017; p. 14, ISBN 978-92-9213-894-3.

10. Norton, B.A.; Coutts, A.M.; Livesley, S.J.; Harris, R.J.; Hunter, A.M.; Williams, N.S.G. Planning for cooler cities: A framework to prioritise green infrastructure to mitigate high temperatures in urban landscapes. *Landsc. Urban Plan.* **2016**, *134*, 127–138. [CrossRef]
11. Wentworth, J. Urban Green Infrastructure. 2013. Available online: http://researchbriefings.parliament.uk/ResearchBriefing/Summary/POST-PN-448 (accessed on 20 April 2020).
12. Chen, L.; Liu, C.; Zhang, L.; Zou, R.; Zhiqiang, Z. Variation in Tree Species Ability to Capture and Retain Airborne Fine Particulate Matter (PM2.5). *Sci. Rep.* **2017**, *7*, 3206. [CrossRef] [PubMed]
13. Gkatsopoulos, P. A Methodology for Calculating Cooling from Vegetation Evapotranspiration for Use in Urban Space Microclimate Simulations. *Procedia Environ. Sci.* **2017**, *38*, 477–484. [CrossRef]
14. Yu, Z.; Xu, S.; Zhang, Y.; Jørgensen, G.; Vejre, H. Strong contributions of local background climate to the cooling effect of urban green vegetation. *Sci. Rep.* **2018**, *8*, 6798. [CrossRef]
15. Abhijith, K.V.; Kumar, P.; Gallagher, J.; McNabola, A.; Baldauf, R.; Pilla, F.; Broderick, B.; Di Sabatino, S.; Pulvirenti, B. Air pollution abatement performances of green infrastructure in open road and built-up street canyon environments—A review. *Atmos. Environ.* **2017**, *162*, 71–86. [CrossRef]
16. Donateo, A.; Rinaldi, M.; Paglione, M.; Villani, M.G.; Russo, F.; Carbone, C.; Zanca, N.; Pappaccogli, G.; Grasso, F.M.; Busetto, M.; et al. An evaluation of the performance of a green panel in improving air quality, the case study in a street canyon in Modena, Italy. *Atmos. Environ.* **2021**, *247*, 118189. [CrossRef]
17. Green City Solutions. Available online: https://greencitysolutions.de/en/ (accessed on 16 December 2020).
18. Raupach, M.R.; Woods, N.; Dorr, G.; Leys, J.F.; Cleugh, H.A. The entrapment of particles by windbreaks. *Atmos. Environ.* **2001**, *35*, 3373–3383. [CrossRef]
19. Bailey, B.N.; Stoll, R.; Pardyjak, E.R.; Mahaffee, W.F. Effect of vegetative canopy architecture on vertical transport of massless particles. *Atmos. Environ.* **2014**, *95*, 480–489. [CrossRef]
20. Chen, H.; Zou, Q.-P. Eulerian–Lagrangian flow-vegetation interaction model using immersed boundary method and OpenFOAM. *Adv. Water Resour.* **2019**, *126*, 176–192. [CrossRef]
21. Huai, W.; Yang, L.; Wang, W.-J.; Guo, Y.; Wang, T.; Cheng, Y.-G. Predicting the vertical low suspended sediment concentration in vegetated flow using a random displacement model. *J. Hydrol.* **2019**, *578*, 124101. [CrossRef]
22. Liu, D.; Zheng, Y.; Chen, Q. Simulation of flow around rigid vegetation stems with a fast method of high accuracy. *J. Fluids Struct.* **2016**, *63*, 1–15. [CrossRef]
23. Lu, J.; Dai, H.C. Three dimensional numerical modeling of flows and scalar transport in a vegetated channel. *J. Hydro Environ. Res.* **2017**, *16*, 27–33. [CrossRef]
24. Martin, P.H. Exchanges between structured canopies and their physical environment: A simple analytical solution for a generic configuration. *Ecol. Model.* **1999**, *122*, 1–24. [CrossRef]
25. Chang, T.-J.; Kao, H.-M.; Wu, Y.-T.; Huang, W.-H. Transport mechanisms of coarse, fine, and very fine particulate matter in urban street canopies with different building layouts. *J. Air Waste Manag. Assoc.* **2009**, *59*, 196–206. [CrossRef]
26. Lonati, G.; Pepe, N.; Pirovano, G.; Balzarini, A.; Toppetti, A.; Riva, G.M. Combined Eulerian-Lagrangian Hybrid Modelling System for PM2.5 and Elemental Carbon Source Apportionment at the Urban Scale in Milan. *Atmosphere* **2020**, *11*, 1078. [CrossRef]
27. Pepe, N.; Pirovano, G.; Balzarini, A.; Toppetti, A.; Riva, G.M.; Lonati, G. Enhanced Air Quality Modelling through AUSTAL2000 Model in Milan Urban Area. In *IOP Conference Series: Earth and Environmental Science*; IOP Publishing: Bristol, UK, 2019; p. 012012. [CrossRef]
28. Wilson, J.D.; Yee, E.; Ek, N.; d'Amours, R. Lagrangian simulation of wind transport in the urban environment. *Q. J. R. Meteorol. Soc.* **2009**, *135*, 1586–1602. [CrossRef]
29. Anfossi, D.; Tinarelli, G.; Trini Castelli, S.; Nibart, M.; Olry, C.; Commanay, J. A New Lagrangian Particle Model for the Simulation of Dense Gas Dispersion. *Atmos. Environ.* **2010**, *44*, 753–762. [CrossRef]
30. Nibart, M.; Armand, P.; Duchenne, C.; Olry, C.; Albergel, A.; Moussafir, J.; Oldrini, O. Flow and Dispersion Modelling in a Complex Urban District Taking account of the Underground Roads Connections. In Proceedings of the 17th International Conference on Harmonisation within Atmospheric Dispersion Modelling for Regulatory Purposes, Budapest, Hungary, 9–12 May 2016.
31. Oldrini, O.; Armand, P.; Duchenne, C.; Olry, C.; Moussafir, J.; Tinarelli, G. Description and Preliminary Validation of the PMSS Fast Response Parallel Atmospheric Flow and Dispersion Solver in Complex Built-Up Areas. *Environ. Fluid Mech.* **2017**, *17*, 997–1014. [CrossRef]
32. Tinarelli, G.; Mortarini, L.; Castelli, S.T.; Carlino, G.; Moussafir, J.; Olry, C.; Armand, P.; Anfossi, D. Review and Validation of Microspray, a Lagrangian Particle Model of Turbulent Dispersion. *Geophys. Monogr. Ser.* **2012**, *200*, 311–327. [CrossRef]
33. Tinarelli, G.; Gomez, F. *PSPRAY General Description and User's Guide, 2017. Version Code 3.7.3.*; ARIANET/ARIA Technologies. Unpublished work, 2017.
34. Castelli, S.T.; Tinarelli, G.; Reisin, T.G. Comparison of Atmospheric Modelling Systems Simulating the Flow, Turbulence and Dispersion at the Microscale within Obstacles. *Environ. Fluid Mech.* **2017**, *17*, 879–901. [CrossRef]
35. Bigi, A.; Veratti, G.; Fabbi, S.; Po, L.; Ghermandi, G. Forecast of the impact by local emissions at an urban micro scale by the combination of Lagrangian modelling and low cost sensing technology: The TRAFAIR project. In Proceedings of the 19th International Conference on Harmonisation within Atmospheric Dispersion Modelling for Regulatory Purposes, Harmo, Bruges, Belgium, 3–6 June 2019.

36. Fabbi, S.; Asaro, S.; Bigi, A.; Teggi, S.; Ghermandi, G. Impact of vehicular emissions in an urban area of the Po valley by microscale simulation with the GRAL dispersion model. In *IOP Conference Series: Earth and Environmental Science*; IOP Publishing: Bristol, UK, 2019; Volume 296, p. 012006.
37. Ghermandi, G.; Fabbi, S.; Bigi, A.; Veratti, G.; Despini, F.; Teggi, S.; Barbieri, C.; Torreggiani, L. Impact assessment of vehicular exhaust emissions by microscale simulation using automatic traffic flow measurements. *Atmos. Pollut. Res.* **2019**, *10*, 1473–1481. [CrossRef]
38. Ghermandi, G.; Fabbi, S.; Veratti, G.; Bigi, A.; Teggi, S. Estimate of Secondary NO_2 Levels at Two Urban Traffic Sites Using Observations and Modelling. *Sustainability* **2020**, *12*, 7897. [CrossRef]
39. Gomez, F.; Ribstein, B.; Makké, L.; Armand, P.; Moussafir, J.; Nibart, M. Simulation of a dense gas chlorine release with a Lagrangian particle dispersion model (LPDM). *Atmos. Environ.* **2021**, *244*, 117791. [CrossRef]
40. Oldrini, O.; Armand, P. Validation and Sensitivity Study of the PMSS Modelling System for Puff Releases in the Joint Urban 2003 Field Experiment. *Bound. Layer Meteorol.* **2019**, *171*, 513–535. [CrossRef]
41. Oldrini, O.; Armand, P.; Duchenne, C.; Perdriel, S. Parallelization Performances of PMSS Flow and Dispersion Modeling System over a Huge Urban Area. *Atmosphere* **2019**, *10*, 404. [CrossRef]
42. Tognet, F.; Turmeau, C.; Ha, T.L.; Tarnaud, E.; Rouïl, L.; Bessagnet, B.; Robine, E.; Morel, Y. Numerical modelling of microorganisms dispersion in Urban area: Application to Legionella, HARMO 2011. In Proceedings of the 14th International Conference on Harmonisation within Atmospheric Dispersion Modelling for Regulatory Purposes, Kos, Greece, 2–6 October 2011; pp. 746–750.
43. Veratti, G.; Fabbi, S.; Bigi, A.; Lupascu, A.; Tinarelli, G.; Teggi, S.; Brusasca, G.; Butler, T.; Ghermandi, G. Towards the coupling of a chemical transport model with a micro-scale Lagrangian modelling system for evaluation of urban NOx levels in a European hotspot. *Atmos. Environ.* **2020**, 223. [CrossRef]
44. Bailey, B.N.; Stoll, R.; Pardyjak, E.R. A Theoretically Consistent Framework for Modelling Lagrangian Particle Deposition in Plant Canopies. *Bound. Layer Meteorol.* **2018**, *167*, 509–520. [CrossRef]
45. Bailey, B.N.; Stoll, R. Turbulence in Sparse, Organized Vegetative Canopies: A Large-Eddy Simulation Study. *Bound. Layer Meteorol.* **2013**, *147*, 369–400. [CrossRef]
46. Mack, A.; Duyzer, J. The Impact of City Trees on Air Quality in the Valkenburgerstraat (Amsterdam), TNO Report, TNO 2019 R10497. Available online: https://www.oudestadt.nl/wp-content/uploads/2019/06/TNO%202018%20R10613%20Rapport%20Luchtkwaliteit_6%20juni_finaal_tcm318-398657.pdf (accessed on 20 April 2020).
47. Zhang, Z.; Chen, Q. Comparison of the Eulerian and Lagrangian Methods for Predicting Particle Transport in Enclosed Spaces. *Atmos. Environ.* **2007**, *1*, 5236–5248. [CrossRef]
48. Béghein, C.; Jiang, Y.; Chen, Q.Y. Using Large Eddy Simulation to Study Particle Motions in a Room. *Indoor Air* **2005**, *15*, 281–290. [CrossRef]
49. Lu, W.; Howarth, A.T.; Adam, N.; Rifat, S.B. Modelling and Measurement of Airflow and Aerosol Particle Distribution in a Ventilated Two-Zone Chamber. *Build. Environ.* **1996**, *31*, 417–423. [CrossRef]
50. Zhang, Z.; Chen, Q. Numerical Analysis of Particle Behaviors in Indoor Air using Lagrangian Method. In *Air Distribution in Rooms*; Roomvent: Coimbra, Portugal, 2004; p. 134, ISBN 9729797323.
51. Moreira, D.; Vilheno, M. *Air Pollution and Turbulence: Modeling and Applications*; CRC Press, Taylor and Francis Group: Milton Park, UK, 2009; p. 354, ISBN 978-1-4398-1144-3.
52. Iannone, F.; Ambrosino, F.; Bracco, G.; De Rosa, M.; Funel, A.; Guarnieri, G.; Migliori, S.; Palombi, F.; Ponti, G.; Santomauro, G.; et al. CRESCO ENEA HPC clusters: A working example of a multifabric GPFS Spectrum Scale layout. In Proceedings of the 2019 International Conference on High Performance Computing Simulation (HPCS), Dublin, Ireland, 15–19 July 2019; pp. 1051–1052. [CrossRef]
53. Buccolieri, R.; Santiago, J.L.; Rivas, E.; Sanchez, B. Review on urban tree modelling in CFD simulations: Aerodynamic, deposition and thermal effects. *Urban For. Urban Green.* **2018**, *31*, 212–220. [CrossRef]
54. ARIANET. Available online: http://www.aria-net.it/ (accessed on 20 April 2020).
55. Cotton, W.R.; Pielke, R.A.; Walko, R.L.; Liston, G.E.; Tremback, C.J.; Jiang, H.; McAnelly, R.L.; Harrington, J.Y.; Nicholls, M.E.; Carrio, G.G.; et al. RAMS 2001: Current Status and Future Directions. *Meteorol. Atmos. Phys.* **2003**, *82*, 5–29. [CrossRef]
56. D'Elia, I.; Piersanti, A.; Briganti, G.; Cappelletti, A.; Ciancarella, L.; Peschi, E. Evaluation of Mitigation Measures for Air Quality in Italy in 2020 and 2030. *Atmos. Pollut. Res.* **2018**, *9*, 977–988. [CrossRef]
57. Mircea, M.; Ciancarella, L.; Briganti, G.; Calori, G.; Cappelletti, A.; Cionni, I.; Costa, M.; Cremona, G.; D'Isidoro, M.; Finardi, S.; et al. Assessment of the AMS-MINNI System Capabilities to Simulate Air Quality Over Italy for the Calendar Year 2005. *Atmos. Environ.* **2014**, *84*, 178–188. [CrossRef]
58. Mircea, M.; Grigoras, G.; D'Isidoro, M.; Righini, G.; Adani, M.; Briganti, G.; Ciancarella, L.; Cappelletti, A.; Calori, G.; Cionni, I.; et al. Impact of Grid Resolution on Aerosol Predictions: A Case Study over Italy. *Aerosol Air Qual. Res.* **2016**, *16*, 1253–1267. [CrossRef]
59. Adani, M.; Piersanti, A.; Ciancarella, L.; D'Isidoro, M.; Villani, M.G.; Vitali, L. Preliminary Tests on the Sensitivity of the FORAIR_IT Air Quality Forecasting System to Different Meteorological Drivers. *Atmosphere* **2020**, *11*, 574. [CrossRef]
60. FORAIT-IT. Available online: http://www.afs.enea.it/project/ha_forecast/ (accessed on 20 April 2020).
61. Litschke, T.; Kuttler, W. On the Reduction of Urban Particle Concentration by Vegetation–A Review. *Meteorol. Z.* **2008**, *17*, 229–240. [CrossRef]

62. GREENCITY SOLUTIONS. *Deliverable Proof–Reports Resulting from the Finalisation of a Project Task, Work Package, Project Stage, Project as a Whole-EIT-BP17, Work Package 4: On Site Data Collection, Deliverable 1 (2018): Results of the Research Field Campaign 3*; Green City Solutions GmbH: Bestensee, Germany, 2018.
63. Splittgerber, V.; Saenger, P. The City Tree: A vertical plant wall. *WIT Trans. Ecol. Environ.* **2015**, *198*, 295–304. [CrossRef]
64. Nanni, A.; Radice, P.; Smith, P. *TREFIC (Traffic Emission Factors Improved Calculation) User's Guide, Report ARIANET R2009.19*; Unpublished work, 2009.
65. Emisia, S.A. COPER-COmputer Programme to Calculate Emissions from Road Transport. 2018. Available online: https://www.emisia.com/utilities/copert/documentation/ (accessed on 11 June 2020).
66. ARPAE. Arpae Emilia-Romagna. Available online: https://www.arpae.it/ (accessed on 20 April 2020).
67. FAIRMODE. Forum for Air Quality Modelling in Europe. Available online: https://fairmode.jrc.ec.europa.eu/ (accessed on 20 April 2020).
68. Thunis, P.; Cuvelier, C. DELTA Version 5.4, Concepts/User's Guide/Diagrams, Joint Research Centre, Ispra (VA). April 2016. Available online: https://fairmode.jrc.ec.europa.eu/Document/fairmode/WG1/DELTA_UserGuide_V5_4.pdf (accessed on 9 June 2020).
69. Carslaw, D.C.; Ropkins, K. Openair—An R Package for Air Quality Data Analysis. *Environ. Model. Softw.* **2012**, *27*, 52–61. [CrossRef]
70. Lemon, J.; Bolker, B.; Oom, S.; Klein, E.; Rowlingson, B.; Wickham, H.; Tyagi, A.; Eterradossi, O.; Grothendieck, G.; Toews, M.; et al. R Cran Package 'Plotrix'. Available online: https://cran.r-project.org/web/packages/plotrix/plotrix.pdf (accessed on 11 June 2020).
71. Pettit, T.; Torpy, F.R.; Surawski, N.C.; Fleck, R.; Irga, P.J. Effective reduction of roadside air pollution with botanical biofiltration. *J. Hazard. Mater.* **2021**, *414*, 125566. [CrossRef]
72. Rotach, M. Turbulence within and above an Urban Canopy. Ph.D. Thesis, Eidgenössischen Technischen Hochschule Zürich, Zürich, Switzerland, 1991. [CrossRef]
73. Kushta, J.; Georgiou, G.K.; Proestos, Y.; Christoudias, T.; Thunis, P.; Savvides, C.; Papadopoulos, C.; Lelieveld, J. Evaluation of EU Air Quality Standards through Modeling and the FAIRMODE Benchmarking Methodology. *Air Qual. Atmos. Health* **2019**, *12*, 73. [CrossRef] [PubMed]
74. Pisoni, E.; Guerreiro, C.; Lopez-Aparicio, S.; Guevara, M.; Tarrason, L.; Janssen, S.; Thunis, P.; Pfäfflin, F.; Piersanti, A.; Briganti, G.; et al. Supporting the Improvement of Air Quality Management Practices: The "FAIRMODE Pilot" Activity. *J. Environ. Manag.* **2019**, *245*, 122–130. [CrossRef]
75. Carslaw, D. Openair: Tools for the Analysis of Air Pollution Data. 2020. Available online: https://cran.r-project.org/web/packages/openair/index.html (accessed on 11 June 2020).

atmosphere

MDPI

Article

Toward Development of a Framework for Prediction System of Local-Scale Atmospheric Dispersion Based on a Coupling of LES-Database and On-Site Meteorological Observation

Hiromasa Nakayama *, Toshiya Yoshida, Hiroaki Terada and Masanao Kadowaki

Japan Atomic Energy Agency, 2-4, Shirakata, Tokai-mura, Naka-gun, Ibaraki 319-1195, Japan; yoshida.toshiya@jaea.go.jp (T.Y.); terada.hiroaki@jaea.go.jp (H.T.); kadowaki.masanao@jaea.go.jp (M.K.)

* Correspondence: nakayama.hiromasa@jaea.go.jp

Abstract: An accurate analysis of local-scale atmospheric dispersion of radioactive materials is important for safety and consequence assessments and emergency responses to accidental release from nuclear facilities. It is necessary to predict the three-dimensional distribution of the plume in consideration of turbulent effects induced by individual buildings and meteorological conditions. In this study, first, we conducted with meteorological observations by a Doppler LiDAR and simple plume release experiments by a mist-spraying system at the site of Japan Atomic Energy Agency. Then, we developed a framework for prediction system of local-scale atmospheric dispersion based on a coupling of large-eddy simulation (LES) database and on-site meteorological observation. The LES-database was also created by pre-calculating high-resolution turbulent flows in the target site at mean wind directions of class interval 10°. We provided the meteorological observed data with the LES-database in consideration of building conditions and calculated the three-dimensional distribution of the plume with a Lagrangian dispersion model. Compared to the instantaneous shots of the plume taken by a digital camera, it was shown that the mist plume transport direction was accurately simulated. It was concluded that our proposed framework for prediction system based on a coupling of LES-database and on-site meteorological observation is effective.

Keywords: large-eddy simulation; database; on-site meteorological observation; water mist dispersion; lagrangian dispersion model

Citation: Nakayama, H.; Yoshida, T.; Terada, H.; Kadowaki, M. Toward Development of a Framework for Prediction System of Local-Scale Atmospheric Dispersion Based on a Coupling of LES-Database and On-Site Meteorological Observation. *Atmosphere* **2021**, *12*, 899. https://doi.org/10.3390/atmos12070899

Academic Editor: Patrick Armand

Received: 11 June 2021
Accepted: 9 July 2021
Published: 13 July 2021

1. Introduction

In an emergency response to accidental release of radioactive materials from nuclear facilities on a local-scale, it is important to accurately and quickly predict the air concentrations at the evaluation point for internal doses and surface concentrations for external doses to evaluate the radiological consequences in consideration of the influence of turbulent effects induced by individual buildings and meteorological disturbances. For investigating plume dispersion over complex surface geometry in distances of up to several kilometers from the emission source, computational fluid dynamics (CFD) models are useful.

In principle, there are two approaches in CFD models: Reynolds-Averaged Navier–Stokes (RANS) and Large-Eddy Simulation (LES) models. RANS-based CFD models calculate a mean wind flow to deliver an ensemble- or time-averaged solution, and all turbulent motions are modeled using turbulence parameterization. The main advantage of the RANS model is its efficiency in simulating a mean flow field at a relatively low computational cost. However, it was reported that lateral dispersion behaviors of a plume are not reproduced well [1–3]. Recently, LES-based CFD models have also become useful tools. The basic idea of LES models resolves grid-scale turbulent motions and parameterizes only subgrid-scale motions. The advantage is that they can well capture plume dispersion behaviors in complex turbulent flows such as impinging, separated, and recirculating flows around buildings [2–7]. However, LES models also have a significant disadvantage of calculation time for emergency response purposes. To solve the trade-off problem of calculation time and prediction accuracy, we developed a combined LES-database/RANS model in which dispersion fields are simulated by RANS with LES pre-calculated wind velocity data and applied to plume dispersion in an actual urban central district [8]. It was shown that the combined model accurately reproduces horizontal concentration distributions obtained from a wind tunnel experiment for a simple street canyon case [5]. In addition, for a real urban area case, it was found that the results of the combined model reasonably agree with those of the LES used alone, although the underestimation in low concentration areas was observed. We concluded from these results that the combined model provides accurate results with a reasonable calculation time 1/40 times shorter than the LES model alone. However, this model evaluation was conducted under an ideal atmospheric condition in which the mean wind speeds and directions are constant. An important issue was remained in incorporating meteorological information into the model as input conditions.

Collier et al. [9] developed an operational dispersion model known as the Nuclear Accident Model (NAME) by combining with meteorological observations (OBS). The NAME model was designed based on a Lagrangian particle model and two mobile Doppler LiDAR systems. The dispersion processes of large numbers of imaginary particles are calculated by the mean flow and the turbulence parameters obtained from the dual-Doppler LiDAR dataset. The NAME is used for many applications such as nuclear accidents, pollutant episode studies, source term estimation, and air quality forecasting over a wide region of several 100 km^2. Recently, the on-site measurement coupled to CFD approach has been applied to wind resource assessment by several researchers [10–12]. For example, Radunz et al. [12] developed a framework based on a combination of on-site OBS and CFD and showed a fast and comprehensive solution to produce a wind resource map and estimate energy yield over a wide region of 144 km^2 by prescribing accurate CFD inflow conditions obtained from on-site OBS. These studies indicate the usefulness of a coupling of on-site OBS and CFD.

In the case of nuclear accidents, radioactive materials are assumed to be released into the atmosphere directly from a nuclear reactor building rather than an exhaust stack in many cases. In such a situation, plume transport and dispersion are highly influenced by individual buildings and structures, which results in inhomogeneous distributions of air concentrations. In this study, we propose a framework for prediction system of local-scale atmospheric dispersion based on a coupling of LES-database and on-site OBS that takes into account both individual buildings and real meteorological conditions. Toward development of the framework, first, we performed LESs of turbulent flows over a target site and created the dataset of mean and turbulent flows for 36 mean wind directions at 10° class interval. Then, we conducted with meteorological observations by a Doppler LiDAR and simple dispersion experiments using a mist-spraying system for representing a plume at the site of Japan Atomic Energy Agency (JAEA). Here, the dispersion behaviors of the water mist plume were observed using a digital camera. Our objective is to investigate the effectiveness of the coupling of the LES-database and on-site OBS by comparing the coupling simulation results with the transport direction of the water mist plume.

2. Field Experiments

Meteorological observations were conducted at the site of JAEA Nuclear Science Research Institute (NSRI), Tokai-mura, Ibaraki prefecture, Japan during the period from 16 November to 23 December 2020. A mist plume was continuously released by a mist-spraying system for 10 min several times during the period from 1049 JST to 1132 JST 26 November and from 1030 JST to 1122 JST 1 December. The target site and the locations of the experimental devices are shown in Figure 1. A Doppler LiDAR (Streamline Pro, HALO Photonics Ltd.) and an ultrasonic anemometer (WindMaster II, Gill Instruments Ltd.) were placed on the building rooftop at 12-m height. The laser beam was intensively directed toward the position of the mist-spraying system. The Doppler LiDAR repeatedly conducted a 3-min scan sequence, which included four Plan Position Indicator (PPI) scans at elevation angles of 1.0°, 7.6°, 14.9° and 21.8°, and six Range Height Indicator (RHI) scans at azimuth angles with a 15° interval. The detection range was from 15 m to 3000 m, with a range resolution of 35 m in the radial direction. However, the missing rate of the Doppler LiDAR was extremely high at the positions 300 m away from there. Nakano et al. [13] measured wind direction and velocity data by a Doppler LiDAR (Windcube WLS7, Leosphere Ltd.) at the site located approximately 2 km south of the NSRI of JAEA for one-year starting from 1st February 2012 and showed that the missing rate of the Doppler LiDAR was high especially during the period from late autumn to winter. The tendency of the high missing rate during our observation period is the same as their one-year observation [13]. Figure 2 shows an example of spatial distributions of the observed radial wind speeds at elevation angles of 1.0°, 7.6°, 14.9°, and 21.8°. Those were linearly interpolated on grids of 1 m × 1 m in the horizontal direction and were used for plume dispersion simulations in the computational domain of 150 m × 150 m. The details are described in the next section.

Figure 3 shows photograph of the mist plume generated by the mist-spraying system (TIC Corporation Ltd.). Hashimoto et al. [14] and Onogi et al. [15] proposed field particle image velocimetry (PIV) imaging technique using a mist-spraying system with digital still cameras and successfully captured the turbulent eddy structures in the surface boundary layer. They emphasized that the field PIV technique using a mist-spraying system has significant advantages in safety and environment. The mist-spraying system used here was composed of a 50-L pump and a fan. This fan had a diameter of 45 cm and was equipped with 10 spray nozzles (KX47S-01) at an interval of 5 cm. The water mist was discharged vertically upward from each nozzle at a rate of 2.7 L per minute. The photograph of the mist plume was taken by the digital camera (SONY HDR-CX680-R) on the ground directed toward the mist-spraying system (see Figure 1). However, the fan was not operated in order to prevent the influence on the ambient flow.

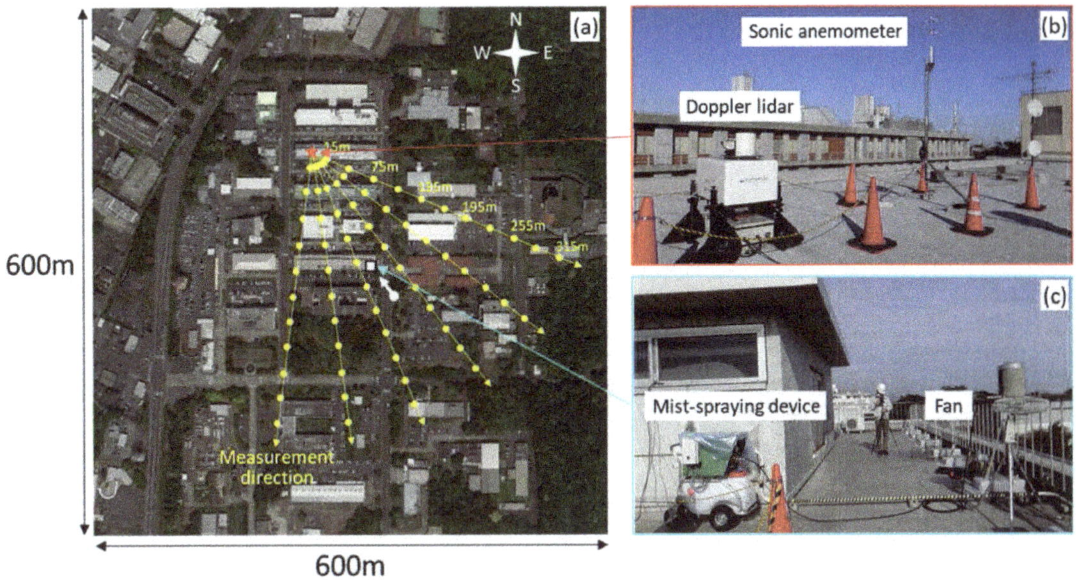

Figure 1. Locations of the meteorological observations and the mist-spraying system. (**a**) Study site in JAEA. (**b**) Doppler LiDAR and sonic anemometer. (**c**) Mist-spraying system. The photograph (**a**) is reproduced by Google^{TM} earth graphic. The red star depicts a position of the Doppler LiDAR and sonic anemometer placed at the building rooftop. The yellow circles are measurement points of wind velocity by the Doppler LiDAR at elevation angle of 1.0°. The white square depicts a position of the mist-spraying system placed at the building rooftop. The white arrow indicates a position of the digital camera on the ground directed toward the mist-spraying system.

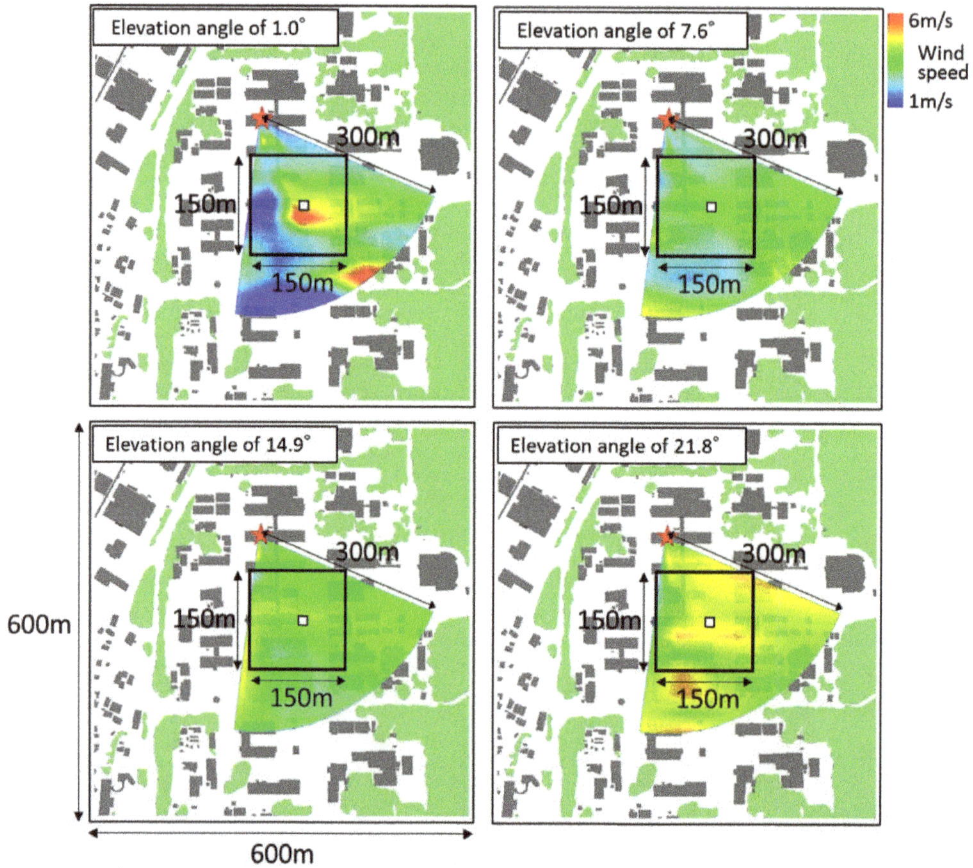

Figure 2. Spatial distributions of radial wind speeds at elevation angles of 1.0°, 7.6°, 14.9°, and 21.8° on 1122 JST 26 November 2020. The square indicates the computational domain used for calculating plume dispersion. The symbols are the same as in Figure 1. The gray and green areas indicate buildings and tree canopy, respectively. The gray and green areas are buildings and tree canopy, respectively.

Figure 3. Photograph of the mist plume generated by the mist-spraying system.

3. LES-Database

3.1. LES Model

The CFD model used here is the LOHDIM-LES (LOcal-scale High-resolution atmospheric DIspersion Model using LES) developed by JAEA [16–18]. The governing equations are the filtered continuity equation, the Navier–Stokes equation in Boussinesq-approximated form, and the transport equations of temperature and concentrations. The subgrid-scale turbulent effect is represented by the Smagorinsky model [19]. The subgrid-scale scalar fluxes are also parameterized by an eddy viscosity model. Buildings and structures are explicitly represented by the use of a digital surface model dataset. The turbulent effects of buildings are represented by the immersed boundary method [20]. The tree canopy effect is represented by a conventional drag force composed of the drag coefficient, the leaf area index, and wind velocities. The performance of LOHDIM-LES was evaluated by conducting detailed simulations of turbulent flows and plume dispersion over a flat terrain and a two-dimensional hill, around an isolated building, within building arrays with different obstacle densities, and within a central district of an actual urban area under ideal and realistic meteorological conditions.

3.2. Computaiotnal Conditions

Figure 4 shows the computational domain for creating the LES-database of mean and turbulent winds over the site of JAEA. The size of the computational domain is 1.2 km × 1.2 km in the horizontal direction at a depth of 200 m. The total mesh number is 1200 × 1200 × 88 nodes. The grid spacing is 1 m in the horizontal direction and 1–10 m stretched in the vertical direction. The buildings and tree canopy were explicitly resolved using a digital surface model dataset in the domain of 600 m × 600 m. The driver sections with a length of 300 m were set, and roughness blocks were placed to efficiently generate turbulent boundary layer (TBL) flows.

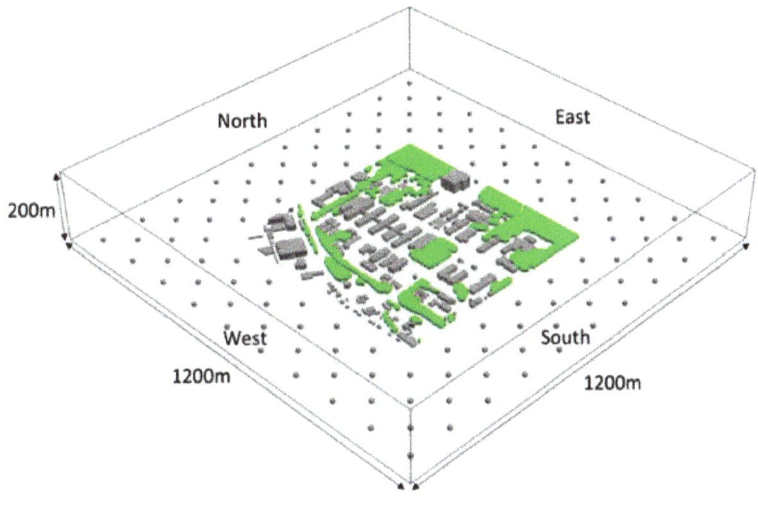

Figure 4. Computational domain for creating the LES-database of mean and turbulent winds over the target site.

We pre-calculated LESs of TBL flows under a neutral stability condition and created a dataset of mean and turbulence velocities for 36 different mean wind directions at class interval 10°. The 10° class interval is enough to reasonably estimate spatial distributions of plume concentrations using the database for changing meteorological conditions [21]. A vertical profile with a power law exponent of 0.14 with a mean wind speed of 15 m/s at 60-m height was imposed at the inflow boundaries. At the same time, time-dependent turbulent inflow data were added to it by the recycling method [22]. At the bottom surface, the Monin-Obukhov similarity theory is applied [23]. The total length of the calculation run is 30 min. The first 20 min are considered to be the spin-up period before turbulent statistics become a statistical steady-state. The mean wind velocities and turbulence standard deviation were computed over the last 10-min period. The calculation time step interval is 0.005 s.

4. Dispersion Simulation Settings

4.1. Coupling of the LES-Database and On-Site OBS

The relationship between the lower atmospheric boundary layer and building morphological characteristics has been studied by many researchers. It is well known that boundary layer flows over buildings are mainly classified into three types: the building canopy layer, the roughness sublayer, and the inertial sublayer [24–26] as shown in Figure 5. In the building canopy layer, the flow patterns are directly determined by building arrangements and show a strong three-dimensionality caused by impinging, separated, and recirculating flows. The depth of the roughness sublayer ranges up to 2 to 5 times the height of the buildings [25]. In the inertial sublayer, the dynamical influence of the surface decreases with height and the flows eventually readjust to the meteorological conditions.

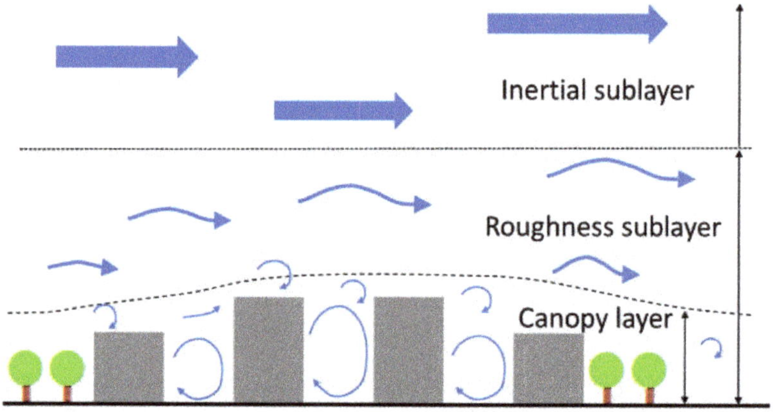

Figure 5. Schematic of TBL flows over buildings. Adopted from Britter and Hanna [27].

Figure 6 shows spatial distributions of the building heights represented by a digital surface model dataset. The averaging building height is 11.1 m. The building height variability defined as the ratio of the standard deviation of building height to the average building height is 5.1 m. The building height at this site is nearly uniform except the structure located at the east side. It is considered from this building morphological characteristics that the height at which the influence of buildings on atmospheric winds becomes fully small is approximately 55 m. Therefore, we took the measuring 6-points averaged winds obtained at 61.2-m height (elevation angle of 21.8°) for a reference meteorological condition.

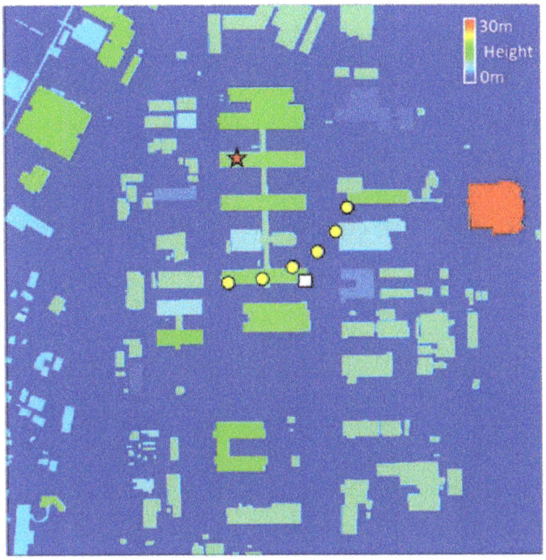

Figure 6. Spatial distributions of building heights in the site of JAEA represented using the digital surface model dataset. The symbols of the red star and the white square are the same as in Figure 1. The yellow circles are measurement points of the Doppler LiDAR at elevation angle of 21.8° used for estimating the reference meteorological condition.

Doppler LiDAR instruments can capture spatial distributions of wind velocities in real meteorological conditions. However, it was pointed out that those might include errors in representing high-frequency turbulent fluctuations [27]. Furthermore, buildings and trees cause obstructions of the laser beam at low elevation angles in certain azimuths [28]. On the other hand, LES models are intrinsically superior in capturing basic flow patterns within building canopy that are governed mainly by the building morphology under ideal meteorological conditions.

In this meteorological observation by the Doppler LiDAR, the radial wind velocities just above the building heights were measured for a case of elevation angle 1.0°. Therefore, first, in estimating a mean flow filed, we prepared the OBS data obtained by the Doppler LiDAR for the region above the building canopy layer and used the LES-database of mean velocities for the region below it.

$$U_{coup}(x,y,z) = U_{OBS}(x,y,h_{OBS}(x,y)) \text{ for } z > h_{OBS}(x,y) \quad (1)$$

$$U_{coup}(x,y,z) = U_{LES}(x,y,z)\,\frac{U_{OBS}(x,y,h_{OBS}(x,y))}{U_{LES}(x,y,h_{OBS}(x,y))} \text{ for } z \le h_{OBS}(x,y) \quad (2)$$

where U_{coup}, U_{OBS}, U_{LES}, and h_{OBS} are the mean wind velocity estimated by a coupling of the LES-database and on-site OBS, the OBS wind velocity linearly interpolated on the LES calculation grids, the LES database of mean wind velocity, and the lowest measurement height of the Doppler LiDAR at the position of (x,y), respectively. U_{LES} is extracted from the LES-database pre-calculated for 36 different mean wind directions at class interval 10° in accordance with the target meteorological condition.

It is well known that the ratio of turbulence standard deviation to mean velocity is not constant and changes with the mean wind velocity for a weak wind condition, while the ratio is almost constant for a strong wind condition regardless of the mean velocity. The LES-database of mean and turbulence velocities was created under a strong neutral wind condition as mentioned in Section 3.2. The turbulence standard deviation can be estimated by multiplying the ratio in the LES-database by the mean wind velocity

U_{coup} for a strong wind condition. However, for a weak wind condition, the turbulence standard deviation can be no longer estimated by the product of them. Therefore, in estimating a turbulent flow field, we adopted Normal Turbulence Model (NTM) proposed by the International Electrotechnical Commission (IEC) in the international standard IEC61400-1 used for determining appropriate locations of wind turbines as shown in the following formulation [29].

$$\sigma_l = I_{ref}(aU + b) \tag{3}$$

where σ_l is the longitudinal turbulence standard deviation and U is the mean wind speed at hub height over a 10-min period. I_{ref} is the expected value of the turbulence intensity at 15 m/s and has three values of 0.12, 0.14, and 0.16 depending on wind turbine classes, respectively. The constants of a and b are 0.75 and 3.8, respectively, for the mean turbulence standard deviation. The measurements were taken at a height of the hub height of typically from 60 m to 80 m. The applicability of NTM IEC was investigated by Ishihara et al. [30].

To investigate the applicability to complex turbulent flow fields such as impinging, separated, and recirculating flows formed at building rooftop, regression analysis for I_{ref} was conducted using the wind velocity data measured by the ultrasonic anemometer. When I_{ref} = 0.16, the coefficient of determination shows a peak and its value is over 0.6 as shown in Figure 7a. It is also found from Figure 7b that the measurement data are generally distributed well along the NTM curve with I_{ref} of 0.16, which indicates that the formulation is applicable to complex turbulent flow fields. Therefore, we applied it to strong three-dimensional turbulent flow fields within and over individual buildings as the following expression.

$$\sigma_{l\_est}(x,y,z) = I_{ref\_l\_LES}(x,y,z)\left(aU_{coup}(x,y,z) + b\right) \tag{4}$$

$$I_{ref\_l\_LES}(x,y,z) = \frac{\sigma_{l\_LES}(x,y,z)}{U_{LES}(x,y,z)} \tag{5}$$

where σ_{l_est} is the estimated longitudinal turbulence standard deviation and $I_{ref\_l\_LES}$ is the longitudinal component of the local turbulence intensity given from the LES-database. Because it is difficult to derive two components of the turbulent fluctuations σ_u and σ_v directly from the longitudinal turbulence standard deviation σ_l, we assumed that σ_l is equal to σ_u for the measuring 6-points averaged wind direction θ ranging from 45° to 135° and from 225° to 315°, and σ_l is equal to σ_v for θ ranging from 0° to 45°, from 135° to 225°, and from 315° to 360°, respectively. When assuming $\sigma_l = \sigma_u$ here, σ_u can be estimated by the following formula.

$$\sigma_{u\_est}(x,y,z) = I_{ref\_LES}(x,y,z)\left(aU_{coup}(x,y,z) + b\right) \tag{6}$$

$$I_{ref\_LES}(x,y,z) = \frac{\sigma_{u\_LES}(x,y,z)}{U_{LES}(x,y,z)} \tag{7}$$

The v- and w-components of the turbulence standard deviation were estimated by the following expression.

$$\sigma_{v\_est}(x,y,z) = \frac{\sigma_{v\_LES}(x,y,z)}{\sigma_{u\_LES}(x,y,z)}\sigma_{u\_est}(x,y,z) \tag{8}$$

$$\sigma_{w\_est}(x,y,z) = \frac{\sigma_{w\_LES}(x,y,z)}{\sigma_{u\_LES}(x,y,z)}\sigma_{u\_est}(x,y,z) \tag{9}$$

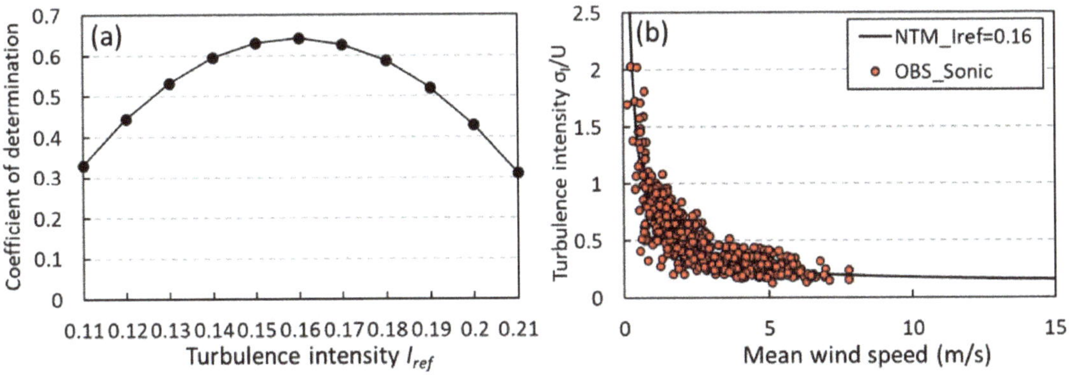

Figure 7. The coefficient of determination for regression analysis (**a**) and variation of the turbulence intensity with the mean wind speed measured by the ultrasonic anemometer placed at the building rooftop (**b**).

4.2. Lagrangian Particle Dispersion Model and Calculation Conditions

We adopted the following expression of a Lagrangian particle dispersion model [31].

$$x_i(t+\Delta t) = x_i(t) + u_{pi}\Delta t \tag{10}$$

$$u_{pi} = U_i + u_i' \tag{11}$$

$$u_i'(t+\Delta t) = a u_i'(t) + b\sigma_{ui}\xi + \delta_{i3}(1-a)t_{Lxi}\frac{\partial \sigma_{ui}^2}{\partial x_i} \tag{12}$$

$$a = exp\left(-\frac{\Delta t}{t_{Lxi}}\right) \tag{13}$$

$$b = \left(1-a^2\right)^{1/2} \tag{14}$$

where x_i, u_{pi}, U_i, u_i', σ_{ui}, ξ, t_{Lxi}, δ_i, t, and Δt are the particle position in the i-direction (east-west direction, i = 1; north-south direction, i = 2; vertical direction, i = 3), velocity of the particle, mean wind velocity, turbulence velocity, standard deviation of the velocity fluctuation, a random number from a Gaussian distribution with zero mean and unit variance, the Lagrangian integral time, the Dirac delta function, time, and calculation time step interval, respectively.

The size of the computational domain is 150 m × 150 m in the horizontal direction at 100-m height as shown in Figure 2. The grid spacing and mesh arrangement are the same as the LES computational conditions as mentioned earlier. The target simulation periods are 10 min from 1122 JST to 1132 JST on 26 November 2020 and from 1112 JST to 1122 JST on 1 December 2020. The plume is released at 1.5-m height from the building rooftop. As described in the previous section, U_i are provided from Equations (1) and (2). Each component of σ_i was given by Equations (6), (8) and (9). The particles are assumed to be reflected at the solid boundaries. The calculation time step interval is 0.05 s. The number of the imaginary particles is 3600.

5. Results

Figure 8 shows horizontal distributions of the Doppler LiDAR-derived wind velocity vectors near the mist plume release height for the two experimental periods of the case A from 1125 to 1131 JST 26 November and the case B from 1116 to 1122 JST 1 December. The wind velocities were interpolated on the grids of 1 m × 1 m. For the case A, first, a western wind blew around the release location and a northwesterly wind was also overserved on the north side from there. Then, a north-northwest wind blew just in the west side and a west wind did in the east side from the release point. At 1131 JST, a west-northwest and northwest wind blew in the whole area. For the case B, first, a northeast wind blew just in the northeast area, southeaste wind did in the southeast area, and an east wind blew in the vicinity of the release point. A vortex weak wind rotating counterclockwise was also observed in the southwest side. Then, an east wind blew in the whole area. The main flow directions above the building canopy changed from 280° to 304° and from 75° to 118° during the periods for the cases A and B, respectively. It is found that the wind direction variability is larger for the case B.

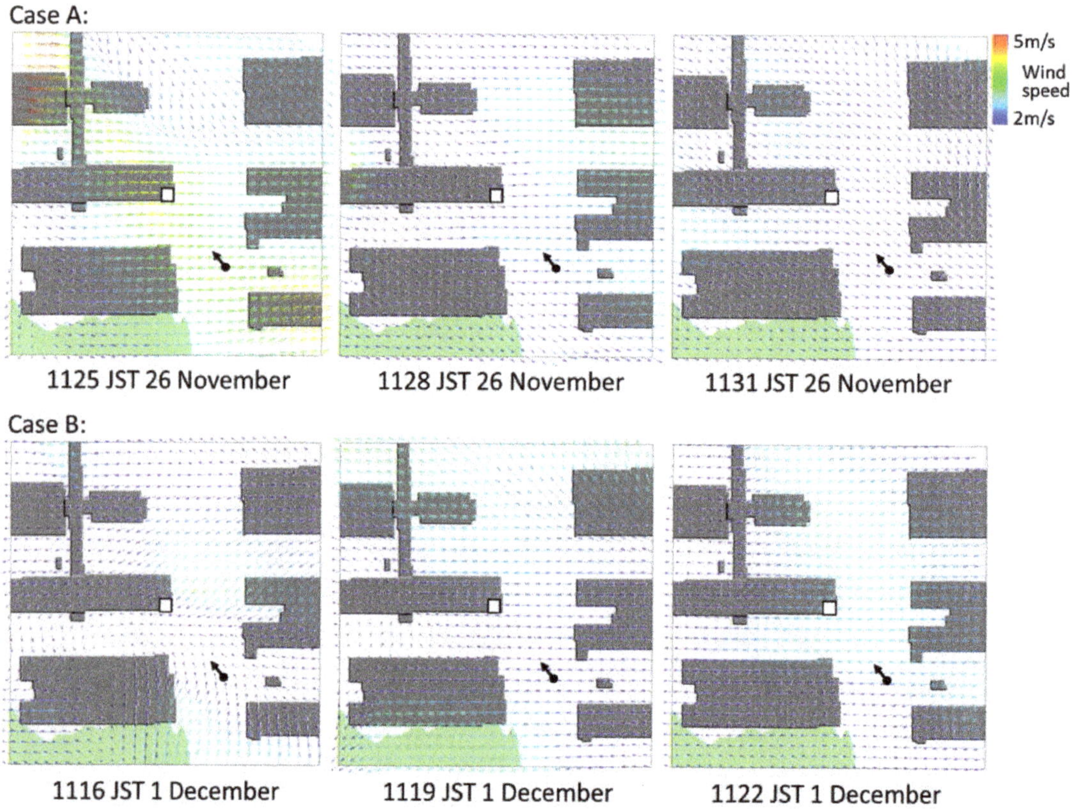

Figure 8. Horizontal distributions of the Doppler LiDAR-derived wind velocity vectors near the mist plume release height. The areas correspond to the square area of 150 m × 150 m shown in Figure 1. The white square depicts a position of the mist-spraying system placed at the building rooftop. The arrow indicates a position of the digital camera on the ground directed toward the mist-spraying system. The case A is the period from 1125 to 1131 JST 26 November. The case B is the period from 1116 to 1122 JST 1 December.

Figure 9 shows instantaneous dispersion shots of the mist plume generated by the mist-spraying system at 3-min interval over the 10-min period after the release for the two cases. Here, the 10-min period was selected as the sampling period for investigating general behaviors of the plume under the influence of turbulent-scale flows. Macdonald and Griffiths [32] and Macdonald et al. [33] investigated plume dispersion behavior over regular arrays of building-like obstacles at a field site. In their field experiments, the 3-min period was selected as the standard sampling period for mean concentration because the time period less than 15–20-min removes the influence of the lateral meandering of a plume due to the meteorological disturbances. For the case A, it is observed that the mist plume was constantly transported in the east from the plume release location. On the other hand, for the case B, the mist plume centroid was actively fluctuated around the general transport direction in the west. This is due to the separated turbulent flows formed at the corner of the building. The horizontal transport direction of the plume generally corresponds to the Doppler LiDAR-derived wind velocity vectors for both cases.

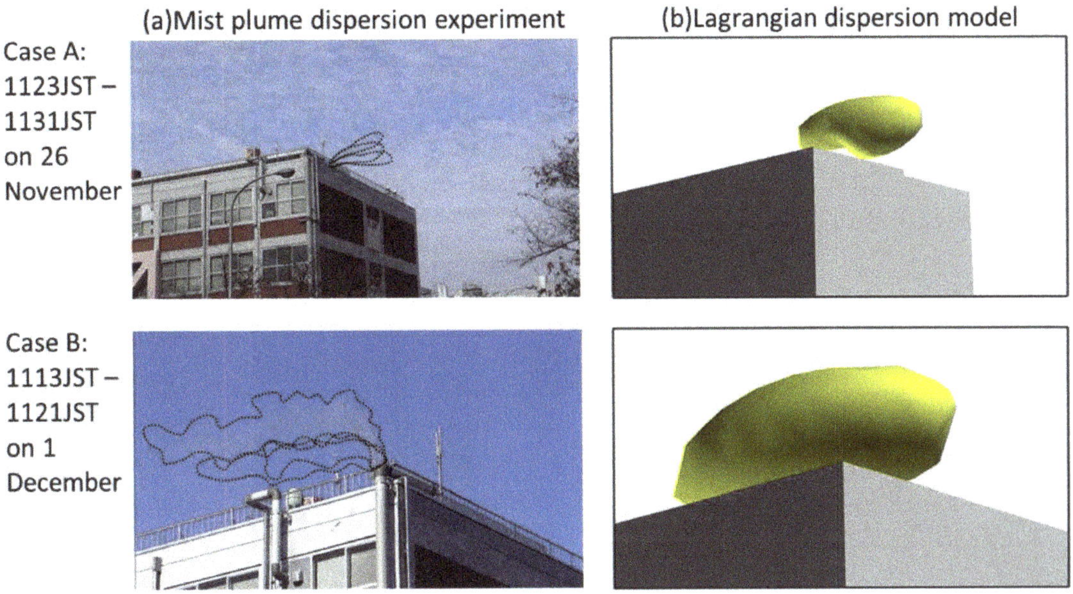

Figure 9. Instantaneous shots of the mist plume dispersion generated by the mist-spraying system at 3-min interval (**a**) and the simulated plume dispersion (**b**). The dashed line indicates the enveloping contour of the mist plume. The photograph was taken by the digital camera directed toward the mist-spraying system as shown in Figure 1. The yellow areas indicate the 50% iso-surface of initial concentration.

Figure 10 shows 3-dimensional distributions of 10-min averaged concentrations of the simulated plume for the cases A and B. Concentration at each mesh was estimated by kernel density estimation [31]. For the case A, the plume is transported in the east and a part of the plume is entrained into the zone of recirculating flow behind the building. For the case B, the plume is transported in the west and a part of the plume is entrained into the gap between the buildings due to the channeling effects. When compared both cases, it is also found that the plume spreads for the case B are larger than the ones for the case A. This is because the initial plume spreads are enhanced by the strong turbulence intensities produced at the building corner.

These results indicate that the general transport direction of the simulated plume is reproduced well by using the Doppler LiDAR-derived wind velocities for the region above the building canopy layer in comparison with the instantaneous shots of the real mist plume shown in Figure 9. The entrainment behavior of the plume into the building wake is also qualitatively represented well by using the LES-database for the region below it. Furthermore, the relative size of the simulated plume between the two cases can be associated with the fact that the fluctuating patterns of the real mist plume centroid, which implies that the local turbulence velocities are reasonably estimated from the empirical formulation of the relationship between mean wind speed and turbulence standard deviation. Therefore, it is concluded that our proposed framework for prediction system of local-scale atmospheric dispersion based on a coupling of LES-database and on-site meteorological observation has an enough potential to simulate plume transport direction by mean winds and reasonably represent dispersion behaviors by turbulence under the influences of individual buildings and meteorological disturbances.

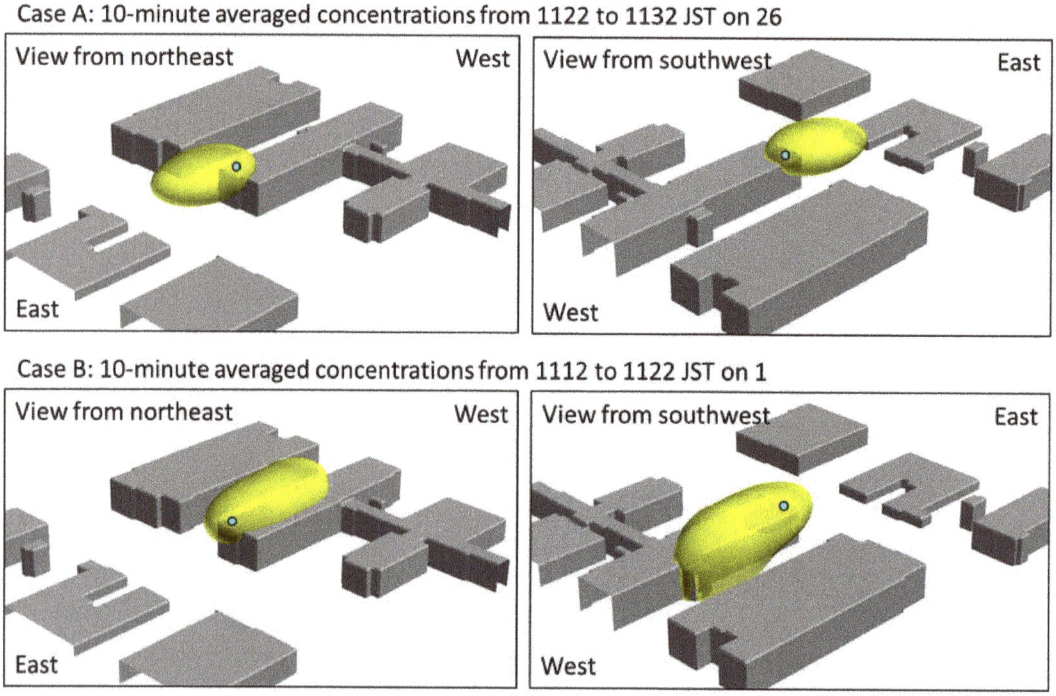

Figure 10. 10-min averaged concentrations of the simulated plume. The yellow areas indicate the 5% iso-surface of initial concentration. The blue circle is the release point.

6. Conclusions

We developed a framework for prediction system of local-scale atmospheric dispersion based on a coupling of LES-database and on-site meteorological observation that takes into account both individual buildings and real meteorological conditions. First, we performed LESs of turbulent flows over a target site and created the dataset of mean and turbulent flows for 36 mean wind directions at 10° class interval. Then, we conducted with meteorological observations by a Doppler LiDAR and simple dispersion experiments using a mist-spraying system at the target site, and observed dispersion behaviors of the water mist plume using a digital camera.

In estimating a mean flow filed, we prepared the OBS data obtained by the Doppler LiDAR for the region above the building canopy layer and used the LES-database of mean velocities for the region below it. In estimating a turbulent flow field, we adopted the empirical formulation of the relationship between mean wind speed and turbulence standard deviation used for determining appropriate locations of wind turbines. Compared to instantaneous shots of dispersion behaviors of the real mist plume, it was shown that the transport direction of the simulated plume was reproduced well by the Doppler LiDAR-derived wind velocities. The entrainment behavior of the plume into the building wake was qualitatively represented well by the LES-database. Furthermore, the plume spreads were also reasonably represented well by the estimated local turbulence velocities from the empirical formulation.

Here, we discuss the feasibility of our proposed framework for prediction system of local-scale atmospheric dispersion a as a practical emergency response system. Since computing time is an essential problem for emergency situations, first result should be provided within a few minutes especially in local-scale emergency response [34]. Our dispersion calculations by this coupling of the LES-database and on-site OBS were executed on a single core of an Intel CPU. The calculation time was approximately 15 s in order to simulate plume concentrations in the target computational domain of 150 m × 150 m × 50 m at 1 m grid. In conducting CFD simulations of turbulent flows and/or plume dispersion in a local-scale, the target site usually has a computational domain size of several kilometers at grid spacing of several meters for the horizontal direction. Considering it, the estimated calculation time is a few minutes. These facts indicate that our proposed coupling of the LES-database and on-site OBS is effective to accurately and quickly predict plume dispersion under the influences of both individual buildings and real meteorological conditions.

For the areas that are not covered by the measurements or for a case that the wind LiDAR is not available, a meso-scale meteorological simulation (MMS) model data should be used. Recently, coupling simulations of CFD and MMS models have been studied and evaluated against the urban tracer field experiments by many researchers [35–37]. We also conducted LESs of plume dispersion in the real urban central district by coupling with the MMS model and showed the reasonable performance of simulating plume dispersion in the built-up area under realistic meteorological conditions [17].

In future work, it is necessary to conduct field dispersion experiments on tracer gas release in a local-scale distances of up to several kilometers. We previously conducted LESs of plume dispersion in various urban areas with a wide range of obstacle density and building height variability, and compared the streamwise variation of mean and r.m.s. concentrations. It was shown that the spatial extent of concentration distribution patterns influenced by complex surface geometry is 1 km from a point source by comparative analysis [38]. Furthermore, it was reported that the assumption of a neutral stability is valid for within an urban area where building-induced turbulence is dominant and the assumption is less valid for the urban wake region because the levels of building-induced turbulence greatly subside [39]. It is necessary to quantitatively evaluate the coupling framework in comparison with the field experimental data at many downstream positions under thermal stability conditions in a local-scale distances of up to several kilometers. When fully evaluating it, our proposed technique is expected to use in source term estimation for contaminant plume dispersion in emergency responses to accidental release from nuclear facilities.

Author Contributions: Conceptualization, H.N.; methodology, H.N.; writing—original draft preparation, H.N.; writing—review and editing, T.Y., H.T. and M.K. All authors have read and agreed to the published version of the manuscript.

Funding: This work was supported by JAEA Nuclear Energy S&T and Human Resource Development Project through concentrating wisdom Grant Number JPJA18P18071754.

Acknowledgments: The computational simulations were performed on the ICEX at Japan Atomic Energy Agency. The authors are thankful to Haruyasu Nagai, Daiki Satoh, Hideyuki Kawamura, Yuki Kamidaira, Tsubasa Ikenoue of JAEA for helping us in conducting the water mist release experiment.

Conflicts of Interest: The authors declare no conflict of interest.

References

1. Murakami, S.; Mochida, A.; Hayashi, Y. Examining the κ-ε model by means of a wind tunnel test and large-eddy simulation of the turbulence structure around a cube. *J. Wind Eng. Ind. Aerodyn.* **1990**, *35*, 87–100. [CrossRef]
2. Tominaga, Y.; Stathopoulos, T. Numerical simulation of dispersion around an isolated cubic building, Model evaluation of RANS and LES. *Build. Environ.* **2010**, *45*, 2231–2239. [CrossRef]
3. Tominaga, Y.; Stathopoulos, T. CFD modeling of pollution dispersion in a street canyon: Comparison between LES and RANS. *J. Wind Eng. Ind. Aerodyn.* **2011**, *99*, 340–348. [CrossRef]
4. Gousseau, P.; Blocken, B.; Stathopoulos, T.; Heijst, G. CFD simulation of near-field pollutant dispersion on a high-resolution grid: A case study by LES and RANS for a building group in downtown Montreal. *Atmos. Environ.* **2011**, *45*, 428–438. [CrossRef]

5. Santiago, J.; Martilli, A. A dynamic urban canopy parameterization for mesoscale models based on computational fluid dynamics Reynolds–averaged Navier–Stokes microscale simulations. *Bound. Layer Meteorol.* **2010**, *137*, 417–439. [CrossRef]
6. Tominaga, Y.; Stathopoulos, T. CFD modeling of pollution dispersion in building array: Evaluation of turbulent scalar flux modeling in RANS model using LES results. *J. Wind Eng. Ind. Aerodyn.* **2012**, *104*, 484–491. [CrossRef]
7. Tominaga, Y.; Stathopoulos, T. CFD simulation of near-field pollutant dispersion in the urban environment: A review of current modeling techniques. *Atmos. Environ.* **2013**, *79*, 716–730. [CrossRef]
8. Yoshida, T.; Nakayama, H. Development of a combined LES/RANS model to predict atmospheric dispersion over urban areas. *Jpn. Soc. Comput. Eng. Sci.* **2020**, 20200013.
9. Collier, C.G.; Davies, F.; Bozier, K.E.; Holt, A.R.; Middleton, D.R.; Pearson, G.N.; Siemen, S.; Willetts, D.V.; Upton, G.J.G.; Young, R.I. Dual-Doppler lidar measurements for improving dispersion models. *Bull. Am. Meteorol. Soc.* **2005**, *86*, 825–838. [CrossRef]
10. Yan, B.; Li, Q. Coupled on-site measurement/CFD based approach for high-resolution wind resource assessment over complex terrains. *Energy Convers. Manag.* **2016**, *117*, 351–366. [CrossRef]
11. Tang, X.Y.; Zhao, S.; Fan, B.; Peinke, J.; Stoevesandt, B. Micro-scale wind resource assessment in complex terrain based on CFD coupled measurement from multiple masts. *Appl. Energy* **2019**, *238*, 806–815. [CrossRef]
12. Radunz, W.C.; Mattuella, J.M.; Petry, A.P. Wind resource mapping and energy estimation in complex terrain: A framework based on field observations and computational fluid dynamics. *Renew. Energy* **2020**, *152*, 494–515. [CrossRef]
13. Nakano, M.; Watanabe, H.; Sumiya, S. *Examination on the Long-Term Practicability of a Doppler Lidar-Data Missing Rate and Comparison of Wind Direction and Speed with the Propeller Type*; Japan Atomic Energy Agency: Tokyo, Japan, 2013. (In Japanese)
14. Hashimoto, K.; Hori, A.; Hara, T.; Onogi, S.; Mouri, H. Dual-camera system for high-speed imaging of particle image velocimetry. *J. Vis.* **2012**, *15*, 193–195. [CrossRef]
15. Onogi, S.; Haginoya, S.; Hori, A.; Yagi, T.; Mouri, H. Field PIV imaging technique with consumer digital still cameras. *J. Wind. Eng.* **2014**, *39*, 63–66. (In Japanese) [CrossRef]
16. Nakayama, H.; Leitl, B.; Harms, F.; Nagai, H. Development of local-scale high-resolution atmospheric dispersion model using large-eddy simulation part 4: Turbulent flows and plume dispersion in an actual urban area. *J. Nucl. Sci. Technol.* **2014**, *51*, 626–638. [CrossRef]
17. Nakayama, H.; Takemi, T.; Nagai, H. Development of local-scale high-resolution atmospheric dispersion model using large-eddy simulation part 5: Detailed simulation of turbulent flows and plume dispersion in an actual urban area under real meteorological conditions. *J. Nucl. Sci. Technol.* **2016**, *53*, 887–908. [CrossRef]
18. Nakayama, H.; Satoh, D.; Nagai, H.; Terada, H. Development of local-scale high-resolution atmospheric dispersion model using large-eddy simulation part 6: Introduction of detailed dose calculation method. *J. Nucl. Sci. Technol.* **2021**, 1–21. [CrossRef]
19. Smagorinsky, J. General circulation experiments with the primitive equations. *Mon. Weather Rev.* **1963**, *91*, 99–164. [CrossRef]
20. Goldstein, D.; Handler, R.; Sirovich, L. Modeling a no-slip flow boundary with an external force field. *J. Comput. Phys.* **1993**, *105*, 354–366. [CrossRef]
21. Nakayama, H.; Takemi, T. Large-eddy simulation studies for predicting plume concentrations around nuclear facilities using an overlapping technique. *Int. J. Environ. Pollut.* **2018**, *64*, 125–144. [CrossRef]
22. Kataoka, H.; Mizuno, M. Numerical flow computation around aeroelastic 3D square cylinder using inflow turbulence. *Wind Struct.* **2002**, *5*, 379–392. [CrossRef]
23. Monin, A.; Obukhov, M. Basic laws of turbulent mixing in the ground layer of the atmosphere. *Tr. Akad. Nauk SSSR Geophiz. Inst.* **1954**, *24*, 163–187.
24. Grimmond, C.S.B.; Oke, T.R. Turbulent heat fluxes in urban areas: Observations and a local-scale urban meteorological parameterization scheme (LUMPS). *J. Appl. Meteorol.* **2002**, *41*, 792–810. [CrossRef]
25. Cheng, H.; Castro, I.P. Near wall flow over urban-like roughness. *Bound. Layer Meteorol.* **2002**, *104*, 229–259. [CrossRef]
26. Britter, R.E.; Hanna, S.R. Flow and dispersion in urban areas. *Annu. Rev. Fluid Mech.* **2003**, *35*, 469–496. [CrossRef]
27. Yoshida, T.; Takemi, T.; Horiguchi, M. Large-eddy-simulation study of the effects of building-height variability on turbulent flows over an actual urban area. *Bound. Layer Meteorol.* **2018**, *168*, 127–153. [CrossRef]
28. Newson, R.K.; Calhoun, R.; Heap, R.; Cregan, E.; Princevac, M. Retrieval of microscale wind and temperature fields from single- and dual-Doppler lidar data. *J. Appl. Meteorol.* **2005**, *44*, 1324–1345. [CrossRef]
29. IEC61400-1. *Wind Turbines Part 1: Design Requirements*, 3rd ed.; International Electrotechnical Commission: Geneva, Switzerland, 2005.
30. Ishihara, T.; Yamaguchi, A.; Sarwar, M.W. A study of the Normal Turbulence Model in IEC 61400-1. *Wind Eng.* **2012**, *36*, 759–766. [CrossRef]
31. Yamada, T.; Bunker, S. Development of a nested grid, second moment turbulence closure model and application to the 1982 ASCOT Brush Creek data simulation. *J. Appl. Meteorol. Climatol.* **1988**, 562–578. [CrossRef]
32. MacDonald, R.W.; Griffiths, R.F.; Cheah, S.C. Field experiments of dispersion through regular arrays of cubic structures. *Atmos. Environ.* **1997**, *31*, 783–795. [CrossRef]
33. Macdonald, R.W.; Griffiths, R.F.; Hall, D.J. A comparison of results from scaled field and wind tunnel modeling of dispersion in arrays of obstacles. *Atmos. Environ.* **1998**, *32*, 3845–3862. [CrossRef]
34. Leitl, B.; Harms, F.; Berbekar, E.; Boris, J.; Patnaik, G.; Obenschain, K.; Fischer, S. Local-scale hazmat dispersion modelling for first responders based on high-resolution computational fluid dynamics-an overview of CT-Analyst Hamburg. *Chem. Eng. Trans.* **2016**, *48*, 43–48.

35. Nelson, M.A.; Brown, M.J.; Halverson, S.A.; Bieringer, P.; Annunzio, A.; Bieberbach, G.; Meech, S. A case study of the weather research and forecasting model applied to the joint urban 2003 tracer field experiment. part 1: Wind and turbulence. *Bound. Layer Meteorol.* **2016**, *158*, 285–309. [CrossRef]
36. Nelson, M.A.; Brown, M.J.; Halverson, S.A.; Bieringer, P.; Annunzio, A.; Bieberbach, G.; Meech, S. A case study of the weather research and forecasting model applied to the joint urban 2003 tracer field experiment. Part 2: Gas Tracer Dispersion. *Bound. Layer Meteorol.* **2016**, *161*, 461–490. [CrossRef]
37. Burman, J.; Jonsson, L.; Rutgersson, A. On possibilities to estimate local concentration variations with CFD-LES in real urban environments. *Environ. Fluid Mech.* **2019**, *19*, 719–750. [CrossRef]
38. Nakayama, H.; Jurcakova, K.; Nagai, H. Large-Eddy Simulation of plume dispersion within various actual urban areas. *Adv. Sci. Res.* **2010**, *10*, 33–41. [CrossRef]
39. Lundquist, J.K.; Chan, S.T. Consequences of urban stability conditions for computational fluid dynamics simulations of urban dispersion. *J. Appl. Meteorol. Climatol.* **2007**, *46*, 1080–1097. [CrossRef]

atmosphere

Article

Large-Eddy Simulation of Plume Dispersion in the Central District of Oklahoma City by Coupling with a Mesoscale Meteorological Simulation Model and Observation

Hiromasa Nakayama [1,*], Tetsuya Takemi [2] and Toshiya Yoshida [1]

1 Japan Atomic Energy Agency, 2-4, Shirakatashirane, Tokai-mura, Naka-gun, Ibaraki 319-1195, Japan; yoshida.toshiya@jaea.go.jp
2 Disaster Prevention Research Institute, Kyoto University, Kyoto 611-0011, Japan; takemi@storm.dpri.kyoto-u.ac.jp
* Correspondence: nakayama.hiromasa@jaea.go.jp

Abstract: Contaminant gas dispersion within an urban area resulting from accidental or intentional release is of great concern to public health and social security. When estimating plume dispersion in a built-up urban area under real meteorological conditions by computational fluid dynamics (CFD), a crucial issue is how to prescribe the input conditions. There are typically two approaches: using the outputs of a mesoscale meteorological simulation (MMS) model and meteorological observations (OBS). However, the influences of the different approaches on the simulation results have not been fully demonstrated. In this study, we conducted large-eddy simulations (LESs) of plume dispersion in the urban central district of Oklahoma City under real meteorological conditions by coupling with a MMS model and OBS obtained at a single stationary point, and evaluated the two different coupling simulations in comparison with the field experiments. The LES–MMS coupling showed better performance than the LES–OBS one. The latter one also showed a reasonable performance comparable to the acceptance criteria on the model prediction within a factor of two of the experimental data. These facts indicate that the approach using observations at a single stationary point still has enough potential to drive CFD models for plume dispersion under real meteorological conditions.

Keywords: large-eddy simulation; plume dispersion; urban area; coupling simulation; mesoscale meteorological simulation model; meteorological observation

Citation: Nakayama, H.; Takemi, T.; Yoshida, T. Large-Eddy Simulation of Plume Dispersion in the Central District of Oklahoma City by Coupling with a Mesoscale Meteorological Simulation Model and Observation. *Atmosphere* **2021**, *12*, 889. https://doi.org/10.3390/atmos12070889

Academic Editor: Patrick Armand

Received: 24 May 2021
Accepted: 6 July 2021
Published: 8 July 2021

1. Introduction

Contaminant gas dispersion within an urban area resulting from accidental release or intentional release of CBRN (chemical, biological, radiological, or nuclear) agent is of great concern to public health and social security. According to the special issue on *CBRN Terrorism & Defense* in *Journal of Japan Society for Safety Engineering*, Inoue [1] provided an overview of CBRN agents in the past and its emergency preparedness, and listed sarin, chlorine, ammonium, dioxin, nitrogen dioxide, and sulfur dioxide, etc. as substances of very high concern in chemical agents. In UNSCEAR 2013 [2], in the event of accidental release of radionuclides into the atmosphere, measurements were largely focused on the radionuclides ^{131}I, ^{134}Cs and ^{137}Cs, because these are considered to be the most significant contributors to exposure. According to the document of WHO First Global Conference on Air Pollution and Health [3], for almost all pollutants, both short-term and long-term exposure can damage health. Additionally, in WHO Air quality guidelines for particulate matter, ozone, nitrogen dioxide and sulfur dioxide [4], for example, the guideline values are 40 μg/m^3 over annual mean and 200 μg/m^3 over 1-h mean for nitrogen dioxide, and 20 μg/m^3 over 24-h mean and 500 μg/m^3 over 10-min mean for sulfur dioxide, respectively. For the assessment of human health hazards from accidental or intentional release, the existence of high-concentration peaks in a plume should be considered in accordance with the exposure time period.

It has long been known that the root mean square (r.m.s) of concentration fluctuations is almost comparable to the mean concentrations on the mean plume axis in the atmosphere [5]. In such a situation, it is necessary to accurately capture not only mean but fluctuating concentrations of a plume within an urban environment by considering the effects of the individual buildings. The usual methodology used to treat this problem is to firstly obtain information about the mean and r.m.s concentrations, the intermittency factor, and the maximum expected concentrations with a given confidence [6,7].

Urban surface geometries are highly inhomogeneous and complex since the ground is covered with many buildings with highly variable heights and shapes. The lower part of atmospheric boundary layer is strongly influenced by the individual roughness obstacles that induce strong three-dimensionality of the flow [8]. In order to understand contaminant gas transport process in complex surface geometries, a computational fluid dynamics (CFD) model is widely used. CFD models are generally designed based on two different practically relevant approaches, Reynolds-averaged Navier–Stokes (RANS) and large-eddy simulation (LES) models. The former computes mean fields of wind flow and plume dispersion, delivering an ensemble- or time-averaged solution, and all turbulent motions are modeled using a turbulence model. The latter resolves large-scale turbulent motions and models only small-scale motions.

When conducting CFD simulations of plume dispersion in urban central districts under real meteorological conditions, a crucial issue is how to prescribe the input conditions for determining turbulent flow and plume dispersion fields. There are typically two approaches to reproduce realistic atmospheric conditions: one is coupling with a mesoscale meteorological simulation (MMS) model, and the other is coupling with meteorological observations (OBS). For example, Warner et al. [9] reproduced realistic atmospheric conditions by prescribing the different OBS data obtained by the URBAN 2000 field experiments conducted in Salt Lake City at the RANS-based CFD model and evaluated the model predictions for plume concentrations within urban environments. They reported that the inputs obtained close to the plume source led to the worst performance due to the significant variability induced by the urban buildings. Tewari et al. [10] evaluated the following two different approaches of supplying initial and boundary conditions: (i) coupling with observation obtained from a single stationary point and (ii) coupling with the MMS model during the URBAN 2000 field experiments. They mentioned that the latter approach shows better performance in comparison to the former one in comparison with the OBS data of mean concentrations.

Recently, many model evaluation studies using meteorological and tracer gas data by the Joint Urban 2003 (JU2003) field experiments have been reported [11–15]. This field campaign was conducted in the central district of Oklahoma City from 28 June through 31 July 2003 [16]. For example, Wyszogrodzki et al. [11] performed coupling simulations of LES and MMS models for the Joint Urban 2003 (JU2003) field experiments conducted in the central district of Oklahoma City and concluded that the accuracy of the model performance of concentrations heavily depends on the quality of the MMS data which accounts for transient effects within complex urban canopy. Nelson et al. [12,13] conducted building-resolving detailed simulations of plume dispersion using WRF model data and various meteorological observations. Burman et al. [14] conducted LESs of turbulent flows in Oklahoma City and investigated the accuracy of three typical LES subgrid models in comparison to the field experimental data. Li et al. [15] proposed a new scheme which employs the turbulence-reconstruction method for the simulation of microscale flow and dispersion in which both the mesoscale field and small-scale turbulence are specified at the boundary of a microscale model. We also conducted LESs of plume dispersion in the urban central district of Oklahoma City by coupling with the MMS model and showed that the prediction accuracy of the high-resolution simulations highly depends on the quality and reproducibility of the wind speed and direction prescribed by the MMS model [17].

However, these model evaluation studies have focused on the influence of the coupling simulations, mainly on the accuracy of the mean concentrations. Thus, the influence of the different coupling approaches on the mean and peak concentrations of a plume in the complex surface geometries has not been fully demonstrated. As mentioned earlier, when the accidental or intentional release of harmful substances into the atmosphere occurs, it is important to estimate not only the mean but the behavior of the concentration fluctuations for consequence assessment and countermeasures. In this study, we conducted LESs of turbulent flows and plume dispersion in an urban central district under realistic atmospheric conditions by coupling with the MMS model and OBS. Our objective is to evaluate mean and peak concentrations obtained by the two different coupling simulations in comparison with the experimental field data.

2. Configuration of the Numerical Experiments

2.1. Brief Outline of JU2003 Field Experimental Dataset

The field experiments of the JU2003 were conducted in the central district of Oklahoma City from 28 June through 31 July 2003 [16]. It is found from Figure 1a that there are residential areas covered with many trees around the central district of Oklahoma City. The meteorological stations of the OBS1 and OBS2 were located in the residential area and central district, respectively. The concentration measurements used for the model evaluation study are shown in Figure 1b. The average building height, maximum building height, building height variability, and building frontal area index of the urban central district are 32 m, 152 m, 1.0, and 0.15, respectively, as described in Nakayama et al. [17]. Here, the building height variability is the ratio of standard deviation of building height to average building height. The building frontal area index is the ratio of total frontal area of roughness elements to total surface area. It seems from this fact that the surface geometry of the residential area is nearly homogeneous and covered with low-rise buildings, although that of the urban central district is sparsely built-up, having buildings with very variable heights.

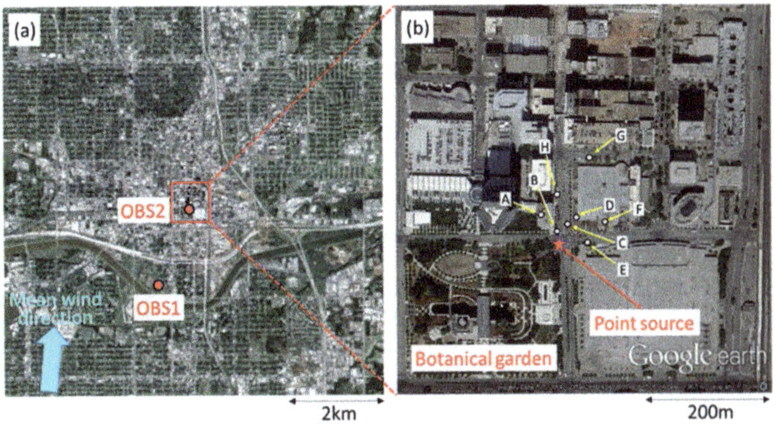

Figure 1. Locations of meteorological stations (**a**) and concentration measurement in the central business district of Oklahoma City (**b**). The star mark depicts a point source position which corresponds to the meteorological station OBS2. Time series of concentration fluctuations obtained at points A–H are used for the model evaluation. The photograph is reproduced by GoogleTM earth graphic.

The wind velocities and potential temperature were measured by Sodar and Rawinsonde at the OBS1, while the wind velocities were measured by minisodar at the OBS2. These OBS data were obtained with 15-min interval. Sulfur hexafluoride (SF6) tracer was released from a ground-level as puff and 30-min continuous releases during the 10 main IOPs (intensive observation periods). The first six IOPs and the last four ones were conducted in the daytime and the nighttime, respectively. During each IOP, three 30-min continuous and four instantaneous releases were typically carried out.

2.2. MesoScale Meteorological Simulation

The Weather Research and Forecasting (WRF), the Advanced Research WRF Version 3.3.1 [18] was adopted as MMS. Two-way nesting was used to resolve the Oklahoma City region at a fine grid spacing by setting three computational domains (with the top being at the 50-hPa level) as shown in Figure 2. The three domains cover 2700 km, 600 km, and 150 km square areas with 4.5 km, 1.5 km, and 0.5 km grids, respectively. The number of vertical levels is 53, with 12 levels in the lowest 1-km depth. The terrain data used here were the global 30-s data (GTOPO30) from the U.S. Geological Survey (USGS). Six-hourly Final Analysis (FNL) data of the U.S. National Centers for Environmental Prediction (NCEP) were used determine the initial and boundary conditions for the atmospheric and surface variables. A physics parameterization closely relevant to the simulation of wind fields was a planetary boundary layer (PBL) mixing parameterization. A Mellor–Yamada Level 2.5 scheme [19] was used for boundary layer physics.

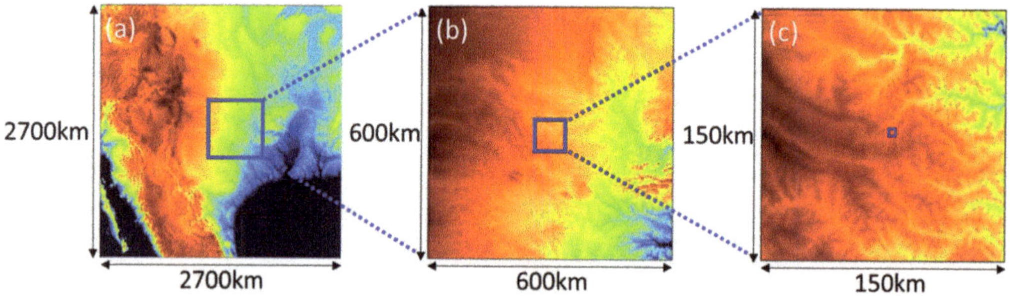

Figure 2. Computational areas of the MMS model. The MMS model is configured with three nested domains covering areas of (**a**) 2700 km × 2700 km at 4.5 km grid, (**b**) 600 km × 600 km at 1.5 km grid, and (**c**) 150 km × 150 km at 500 m grid [17].

2.3. LES-Based CFD Model

The LES-based CFD model used in this study is the LOHDIM-LES [17]. The governing equations are the filtered continuity equation, the Navier–Stokes equation in Boussinesq-approximated form, and the transport equations of temperature and concentrations. The subgrid-scale turbulent effect is represented by the Lagrangian dynamic Smagorinsky model [20]. The subgrid-scale scalar fluxes are also parameterized by an eddy viscosity model. The turbulent Prandtl and Schmidt numbers are 0.5.

Buildings and structures in the central district of Oklahoma City were explicitly represented by the use of a digital surface model dataset. The turbulent effects are represented by the immersed boundary method [21]. The model is configured using two nested domains with one-way as shown in Figure 3: one is an outer domain for generating atmospheric boundary layer flow and the other is an inner domain for detailed simulations of plume dispersion in the central business district. The outer and inner domains cover areas of 8 km by 8 km by 2.0 km with the grid spacing of 20 m by 20 m by 2–20 m stretched and 1.6 km by 1.6 km by 0.75 km with the grid spacing of 4 m by 4 m by 2–20 m stretched in the streamwise, spanwise and vertical directions.

According to the guidelines for CFD simulations of wind environment in urban areas [22,23], the minimum grid resolution should be 10 grid points to accurately reproduce separating flows around buildings. Individual urban buildings in the LES inner domain are basically resolved by 10 grid points, which are expected to capture the basic patterns of turbulent flows.

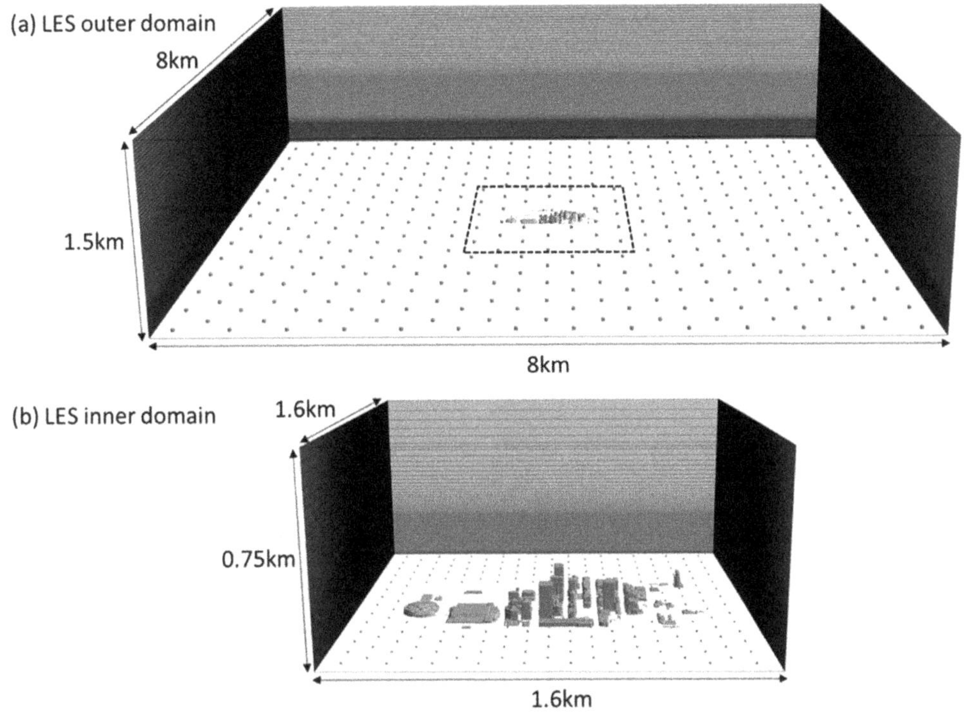

Figure 3. LES computational domains and mesh arrangement. The LES model is configured with two nested domains: (**a**) the outer domain of 8 km × 8 km at 20 m grid for generating atmospheric boundary layer flow and (**b**) the inner domain of 1.6 km × 1.6 km at 4 m grid for detailed simulations of plume dispersion in the central business district [17]. The dashed line in (**a**) the outer domain indicates (**b**) the inner domain.

3. Coupling with the MMS Model and OBS

The target events are the three 30-min continuous plume release from the Botanical Garden at 0900, 1100, and 1300 Central Daylight Time (CDT) on 16 July in IOP6 for cases1, 2, and 3, respectively. In this study, we compared the concentration data measured by fast-response tracer analyzers with a time response of approximately 1-Hz by Lawrence Livermore National Laboratory.

Figure 4 shows schematic of the LES–MMS and LES–OBS coupling cases. For the LES–MMS coupling case, the MMS model data of wind velocities (U_{MMS}) and potential temperature (θ_{OBS}) of the innermost domain shown in Figure 2 were used as the input conditions of the LES outer domain. First, the atmospheric conditions of IOP6 were reproduced by the MMS model from 1900 CDT 14 July to 1900 CDT 16 July 2003. The first and second inner domains were initialized at 1900 CDT 15 July and at 0700 CDT 16 July, respectively. Then, the boundaries of the inflow and ground surface were determined by the MMS outputs (with 1-min interval and 500-m resolution) linearly interpolated on the grids of the LES outer domain with 1-min interval. On the other hand, for the LES–OBS coupling case, the meteorological observation data obtained at a single stationary point of the OBS1 with 15-min interval from 0700 CDT to 1400 CDT were used as the input conditions of the LES outer domain. Since the dominant mean wind directions were nearly south and the OBS1 was located over the homogeneous ground surface at the upstream position of the urban central district for the target period as shown in Figure 1, the vertical profile of the OBS data obtained at a single stationary point was given at the inflow boundaries under the assumption of horizontal homogeneity. Due to the limitation of the observation range, the upper winds above 300-m height were given by the MMS model data under the condition that the OBS data gradually match with them for the vertical direction. The 15-min averaged OBS data were linearly interpolated on the temporal and spatial resolutions of the LES outer domain and were given as inputs. Furthermore, the input conditions of the LES inner domain were determined by the outputs of the LES outer domain linearly interpolated on the grids of the inner domain with 3-s interval in both coupling cases.

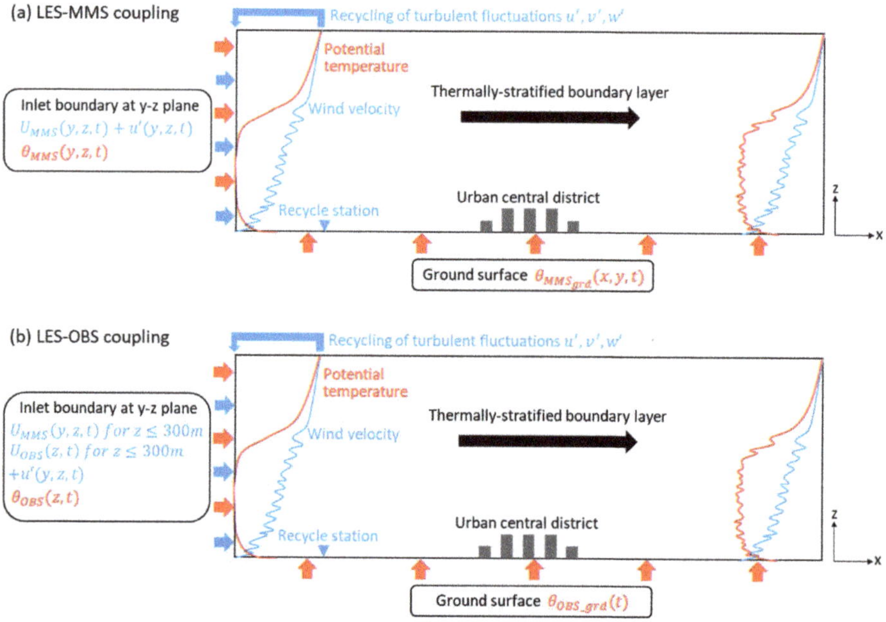

Figure 4. Schematic of (**a**) LES–MMS and (**b**) LES–OBS coupling cases.

Inflow and outflow boundaries of the LES outer domain are automatically determined in accordance with mean wind direction in the meteorological field. For example, when the mean wind direction lies in the range from 0° to 90°, the vertical planes on the north and east sides are set to inflow boundaries, and those on the south and west sides are set to outflow boundaries. The details were described in Nakayama and Takemi [24]. Each component of the turbulent fluctuations is generated by the recycling method [25]. The following are the formulations at the inflow boundary of the *y*–*z* vertical plane.

$$u'(y,z,t)_{inlt} = \phi(z)\left(u_{recy}(y,z,t) - [u(z,t)]\right), \tag{1}$$

$$v'(y,z,t)_{inlt} = \phi(z)\left(v_{recy}(y,z,t) - [v(z,t)]\right), \tag{2}$$

$$w'(y,z,t)_{inlt} = \phi(z)\left(w_{recy}(y,z,t) - [w(z,t)]\right), \tag{3}$$

where u', v', w', and $\phi(z)$ are each components of the turbulent fluctuations and a damping function to control the unwanted development of turbulent fluctuations in the upper part of the simulated boundary layer, respectively. The suffixes of *inlt* and *recy* indicate the inlet boundary and the recycle station, respectively. $[u]$, $[v]$, and $[w]$ are horizontally averaged winds over the driver region ranging from the inlet boundary to the recycle station, respectively. At the bottom surface, the Monin–Obukhov similarity theory [26] is applied. At the upper boundary, a free-slip condition for the horizontal velocity components and a zero-speed condition for the vertical velocity component are imposed. At the outlet boundary, a free-slip condition is applied for each component of wind velocity.

Table 1 shows input conditions of the two different coupling cases. Here, subscripts i, j, and k stand for coordinates (horizontal directions; x, y and vertical direction; z). u, u_*, and θ are the LES calculated data of the instantaneous wind velocity, the friction velocity, and the potential temperature, respectively. τ_{i3}, q_3, ψ_m, ψ_h, z_0, z_{0T}, κ, and t are the local instantaneous local wall stress, the surface heat flux, the stability corrections for momentum and heat, roughness length for momentum and heat, the von Karman constant, and the time, respectively. The suffixes of ws and grd are the wind speed and ground, respectively. ψ_m and ψ_h are estimated by the formulation of Businger et al. [27]. z_{0T} is estimated by the formulation of Moriwaki and Kanda [28].

Table 1. Input conditions of the LES–MMS and LES–OBS coupling cases.

Input Data		LES–MMS Coupling Case	LES–OBS Coupling Case
Wind velocity field	Inlet	$U_{MMS}(y,z,t)+u'(y,z,t)$	$U_{OBS}(z,t)+u'(y,z,t)$
	Ground surface	$\tau_{i3}=-\left(\frac{u_{ws\_grd}(x,y,t)\kappa}{ln(z/z_0)-\psi_m}\right)^2\frac{u_{i\_grd}(x,y,t)}{u_{ws\_grd}(x,y,t)}$	
Potential temperature field	Inlet	$\theta_{MMS}(y,z,t)$	$\theta_{OBS}(z,t)$
	Ground surface	$q_3=\frac{u_*\kappa\left(\theta_{MMS\_grd}(x,y,t)-\theta_{grd}(x,y,t)\right)}{ln(z/z_{0T})-\psi_h}$	$q_3=\frac{u_*\kappa\left(\theta_{OBS\_grd}(t)-\theta_{grd}(x,y,t)\right)}{ln(z/z_{0T})-\psi_h}$

The simulation periods of the LES outer and inner domains are from 0700 to 0940 and from 0800 to 0940 for case1, from 0900 to 1140 and from 1000 to 1140 for case2, and from 1100 to 1340 and from 1200 to 1340 for case3, respectively. The calculation time step intervals are 0.2- and 0.02-s in the LES outer and inner domains, respectively.

4. Results

4.1. Flow Field

Figure 5 shows vertical profiles of wind speed, wind direction, and potential temperature of the OBS, MMS, LES–MMS, and LES–OBS cases at each plume release time at the OBS1 position. The wind speed of the OBS increased with height at 0900 CDT. With the development of daytime convective boundary layer (CBL), the OBS wind speed gradually became nearly homogeneous up to 300-m height due to the turbulent mixing motions. According to Wyszogrodzki et al. [11], the low-level jet was observed during the early morning in IOP6. The OBS data of the sharp peak in the range of 300–400 m height at 1100 and 1300 CDT were due to the residual low-level jet. As the MMS model did not capture the rapid change of the wind speed, nor was it simulated by the LES–MMS case, the LES–OBS case did not reproduce the low-level jet either, since the upper winds above 300-m height were given by the MMS model. Both cases showed nearly constant vertical profiles of the wind direction at 0900 and 1100 CDT although they did not reproduce the rapid change of the OBS wind directions across 200-m height at 1300 CDT. The potential temperature profiles of the LES–MMS case were similar to the OBS data and those of the LES–OBS were consistent with the OBS at each time point.

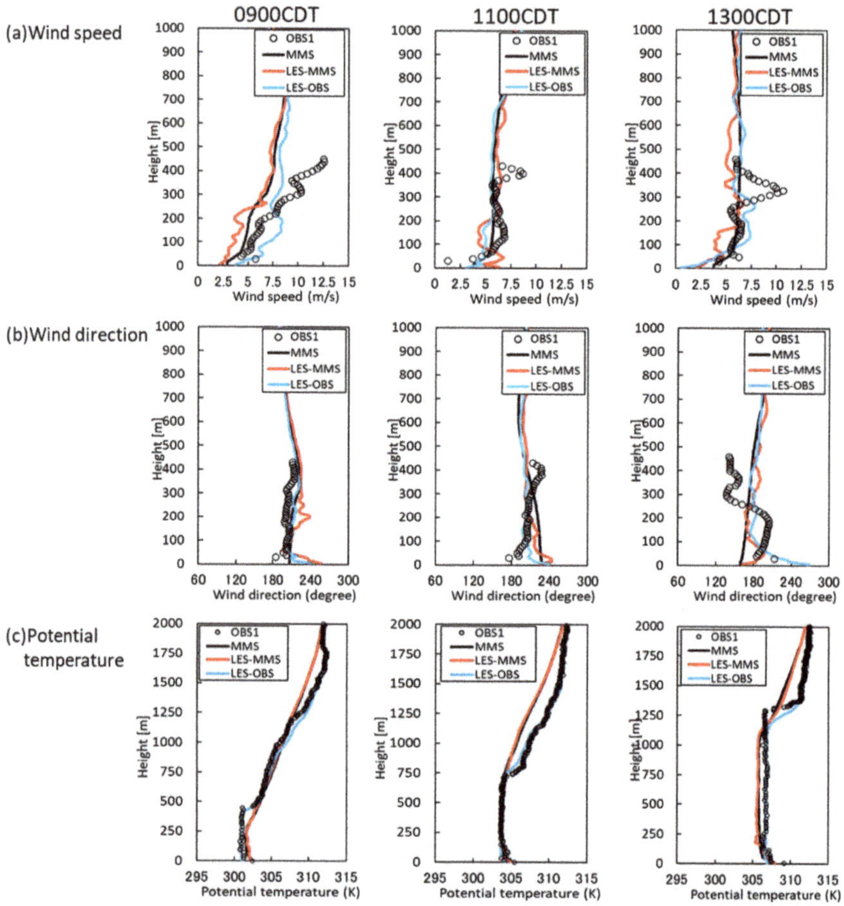

Figure 5. Vertical profiles of wind speed (**a**), wind direction (**b**), and potential temperature (**c**) of the OBS, MMS, LES–MMS, and LES–OBS cases at each plume release time at the position of the OBS1.

Figure 6 shows the vertical profiles of the wind speed and wind direction at the plume release point. The OBS wind speed rapidly decreases near the ground surface. The OBS wind direction also rapidly varies from the ground surface to 30-m height, especially at 0900 and 1100 CDT. According to the study on turbulent boundary layer flows over urbanlike roughness by Cheng and Castro [8], the flow patterns are directly determined by building arrangements and show a strong three-dimensionality caused by impinging, separated, and recirculating flows in the building canopy layer, while the dynamical influence of the surface decreases with height and the flows eventually readjust to the meteorological conditions. It seems that these rapid changes are due to the influence of various obstacles such as buildings and trees in the Botanical Garden. As these were not explicitly resolved in the MMS models, the LES–MMS case did not reproduce the local variations and rapid decrease in the wind speed and wind direction near the ground surface. The LES–OBS case did not capture such a tendency either. Figure 7 shows the time series data of the wind speed and wind direction of the OBS and the 10-min averaged LES data. These data of both cases are different from the OBS data, although the LES–OBS case was consistent with the OBS data of the wind speed in the morning.

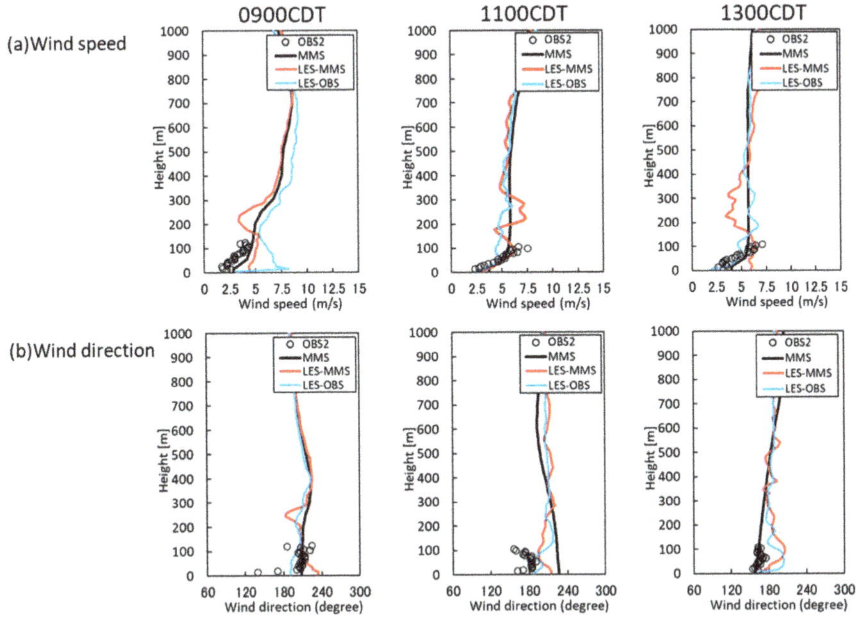

Figure 6. Vertical profiles of the wind speed (**a**) and wind direction (**b**) of the OBS, MMS, LES–MMS, and LES–OBS cases at the plume release point.

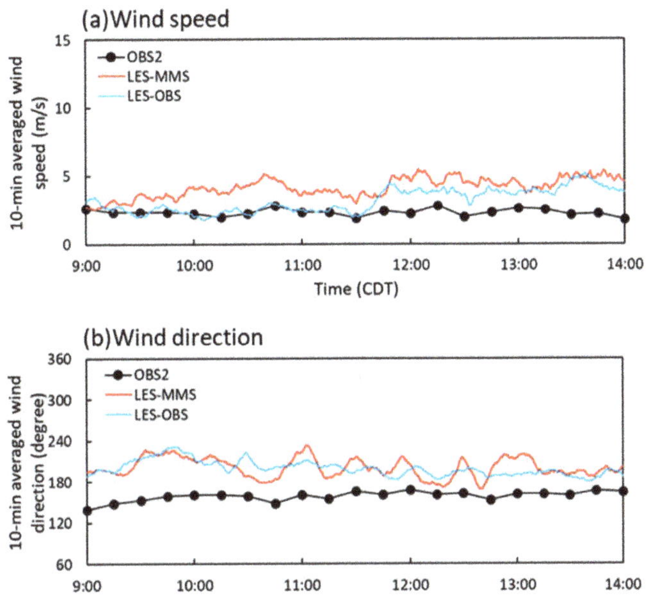

Figure 7. Time series of the wind speed (**a**) and wind direction (**b**) at 15-m height.

It is found from these results that there are quantitatively differences in the vertical profiles and time series of the wind velocities between the OBS and both LES coupling cases. However, the tendencies of the development of CBL were simulated well by both

LES coupling cases. It is also considered from these results that the OBS data obtained at a single stationary point used for the LES–OBS case are reasonable to prescribe the input condition under the assumption of horizontal homogeneity.

There have been many studies on incorporating aerodynamic effects of roughness elements on urban boundary layer flows into MMS models. For example, Wyszogrodzki et al. [11] compared performance of a single-layer urban canopy model (SUCM) and a multilayer urban canopy model (MUCM) used in WRF model. They concluded that WRF–MUCM reproduces the observed mean near-surface and boundary-layer winds and temperature fields during daytime conditions, and provides statistics during the nighttime more accurately than WRF–SUCM. In order to represent urban aerodynamic effects of urban surfaces more accurately by MMS model, such urban canopy models should be used.

4.2. Concentration Field

Figure 8 shows horizontal distributions of the 30-min averaged concentrations near the ground surface after each plume release time for LES–MMS and LES–OBS cases. The horizontal spread of the plume becomes larger with the development of the daytime CBL, and the general distribution patterns were similar between both LES coupling cases. The high-concentration regions in the LES–OBS case were formed mainly around the plume release point and the mean concentrations rapidly decreased with downwind distance, while those in the LES–MMS case were formed along the main street parallel to the mean wind direction at 1300CDT.

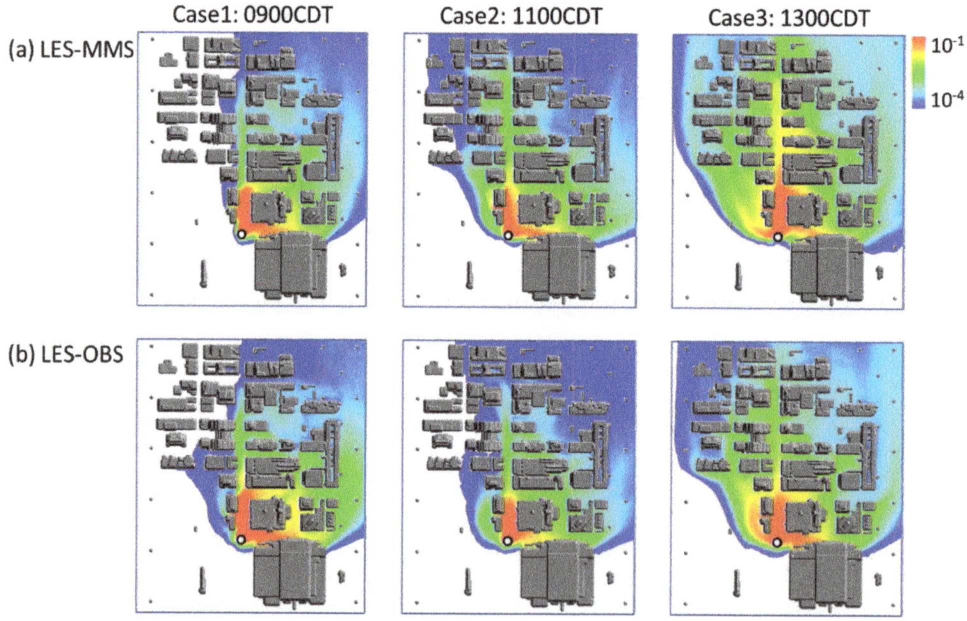

Figure 8. Horizontal distributions of the 30-min averaged concentrations near the ground surface after each plume release time for cases of LES-MMS (**a**) and LES-OBS (**b**) cases. The circle mark depicts the plume source point.

Nelson et al. [13] conducted building-resolving detailed simulations of plume dispersion using WRF model data and various meteorological observations, and evaluated them against the Joint Urban 2003 tracer gas measurements. They mentioned that local intermediate-scale variabilities (i.e., between mesoscale and microscale) reproduced by using only an OBS single stationary point as input were represented as if they were sudden changes in the mesoscale wind field, which significantly overpredicted the lateral spread of the plume. There is a possibility that the larger horizontal spread of the plume for the LES–OBS case is due to the use of a single measurement location.

Figure 9 compares time series of concentration fluctuations obtained by the LES–MMS and LES–OBS cases with the field experimental data at the positions of C, E, and G for cases 1, 2, and 3. At point C, instantaneous high concentrations of the field experiments intermittently occur for each case. Those obtained by both coupling cases were less intermittent and were smoother than the experimental values. Thus, the concentration fluctuating patterns were considerably different from the experimental data. According to the LES sensitivity analysis of grid resolution for a point source corresponding to 1.0 and 10 times the real diameter on the plume intermittency by Michioka et al. [29], it was shown that the intermittency of the concentrations was overestimated by the excess subgrid turbulent mixing for a case of a coarse grid. The hose diameter used for continuous gas release in the JU2003 field experiments was approximately 1.6-cm [30], which was far smaller than the grid resolution of 4 m by 4 m set up in this calculation condition. This overestimation of the intermittency is clearly due to the excess action of the subgrid turbulent mixing by a coarse grid resolution. At point E, instantaneous high concentrations frequently occurred in the field experiments for each case. These tendencies were simulated well by both coupling cases. At point G, the concentrations fluctuated with a large time scale and showed high peaks in the experiments, especially for case 1. Both coupling cases showed sharper peak concentrations than the experimental ones. The fluctuating concentration patterns were considerably different from the experimental data.

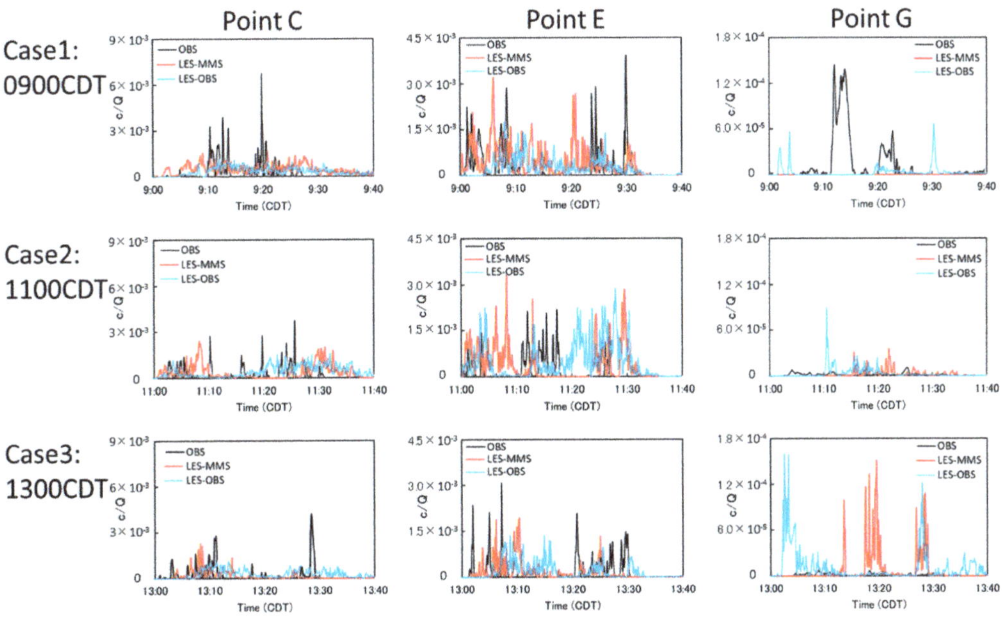

Figure 9. Time series of concentrations fluctuations of the OBS, the LES–MMS and LES–OBS cases.

Figure 10 shows a scatter plot of the 30-min averaged and maximum concentrations during the plume release duration time. Here, FAC2, whose value is within 0.5–2.0, is the ratio of the LES data to the OBS data. From this definition, the best results have a value of 1.0. Statistical performance measures for quantitatively evaluating the predictions of a model with field measurements usually include fractional bias (FB), geometric mean bias (MG), normalized mean square error (NMSE), geometric variance (VG), and the fraction of predictions within a factor of 2 of field measurements (FAC2). According to the model performance evaluation study by Chang and Hanna [31], the FAC2 is the most robust performance measure, because it is not overly influenced by either low or high outliers. The mean concentrations of both LES cases were overestimated in the high concentrations. In addition, those of the LES–OBS were comparatively scattered and were somewhat underestimated in the low concentrations. The maximum concentrations of both LES cases were consistent with the OBS data especially in the high concentrations. However, those of the LES–OBS case were comparatively scattered in the low concentrations. Burman et al. [14] conducted LESs of turbulent flows in Oklahoma City and investigated the accuracy of three typical subgrid models in comparison to the field experimental data. They mentioned that in the urban central district, simulated turbulence is mainly determined by buildings and their configurations and is only weakly affected by LES subgrid model type, while outside and upwind of the central district the turbulence given at the inflow boundaries is very important. It seems that the deviations between measured and simulated concentration data within the built-up areas are caused mainly by geometrical simplifications.

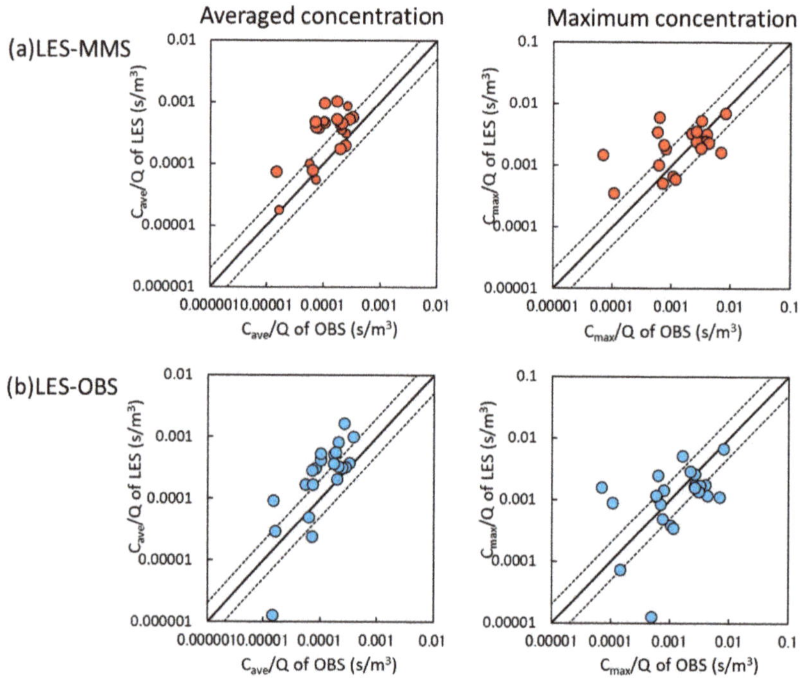

Figure 10. Scatter plot of the 30-min averaged and maximum concentrations in the LES–MMS (**a**) and LES–OBS (**b**) cases. The solid and dashed lines indicate the perfect and FAC2 lines, respectively.

The FAC2 of the mean and maximum concentrations were 0.42 and 0.63 for LES–MMS, and 0.30 and 0.58 for LES–OBS, respectively. Various methodologies have been investigated to evaluate the performance of local-scale atmospheric dispersion models [32–35]. For example, Hanna and Chang [34] suggested that a good model should have FAC2 $\geq$ 30% for urban areas based on a large number of model evaluation exercises by the field experimental data of mean concentrations. FAC2 of the mean concentrations by the LES–OBS case was comparable to the recommended value, whereas that of the LES–MMS case exceeded it. The peak concentrations obtained by both LES cases were distributed around the perfect line in the high concentrations. As there is no universal definition of acceptable criteria for peak concentrations at present, it is difficult to fully evaluate the model performance for them. However, it can be expected that the peak concentrations were reasonably captured depending on different downwind positions.

It is found from these results that the LES–MMS coupling case show good performance. The LES–OBS one also shows reasonable performance. This indicates that it is effective to prescribe the input conditions using the OBS data obtained even at a single stationary point over the homogeneous ground surface with the assumption of horizontal homogeneity.

5. Conclusions

We conducted LESs of plume dispersion in the urban central district under realistic meteorological conditions by coupling with the MMS model and OBS obtained at a single stationary point, and evaluated the two different cases in comparison with the JU2003 field experiments conducted in the central district of Oklahoma City.

For the LES–MMS coupling case, the MMS model data of wind velocities and potential temperature were used as the input conditions of the LES domain. For the LES–OBS coupling case, the OBS data obtained at a single stationary point were used as the input conditions of the LES domain under the assumption of horizontal homogeneity. Although there were quantitative differences in the vertical profiles and time series of the wind velocities between the OBS and both LES coupling cases, the tendencies of the development of CBL were simulated well. This indicates that the OBS data used for the LES–OBS case was reasonable for prescribing the input condition under the assumption of horizontal homogeneity.

The general distribution patterns of the mean concentrations were similar between both LES coupling cases, although the high-concentration regions were a little different in the daytime CBL development. The fluctuating concentration patterns were considerably different from the field experimental data at several measuring points. This is due to coarse grid resolution for a plume release point. Focusing on a scatter plot of the 30-min averaged and maximum concentrations during the plume release duration time, it was shown that the mean concentrations obtained by both LESs were overestimated in the high-concentration regions. Those of the LES–OBS case were comparatively scattered and were a little underestimated, especially in the low concentrations. The maximum concentrations of both LESs were consistent with the experimental data, especially in the high concentrations. The LES–MMS case showed better performance than the LES–OBS one. However, FAC2 of the latter case was comparable to the acceptance criteria on the model prediction within a factor of two of the field experimental data and showed reasonable performance. This indicates that it is promising to prescribe the input conditions using the OBS data obtained at a single stationary point over the homogeneous ground surface under the assumption of horizontal homogeneity.

In order to further improve the simulation accuracy, the following three points should be considered.

1. Appropriately represent aerodynamic effects of urban surfaces with the MMS model;
2. Utilize multiple spatially varying wind field observations;
3. Incorporate data assimilation techniques into the LES model.

Regarding the first point, one conventional approach to parameterize the effects of buildings and structures is to introduce an urban canopy model. Several researchers [35–37] recommended the use of multilayer urban canopy models for appropriately representing urban effects on boundary layer structure. On the second point, Nelson et al. [13] mentioned that using the spatially varying flow fields generated from multiple observation profiles generally provided the best performance from model evaluation against the Joint Urban 2003 tracer gas measurements. Multiple spatially varying meteorological observations are favorable in terms of avoiding unrealistically reproduced temporal and spatial variability by using only a single measurement location as an input. The final point is that it is effective to use a data assimilation technique which is capable of reproducing more realistic simulated states by incorporating observational data into simulation models. In future work, we will apply the data assimilation method using a vibration equation [38] to LESs of plume dispersion in complex urban environments while both nudging the flow fields toward a target mean state and retaining the fluctuating nature of turbulent flows.

Author Contributions: Conceptualization, methodology, writing—original draft preparation, H.N.; writing—review and editing, T.T. and T.Y. All authors have read and agreed to the published version of the manuscript.

Funding: This work was supported by JSPS KAKENHI grant 18H01680 and 21H01591.

Institutional Review Board Statement: Not applicable.

Informed Consent Statement: Not applicable.

Conflicts of Interest: The authors declare no conflict of interest.

References

1. Inoue, T. CBRN Terrorism & Defense. *J. Jpn. Soc. Saf. Eng.* **2016**, *55*, 398–405. (In Japanese)
2. *Report of the United Nations Scientific Committee on the Effects of Atomic Radiation to the General Assembly, 2013 Report Vol. I. Sources, Effects and Risks of Ionizing Radiation*; United Nations Scientific Committee on the Effects of Atomic Radiation: New York, NY, USA, 2013.
3. World Health Organization. *WHO Air Quality Guidelines for Particulate Matter, Ozone, Nitrogen Dioxide and Sulfur Dioxide: Global Update 2005: Summary of Risk Assessment*; WHO: Geneva, Switzerland, 2006.
4. World Health Organization. The invisible killer—Health effects of air pollution. In Proceedings of the First WHO Global Conference on Air Pollution and Health, Geneva, Switzerland, 30 October–1 November 2018; WHO: Geneva, Switzerland, 2018.
5. Hanna, S.R. The exponential probability density function and concentration fluctuations in smoke plumes. *Bound. Layer Meteorol.* **1984**, *29*, 361–375. [CrossRef]
6. Efthimiou, G.C.; Bartzis, J.G.; Koutsourakis, N. Modelling concentration fluctuations and individual exposure in complex urban environments. *J. Wind Eng. Ind. Aerodyn.* **2011**, *99*, 349–356. [CrossRef]
7. Efthimiou, G.C.; Bartzis, J.G. Atmospheric dispersion and individual exposure of hazardous materials. *J. Hazard. Mater.* **2011**, *188*, 375–383. [CrossRef]
8. Cheng, H.; Castro, I.P. Near wall flow over urban-like roughness. *Bound. Layer Meteorol.* **2002**, *104*, 229–259. [CrossRef]
9. Warner, S.; Platt, N.; Jeffry, T.; Heagy, J.F. Comparisons of Transport and Dispersion Model Predictions of the URBAN 2000 Field Experiment. *J. Appl. Meteorol.* **2004**, *43*, 829–846. [CrossRef]
10. Tewari, M.; Kusaka, H.; Chen, F.; Coirier, W.J.; Kim, S.; Wyszogrodzki, A.; Warner, T. Impact of coupling a microscale computational fluid dynamics model with a mesoscale model on urban scale contaminant transport and dispersion. *Atmos. Res.* **2010**, *96*, 656–664. [CrossRef]
11. Wyszogrodzki, A.; Miao, S.; Chen, F. Evaluation of the coupling between mesoscale-WRF and LES-EULAG models for simulating fine-scale urban dispersion. *Atmos. Res.* **2012**, *118*, 324–345. [CrossRef]
12. Nelson, M.A.; Brown, M.J.; Halverson, S.A.; Bieringer, P.; Annunzio, A.; Bieberbach, G.; Meech, S. A case study of the weather research and forecasting model applied to the joint urban 2003 tracer field experiment. Part 1: Wind and turbulence. *Bound. Layer Meteorol.* **2016**, *158*, 285–309. [CrossRef]
13. Nelson, M.A.; Brown, M.J.; Halverson, S.A.; Bieringer, P.; Annunzio, A.; Bieberbach, G.; Meech, S. A Case Study of the Weather Research and Forecasting Model Applied to the Joint Urban 2003 Tracer Field Experiment. Part 2: Gas Tracer Dispersion. *Bound. Layer Meteorol.* **2016**, *161*, 461–490. [CrossRef]
14. Burman, J.; Jonsson, L.; Rutgersson, A. On possibilities to estimate local concentration variations with CFD-LES in real urban environments. *Environ. Fluid Mech.* **2019**, *19*, 719–750. [CrossRef]
15. Li, H.; Cui, G.; Zhang, Z. A New Scheme for the Simulation of Microscale Flow and Dispersion in Urban Areas by Coupling Large-Eddy Simulation with Mesoscale Models. *Bound. Layer Meteorol.* **2018**, *167*, 145–170. [CrossRef]
16. Allwine, K.J.; Flaherty, J.E. *Joint Urban 2003 Study Overview and Instrument Locations*; PNNL-15967; Pacific Northwest National Laboratory: Richland, WA, USA, 2003.
17. Nakayama, H.; Takemi, T.; Nagai, H. Development of local-scale high-resolution atmospheric dispersion model using large-eddy simulation part5: Detailed simulation of turbulent flows and plume dispersion in an actual urban area under real meteorological conditions. *J. Nucl. Sci. Technol.* **2016**, *53*, 887–908. [CrossRef]
18. Skamarock, W.C.; Klemp, J.B.; Dudhia, J.; Gill, D.O.; Barker, D.M.; Duda, M.G.; Huang, X.; Wang, W.; Powers, J.G. *A Description of the Advanced Research WRF Version 3*; NCAR Tech. Note, NCAR/TN-475+STR; National Center for Atmospheric Research: Boulder, CO, USA, 2008; p. 1.
19. Janjic, Z.I. *Nonsingular Implementation of the Mellor-Yamada Level 2.5 Scheme in the NCEP Meso Model*; NCEP Office Note; National Centers for Environmental Prediction: Camp Springs, MD, USA, 2002; Volume 437, p. 61.
20. Meneveau, C.; Lund, T.; Cabot, W. A Lagrangian dynamic sub-grid-scale model of turbulence. *J. Fluid Mech.* **1996**, *319*, 353–385. [CrossRef]
21. Goldstein, D.; Handler, R.; Sirovich, L. Modeling a no-slip flow boundary with an external force field. *J. Comput. Phys.* **1993**, *105*, 354–366. [CrossRef]

22. Franke, J.; Hellsten, A.; Schlunzen, H.; Carissimo, B. *Best Practice Guideline for the CFD Simulation of Flows in the Urban Environment, COST Action 732, Quality Assurance and Improvement of Microscale Meteorological Models*; Meteorological Inst.: Brussels, Belgium, 2007; ISBN 3-00-018312-4.
23. Tominaga, Y.; Mochida, A.; Yoshie, R.; Kataoka, H.; Nozu, T.; Yoshikawa, M.; Shirasawa, T. AIJ guidelines for practical applications of CFD to pedestrian wind environment around buildings. *J. Wind Eng. Ind. Aerodyn.* **2008**, *96*, 1749–1761. [CrossRef]
24. Nakayama, H.; Takemi, T. Large-eddy simulation studies for predicting plume concentrations around nuclear facilities using an overlapping technique. *Int. J. Environ. Pollut.* **2018**, *64*, 125–144. [CrossRef]
25. Kataoka, H.; Mizuno, M. Numerical flow computation around aeroelastic 3D square cylinder using inflow turbulence. *Wind Struct.* **2002**, *5*, 379–392. [CrossRef]
26. Monin, A.; Obukhov, M. Basic laws of turbulent mixing in the ground layer of the atmosphere. *Tr. Akad. Nauk SSSR Geophiz. Inst.* **1954**, *24*, 163–187.
27. Businger, J.A.; Wyngaard, J.C.; Izumi, Y.; Bradley, E.F. Flux-profile relationships in the atmospheric surface layer. *J. Atmos. Sci.* **1971**, *28*, 181–189. [CrossRef]
28. Moriwaki, R.; Kanda, M. Flux-gradient profiles for momentum and heat over an urban surface. *Theor. Appl. Climatol.* **2006**, 127–135. [CrossRef]
29. Michioka, T.; Sato, A.; Sada, K. Large-eddy simulation for the tracer gas concentration fluctuation in atmospheric boundary layer. *Trans. Jpn. Soc. Mech. Eng.* **2003**, *B69*, 868–875. (In Japanese) [CrossRef]
30. Warner, S.; Urban, J.T.; Heagy, J.F. Comparisons of T&D model predictions of the Joint Urban 2003 field experiment. *J. Appl. Meteorol. Climatol.* **2008**, *47*, 1910–1928.
31. Chang, J.C.; Hanna, S.R. Air quality model performance evaluation. *Meteorol. Atmos. Phys.* **2004**, *87*, 167–196. [CrossRef]
32. Hanna, S.R.; Hansen, O.R.; Dharmavaram, S. FLACS CFD air quality model performance evaluation with Kit Fox, MUST, Prairie Grass, and EMU observations. *Atmos. Environ.* **2004**, *38*, 4675–4687. [CrossRef]
33. Hanna, S.R.; Chang, J.C. Acceptance criteria for urban dispersion model evaluation. *Meteorol. Atmos. Phys.* **2012**, *116*, 133–146. [CrossRef]
34. Herring, S.; Huq, P. A review of methodology for evaluating the performance of atmospheric transport and dispersion models and suggested protocol for providing more informative results. *Fluids* **2018**, *3*, 20. [CrossRef]
35. Martilli, A.; Clappier, A.; Rotach, M.W. An urban surface exchange parameterisation for mesoscale models. *Bound. Layer Meteorol.* **2002**, *104*, 261–304. [CrossRef]
36. Kondo, H.; Tokairin, T.; Kikegawa, Y. Calculation of wind in a Tokyo urban area with a mesoscale mode lincluding a multi-layer urban canopy model. *J. Wind Eng. Ind. Aerodyn.* **2008**, *96*, 1655–1666. [CrossRef]
37. Salamanca, F.; Martilli, A.; Tewari, M.; Chen, F. A study of the urban boundary layer using different urban parameterizations and high-resolution urban canopy parameters with WRF. *J. Appl. Meteorol. Climatol.* **2011**, *50*, 1107–1128. [CrossRef]
38. Nakayama, H.; Takemi, T. Development of a data assimilation method using vibration equation for large-eddy simulations of turbulent boundary layer flows. *J. Adv. Model. Earth Syst.* **2020**, *12*. [CrossRef]

atmosphere

MDPI

Article

Large Eddy Simulations of Turbulent and Buoyant Flows in Urban and Complex Terrain Areas Using the Aeolus Model

Akshay A. Gowardhan *, Dana L. McGuffin, Donald D. Lucas, Stephanie J. Neuscamman, Otto Alvarez and Lee G. Glascoe

Lawrence Livermore National Laboratory, Livermore, CA 94551, USA; mcguffin1@llnl.gov (D.L.M.); lucas26@llnl.gov (D.D.L.); neuscamman1@llnl.gov (S.J.N.); alvarez45@llnl.gov (O.A.); glascoe1@llnl.gov (L.G.G.)
* Correspondence: gowardhan1@llnl.gov

Abstract: Fast and accurate predictions of the flow and transport of materials in urban and complex terrain areas are challenging because of the heterogeneity of buildings and land features of different shapes and sizes connected by canyons and channels, which results in complex patterns of turbulence that can enhance material concentrations in certain regions. To address this challenge, we have developed an efficient three-dimensional computational fluid dynamics (CFD) code called Aeolus that is based on first principles for predicting transport and dispersion of materials in complex terrain and urban areas. The model can be run in a very efficient Reynolds average Navier–Stokes (RANS) mode or a detailed large eddy simulation (LES) mode. The RANS version of Aeolus was previously validated against field data for tracer gas and radiological dispersal releases. As a part of this work, we have validated the Aeolus model in LES mode against two different sets of data: (1) turbulence quantities measured in complex terrain at Askervein Hill; and (2) wind and tracer data from the Joint Urban 2003 field campaign for urban topography. As a third set-up, we have applied Aeolus to simulate cloud rise dynamics for buoyant plumes from high-temperature explosions. For all three cases, Aeolus LES predictions compare well to observations and other models. These results indicate that Aeolus LES can be used to accurately simulate turbulent flow and transport for a wide range of applications and scales.

Keywords: urban dispersion; large eddy simulation; complex terrain; fast-response dispersion modeling; computational fluid dynamics

Citation: Gowardhan, A.A.; McGuffin, D.L.; Lucas, D.D.; Neuscamman, S.J.; Alvarez, O.; Glascoe, L.G. Large Eddy Simulations of Turbulent and Buoyant Flows in Urban and Complex Terrain Areas Using the Aeolus Model. *Atmosphere* **2021**, *12*, 1107. https://doi.org/10.3390/atmos12091107

Academic Editor: Patrick Armand

Received: 9 July 2021
Accepted: 23 August 2021
Published: 27 August 2021

1. Introduction

More than half of the world's population lives in urban areas and the danger from an accidental or deliberate release of hazardous materials can be significant. The transport and dispersion of atmospheric contaminants in urban areas is strongly influenced by surrounding buildings, which significantly modify the winds, leading to areas of channeling along the streets, updrafts and downdrafts in the wake of the buildings, and recirculating flow in street canyons [1,2]. In addition, urban areas create highly heterogeneous regions of wind speed and turbulence intensity. Similarly in complex terrain, the local terrain impacts the flow field significantly, producing similar complex effects which can lead to non-intuitive dispersion patterns. There is a great need to have an accurate and efficient capability to predict dispersion and deposition patterns in these complex scenarios.

High-resolution computer models can predict how airborne materials spread around buildings in urban areas and land features in complex terrain. However, the modeling tool must be flexible enough to use for a variety of applications, and should be coupled to many relevant databases, such as terrain, building shapefiles, land-use characteristics, and population. Many fast-response urban dispersion models have been developed to predict transport for these scenarios. Gaussian plume models, which run in seconds on a laptop, have been modified to account for the plume centerline shift that may occur due to channeling in street canyons [3]. Hall et al. (2000) developed a Gaussian puff model called the Urban Dispersion Model (UDM) for use from neighborhood to city scales [4]. Röckle (1990) derived a diagnostic model that computes three-dimensional (3-D) flow around buildings using empirical equations and mass conservation [5]. Most fast-response models rely on empirical algorithms based on idealized building configurations. This makes it difficult to generalize the accuracy of these models for flow fields in highly heterogeneous urban terrain without many validation exercises [6].

Computational fluid dynamics (CFD) models have been used to compute the flow field in urban areas and complex topography. Comparison of these results with field measurements shows that these models work well in most regions [7–13]. These CFD models, however, are computationally very expensive and prohibitive for applications related to toxic releases in cities or at industrial facilities where turnaround time is very important.

Further, before running CFD models, users need to generate detailed grids that account for the 3-D geometry of the surrounding city and terrain. This is usually a time-consuming process, which renders these models useless for an operational response where a quick answer is needed in a small amount of time.

Gowardhan et al. (2011) developed a fast-response CFD model which represents an intermediate model type that produces fast runtimes (in the order of minutes for a several-block problem) and a reasonably accurate solution [9]. Neophytou et al. (2011) also evaluated this model and showed these fast-response CFD models can accurately predict the flow features in complex urban configurations [10].

Based on work carried out by Gowardhan (2008) and Gowardhan et al. (2011), we have developed a new fast-response operational dispersion modeling system called Aeolus, which can predict the flow and transport of airborne contaminants in urban areas and complex terrain [9,11]. The model can be run in a very efficient (~minutes) RANS (Reynolds average Navier–Stokes) mode or a detailed (~hours) LES (large eddy simulation) mode.

In this paper, we describe the Aeolus model and evaluate its performance in large eddy simulation mode. The LES version is validated for two different cases and applied to a third case to showcase the model capabilities over a wide range of applications. We first present validation results for turbulence generated in neutrally stratified flow over complex terrain case using data from Askervien hill campaign [14]. Next, flow and dispersion in an urban area is validated using wind and tracer data from the Joint Urban 2003 Oklahoma City field experiment [15]. Last, we apply the model's capability to simulate the cloud rise dynamics of a high-temperature bubble from a nuclear explosion in the troposphere using data from the Upshot-Knothole Dixie test. This test was a high-altitude air burst with an explosive yield of 11 kilotons and a height of burst (HOB) of 1836 m above ground level (AGL).

2. Introduction to Aeolus Modeling System

Aeolus is an efficient 3-D CFD code based on a finite volume method. It solves the time-dependent incompressible Navier–Stokes equations on a regular cartesian staggered grid using a fractional step method algorithm. It also solves a scalar transport equation for potential temperature, which is coupled to the flow using an anelastic approximation. The model includes a Lagrangian dispersion model for predicting the atmospheric transport and dispersion of tracers and other materials. The RANS version of Aeolus is used as an operational model in the National Atmospheric Release Advisory Center (NARAC) at Lawrence Livermore National Laboratory (LLNL) for quickly simulating the impacts of airborne hazardous materials in urban areas. NARAC uses Aeolus and other operational atmospheric models to provide the United States Department of Energy information and services pertaining to chemical, biological, radiological, and nuclear airborne hazards [16,17]. NARAC can simulate downwind effects from a variety of scenarios, including fires, industrial and transportation accidents, radiation dispersal device explosions, hazardous material spills, sprayers, nuclear power plant accidents, and nuclear detonations.

2.1. Large Eddy Simulation Model

Aeolus can be run in a high-fidelity mode using an LES model. LES resolves the time-dependent turbulent flow field at both small and large scales, allowing better fidelity than alternative approaches such as RANS. The smallest scales of the solution rely on a Smagorinsky model with a constant of C_s = 0.12 to account for unresolved scales in the flow, rather than resolving them directly as in expensive direct numerical simulation (DNS) methods. This makes the computational cost for applying LES to realistic engineering systems with complex geometry or flow configurations practical and attainable using supercomputers. In contrast, direct numerical simulation, which resolves every scale of the solution, is prohibitively expensive for nearly all atmospheric dispersion problems with complex geometry.

2.2. Numerics

The model uses the 3rd order accurate Quadratic Upstream Interpolation for Convective Kinematics (QUICK) scheme [18] for the advective terms and 2nd order central difference for the diffusive terms. The scalar transport equation uses a Bounded QUICK (BQUICK) scheme to obtain a bounded solution while maintaining spatial accuracy and reducing dispersion errors. The law of the wall boundary condition is imposed at the rigid surface by applying a free slip boundary condition at the surface. The tangential shear stress is set equal to ${u_*}^2$. The value of friction velocity u_* is evaluated using a log-law ($u_* = uk/\ln(0.5 \times \Delta z/z_0)$), where u is the magnitude of the tangential velocity and z_0 is the surface roughness, k is the von Karman constant with a value of 0.4, and Δz is the vertical grid resolution.

The pressure Poisson equation is solved using the successive over-relaxation method (SOR via the methodology described above). A free slip condition is used at the top and side boundaries. The following outflow boundary condition is prescribed at the outlet:

$$\frac{\partial \varphi}{\partial t} = U_b \frac{\partial \varphi}{\partial n} \tag{1}$$

where n denotes the direction normal to the boundary face, φ is the advected variable, and U_b is a velocity that is independent of the location of the outflow surface and is selected so that an overall mass balance is maintained. This boundary condition allows the convection of turbulent structures out of the domain and avoids problems with reflection of pressure waves back to the interior of the domain.

2.3. Dispersion Model

To model dispersion within the atmosphere, Aeolus models the 3-D, incompressible, advection–diffusion equation with sources and sinks using a Lagrangian framework [19]:

$$\frac{DC}{Dt} = Q - S \tag{2}$$

where the total derivative represents the advection and diffusion that occurs to species in a Lagrangian reference frame, C is the air concentration of the species, Q is the source term, and S is the sink term, which accounts for removal processes such as deposition.

The equations for the Lagrangian particle displacement due to advection, diffusion, and settling in the three coordinate directions are:

$$dx_i = \tilde{u}_i dt + \frac{\partial \frac{v_T}{Sc}}{dx_i} dt + \left(2\frac{v_T}{Sc}\right)^{1/2} dW_{x_i} \tag{3}$$

where $\tilde{u}_i$ is the wind components in the x, y, and z projection directions, respectively, v_T is the eddy diffusivity, Sc is the Schmidt number, $dW_{x,y,z}$ are three independent normal random variates with zero mean and variance dt, which is the timestep of advection of the Lagrangian particle. The stochastic differential equations above are then integrated in time to calculate an independent trajectory for each Lagrangian particle. The concentration $\widetilde{c}$, at any time t, can then be calculated from the Lagrangian particle locations at time t and the contaminant mass associated with each particle. The model does not apply kernel smoothing and the grid cell concentration depends on the number of particles in the respective grid cell.

2.4. Grid Generation

The model uses a cartesian grid and straightforward masking approach for generating a computational grid. New model grids can be generated in seconds from geographic information system shapefiles (for a few kilometers) and/or building data available from the National Geospatial Intelligence Agency and United States Geological Service (USGS) dataset containing building data for over 100 U.S. cities. Apart from the building dataset, the model also uses the USGS national elevation data at 10 m resolution (NED10) for terrain information which covers the 48 contiguous U.S. states, Hawaii, and portions of Alaska. The built-in datasets and fast grid generation tools are useful for operational applications. Examples of the grids produced for urban areas and complex terrain areas are shown in Figure 1.

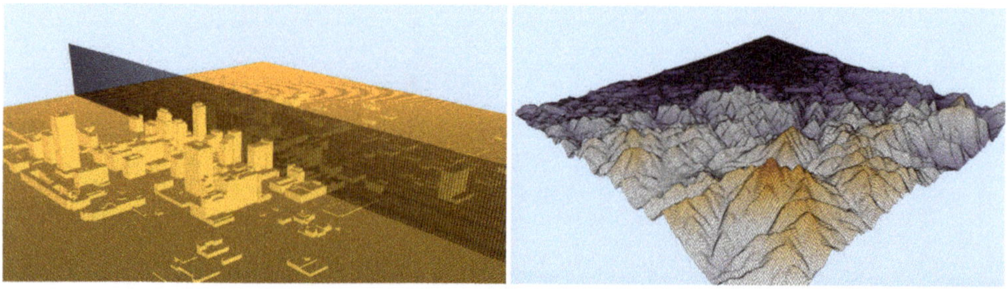

Figure 1. Grids produced by Aeolus modeling system for (**left**) the central business district in Oklahoma City showing the vertical resolution and (**right**) a region with complex terrain.

Apart from the elevation and building databases, the modeling system also integrates data about land-use characteristics, daytime and nighttime population, as well as meteorological fields from operational weather forecast centers and other sources. All the above information provides the model with the required initial and boundary conditions and subsequently helps to reduce the model setup time to minutes. The integrated databases also ensure that relevant products can be produced quickly and be distributed to relevant authorities.

2.5. Inflow Turbulence

Large eddy simulation models often need a precursor simulation to build a turbulent inflow profile. However, this process can be time consuming and difficult to achieve in an operational setup. Following DeLeon and Senocak (2017), we have developed a robust inflow turbulence generator which uses temperature perturbations in cells near the inflow boundary to produce a turbulent profile [20]. The inflow turbulence zone is contained within the five grid cells nearest to the inlet where a mean velocity is prescribed. The buoyancy effect due to the perturbation in the temperature field propagates to the velocity field and produces the requisite turbulent structures.

3. Complex Terrain Validation

The Aeolus model using the large eddy simulation methodology was validated against experimental data for three different applications: flow over complex terrain, flow and dispersion of tracers in urban areas, and cloud rise dynamics for buoyant plumes from high-temperature explosions. This section covers the complex terrain validation, while the following two sections cover the urban area and cloud rise validations.

The Askervein Hill project [14] was a field study conducted in 1982 and 1983 to study the boundary-layer flow over low-profile hills. It was performed under a collaborative effort under the auspices of the International Energy Agency Programme for Research and Development on Wind Energy Conversion Systems. Askervein Hill is a low, isolated elliptic hill on the west coast of the island of South Uist in the Outer Hebrides of Scotland, which peaks at about 116 m above the ground. During these field campaigns, more than 50 towers were deployed and instrumented for wind measurements on and around this low-profile hill, as shown in Figure 2. Towers were placed in two arrays along the major axis of the hill (lines A and AA), in the prevailing wind direction, and one array along the minor axis of the hill (line B). Lines A and AA pass through points hilltop (HT) and center point (CP), respectively, and along an orthogonal line B. A measurement site called reference site (RS) was also placed upstream of the hill to characterize the inflow conditions.

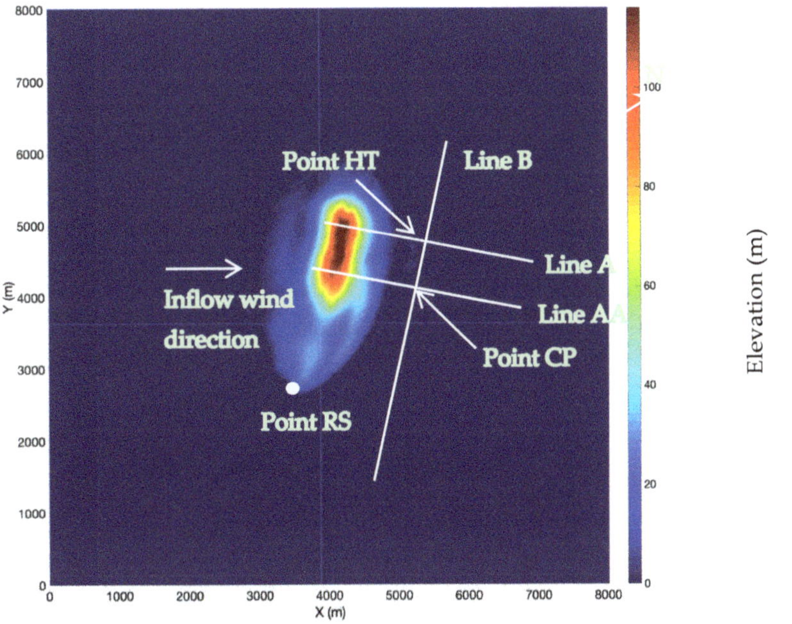

Figure 2. Terrain elevation map of Askervien Hill field study area.

The Aeolus grid was generated by rotating the elevation dataset clockwise by 60 degrees so that the inflow wind direction is orthogonal to the grid as shown in Figure 2, with the vertical extent of 1 km. A uniform Cartesian mesh was created using a grid resolution of Δx, $\Delta y = 20$ m, and $\Delta z = 10$ m resulting in~16 million grid cells. The inflow velocity profile was created using a log-law profile ($u = (u_*/k)\ \ln(z/z_0)$), to fit the observed data from the upstream site RS as shown in Figure 3. The value of surface roughness, z_0, was 0.03 m [14] and the friction velocity u_* was derived using the velocity reading ($u_r = 14$ m/s at $z_r = 60$ m) at the reference site RS.

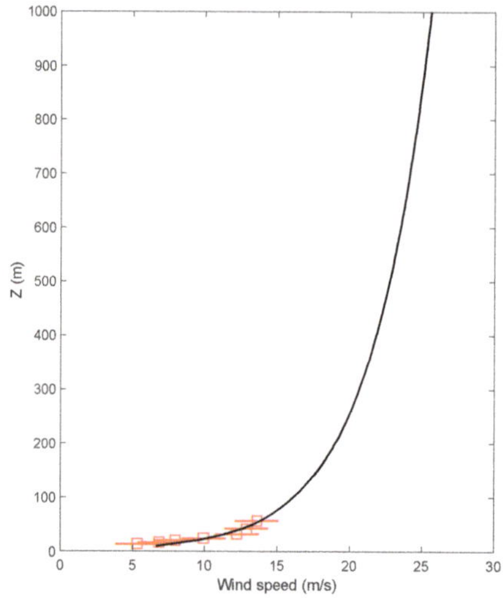

Figure 3. Inflow profile for the Aeolus model (black line) overlaid with field data from site RS (red squares are the time averaged values and the red line represent the variance).

Figure 4 shows the turbulent structures in the velocity magnitude in a vertical slice along lines A and AA. The velocity magnitude increases as it passes over the Askervein Hill top and the flow separates in the lee side of the hill. It can be observed that a larger wake is created behind the plane passing along line A. This larger region of separation occurs due to a steeper drop in elevation along line A and has been observed in other model results, as well as observed data.

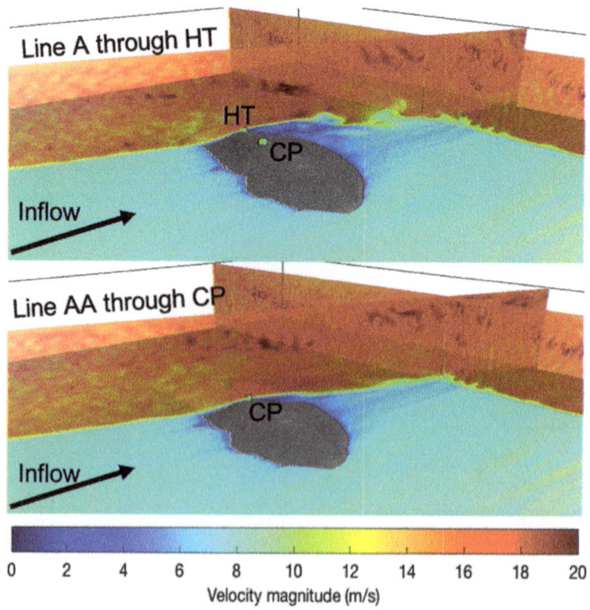

Figure 4. Turbulent structures in the vertical planes passing along line A through point HT (**upper**) and along line AA through point CP (**lower**).

Winds were simulated for 2 h, which took about 6 h of computer time on a quad-core machine. After a 1.5 h spinup period, Aeolus data from the last 30 min of the simulation was time averaged for comparision with the observed data.

Observations along lines A and AA are compared with averaged velocities from the Aeolus simulation. Historically, other model simulations have been compared to the Askervein Hill dataset in terms of fractional speedup ΔS, which is defined as

$$\Delta S = \frac{S(z) - S_{RS}(z)}{S_{RS}(z)} \tag{4}$$

where S is the horizontal wind speed at a specified height above the surface z = 10 m, and S_{RS} is the wind speed at the reference site. The fractional speedup ΔS provides a measure of the influence of the terrain on the wind field based on the upwind undisturbed inflow.

Figure 5 compares the fractional speedup ΔS at 10 m above ground from Aeolus with field data and two other peer-reviewed models, the standard Weather Research and Forecasting (WRF) model and a version of WRF with an immersed boundary method (WRF-IBM) [8]. For results along line A, Aeolus compares reasonably well with the observed data and the other models. It correctly captures the speed-up observed at the top of hill, as well as the separation of the flow on the lee side of hill. Aeolus is also able to predict the slight deceleration in the upwind part of hill reasonably well. Predictions along line AA have been challenging for many models, but here also, Aeolus is able to predict the key features observed in the data.

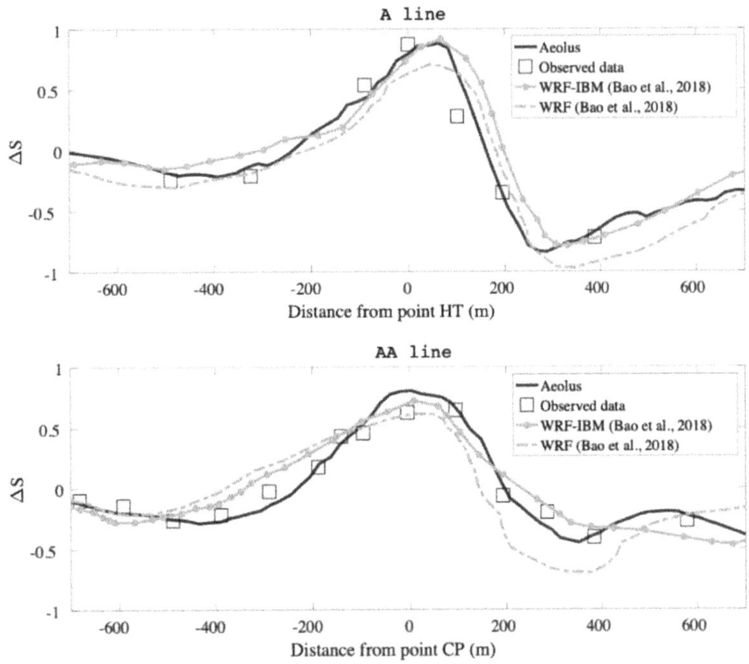

Figure 5. Comparison of fractional speedup predicted by Aeolus along lines A (**top**) and AA (**bottom**) with field campaign data as well as other peer-reviewed models.

This validation study shows us that Aeolus is able to predict key flow features in complex terrain and this makes it an important tool for many applications ranging from wind turbine optimization studies to predicting dispersion patterns in regions with complex terrain. In future work, we plan to validate Aeolus for predicting flow and dispersion pattern in more complicated terrain involving multiple hills and valleys.

4. Urban Area Flow and Dispersion Validation

Aeolus was validated using data from the Joint Urban 2003 field experiment, which was performed in July 2003 in the central business district of Oklahoma City. A large number of meteorological instruments and tracer-gas air samplers were deployed in the urban area. Meteorological measurements were taken at over 160 different locations [15] while tracer measurements were made at over 130 locations [21]. Ten intensive operation periods (IOPs) were conducted for both daytime and nighttime periods, during which most meteorological and gas sampler instruments were activated. During the IOPs, the winds were predominantly from the south. Further details about the experiment, instrument types and locations, and tracer release information can be found in Allwine et al. (2004), Clawson et al. (2005), Flaherty et al. (2007), Nelson et al. (2007), and Brown et al. (2004) [15,21–24].

Aeolus results were compared to field data from a continuous release of sulfur hexafluoride (SF_6) during IOP 8 trial 2. As noted previously, the winds were predominantly from the south for this release. The event was chosen because there was little variation in the inflow wind direction and the edge of the plume was well captured by the gas sampler data. The portable wind detector at the city post office (PWID 15), a propeller anemometer, was used to record the 'wake-free' inflow profile for wind direction and wind speed. It was located about 500 m upstream of the central business district at 50 m above ground on a 35 m rooftop tower, and was free from building effects. A total of 5488 g of SF_6 gas was released continuously for 30 min from the Westin location shown in Figure 6.

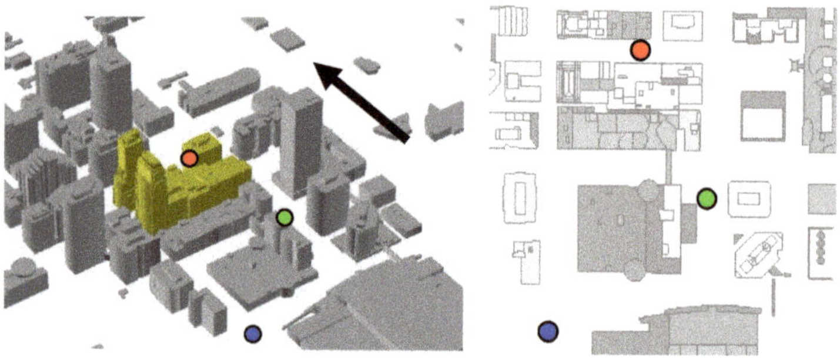

Figure 6. SF_6 release locations in the Oklahoma City central business district during Joint Urban 2003 (• Park Avenue, • Westin, and • Botanical). The northward direction is indicated by the black arrow.

The computational domain is displayed in Figure 1 (left) and was 1.2 km × 1.4 km × 0.21 km in the x, y, and z directions discretized on a regular grid ($\Delta x = \Delta y = 5$ m, $\Delta z = 3$ m). The horizontal grid resolution of 5 m is the minimum grid spacing needed to resolve a typical street canyon. The grid consists of about 4.5 million cells. Time varying input for the simulation was constructed using data from the PWID 15 anemometer. Six log-law profiles using a surface roughness value of $z_0 = 0.1$ m (5 min average) were used in the LES simulation. Figure 7 shows the wind speed and direction measured by the anemometer (dashed lines) and the five minute averaged values used to construct the Aeolus input wind profile (solid line, squares). The averaged log-law profiles were used to create the mean inflow profile while the inflow turbulence gerator perturbed the velcoity field to create physically realistic turbulent features. The source was defined as a sphere of 1 m radius and a release amount of 5488 g was simulated by releasing 0.5 million lagrangian particles over the release duration of 30 min. It was found that 0.5 million particels are sufficient to estimate 30 min averaged concentration at this spatial resolution.

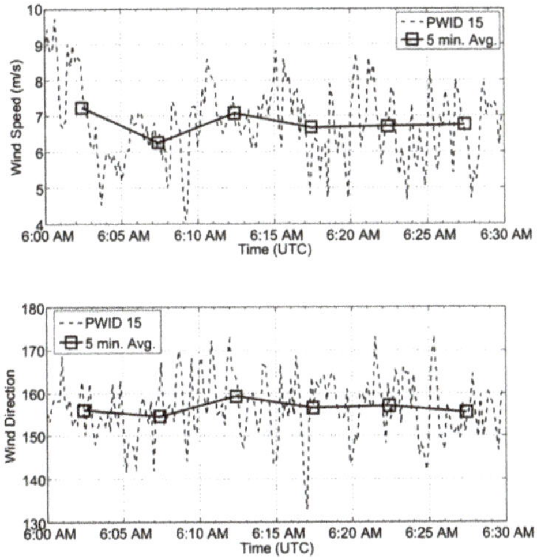

Figure 7. Five minutes of averaged data from PWID 15 were used to build the inflow profiles for the Aeolus LES simulation.

Similarly to the complex terrain case, the flow field was simulated for 1.5 h, with the initial hour used for spinup and the final 30 min used for analysis and comparison with observations. The simulation took about 2 h of computer time to run on a quad-core machine.

Figure 8 shows the velocity vector field for flow around Oklahoma City in the x–y plane at 8 m above ground level (AGL). Simulated wind vectors from the Aeolus model (grey arrows) are overlaid with meteorological observations (black arrows). Longer arrows in the figure indicate higher wind speed. The Aeolus wind speed prediction is also represented by the color shading around the buildings, where warmer colors indicate higher predicted wind speed values. From this figure, it can be observed that the Aeolus model is able to predict the important flow features reasonably well.

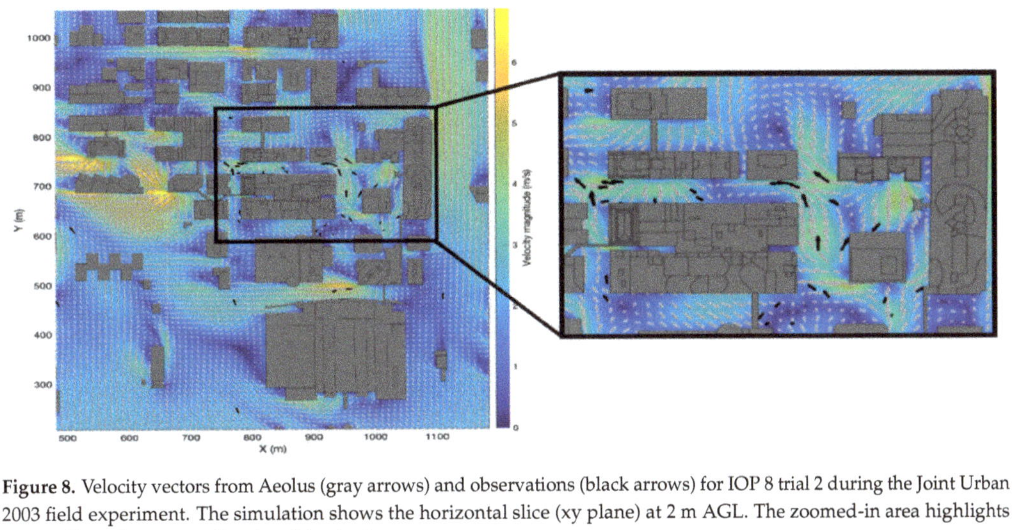

Figure 8. Velocity vectors from Aeolus (gray arrows) and observations (black arrows) for IOP 8 trial 2 during the Joint Urban 2003 field experiment. The simulation shows the horizontal slice (xy plane) at 2 m AGL. The zoomed-in area highlights urban effects that are predicted well by Aeolus.

The model captures the channeling effects along north–south running streets and predicts the high wind speeds measured in these regions. Aeolus also predicts the reverse flow in the street canyons and wake regions in the domain. The model-produced velocities in the intersection areas are in good agreement with the field data.

Figure 9 shows the measured and predicted SF_6 air concentration values at ground level. The colored circles represent the measured air concentrations averaged over the 30 min of the continuous release. The colored contours represent the Aeolus prediction of the 30 min average air concentration, with higher predicted values near the source (red, orange areas). The Aeolus model predictions agree with the experimental results well; the areas of highest concentration and the general downwind plume spreading are captured in the simulation results.

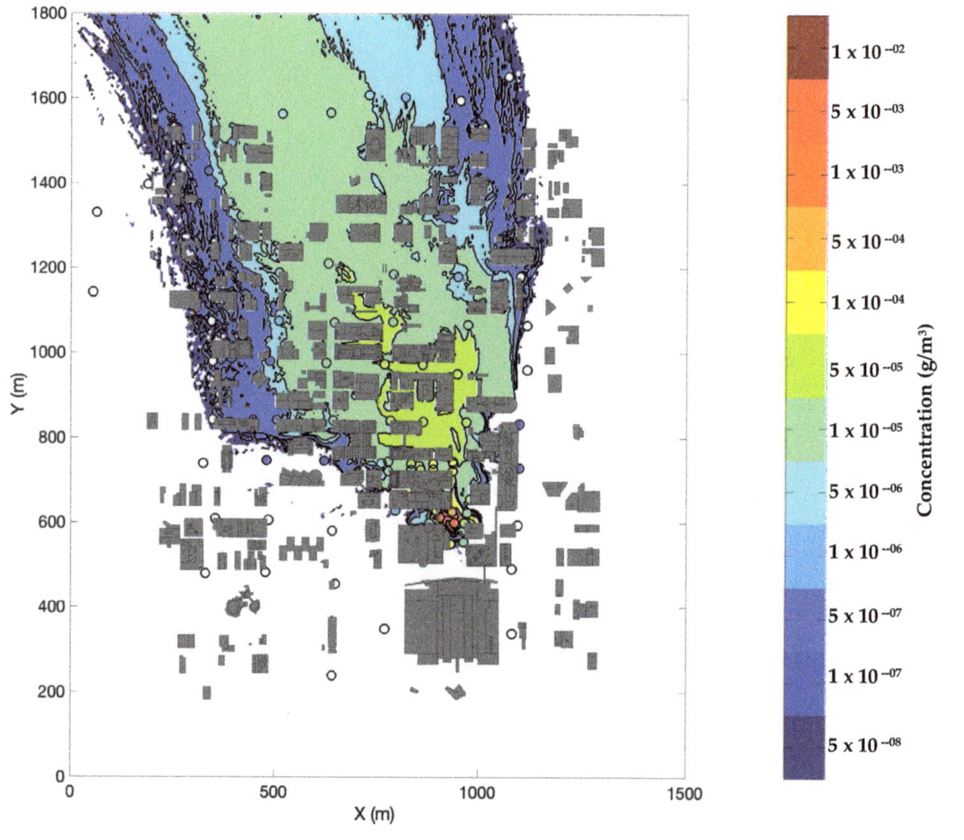

Figure 9. Contours of 30 min averaged SF_6 air concentration (g/m^3) from Aeolus overlaid with 30 min averaged field concentration data (filled circles) for IOP 8 trial 2 during the Joint Urban 2003 field experiment. The simulation shows the horizontal slice (*xy* plane) at 2 m AGL.

Figure 10 displays scatter plots of the paired (point-to-point) values from the Aeolus predictions and the field experiment measurements. Data points (blue circles) that fall on the solid black diagonal represent perfect matching between the predicted and measured values. Points that lie above and below the black line represent values that are over- and under-predicted by Aeolus, respectively, as compared to the measured data. The green, blue, and orange colored diagonal lines represent factors of 2, 5, and 10 model–measurement mismatches, respectively (FAC2, FAC5, and FAC10). The scatter plots show good agreement between predicted and measured values, with most pairs falling within the blue FAC5 lines. Figure 10 also indicates that the number of matched zeros, which show how often the model correctly predicts zero-valued measurements (data below the instrument minimum level of detection, MLOD = $10^{-7.5}$ g/m^3). The number of matched zeros shows that the model is able to correctly predict the spread of the plume. Overall, we found that 48.9%, 84.7%, and 91.5% of the simulated points fall within FAC2, FAC5, and FAC10, respectively, indicating excellent performance for predicting dispersion in complex urban areas which are consistent with the values suggested in Hanna and Chang (2012) [25].

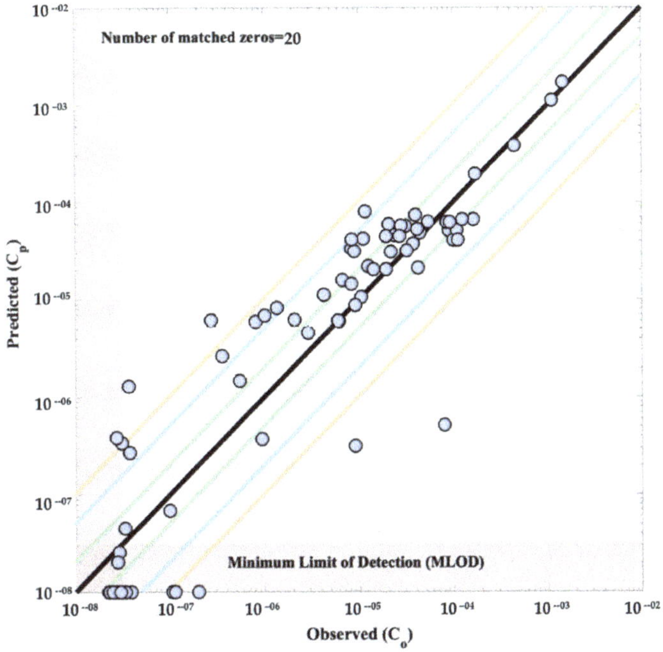

Figure 10. Scatter plot showing points paired in time and space for predicted and observed 30 min averaged SF_6 concentrations (g/m^3) for IOP 8 trial 2 during the Joint Urban 2003 field experiment.

Further quantitative analysis of our results is given in terms of absolute value of fractional bias (|FB|) and the normalized mean square error (NMSE) for concentration. Fractional bias is a normalized value of mean error [26]. |FB| values range from 0 to +2. A perfect agreement between model and measurement would result in FB = 0.

$$\mathrm{FB} = \left(\frac{(C_P^i - C_o^i)}{0.5(C_P^i + C_o^i)} \right) \tag{5}$$

where C_o^i is the ith observation (measurement), and C_P^i is the corresponding model prediction. NMSE captures the overall absolute departure of the modeled results from measurements. Lower values of NMSE indicate better agreement between model and experimental values.

$$\mathrm{NMSE} = \frac{\frac{1}{n}\sum(C_P^i - C_o^i)^2}{\overline{C_o}^2} \tag{6}$$

where n is the number of valid measurement–model data pairs and $\overline{C_o}$ is the mean measurement value. Hanna and Chang (2012) suggest the following limits on the comparison metrics for acceptable performance of an urban model [25]:

- $|FB| \lesssim 0.67$, i.e., the relative mean bias less than a factor of ~2.
- NMSE $\lesssim$ 6, i.e., the random scatter $\lesssim$ 2.4 times the mean.
- FAC2 $\lesssim$ 0.30, i.e., 30% or more of model predicted values are within a factor of two of measured values.

The absolute value of the fractional bias ($|FB|$) was found to be 0.015 and the NMSE for the LES simulation was 0.29, indicating relatively low simulation errors compared to the experimental data. This excellent comparison of the Aeolus model with field measurements in complex urban areas makes it a very useful tool for predicting flow features and dispersion patterns in these scenarios.

5. High-Temperature Nuclear Cloud Rise Dynamics

For the final application, we ran Aeolus to simulate the dynamics of a hot nuclear detonation cloud rising in the troposphere with a specified ambient potential temperature profile θ^a. In this set-up, the source is not only a mass release, but also a large temperature perturbation. Therefore, the initialization requires the temperature, altitude, and size of the fireball formed from the detonation.

The Smagorinsky scheme in Aeolus (see Section 2.1) is useful for simulating turbulence at standard atmospheric conditions, but not for including all the relevant turbulence scales in a buoyant nuclear cloud. Realistically, there is additional mass and energy exchange as ambient air is entrained into the cloud. In Section 5.3, we describe a new parameterization that was added to Aeolus to represent this entrainment process.

5.1. Dixie Event Description

The Dixie test was performed during operation Upshot-Knothole on 6 April, 1953 at 07:30 local time in Nevada (37°5′5″ N, 116°1′5″ W). The device was detonated in the atmosphere at 1.84 km AGL (3.1 km above mean sea level) with an explosive yield of 11 kilotons [27]. These characteristics result in a scaled height of burst of 831 m corresponding to 'regime 1′ in which no soil or dirt is disturbed due to the detonation [28]. During the shot, high-frequency cameras captured the formation and propagation of the shockwave and fireball for comparison with simulations, such as Miranda [29]. Miranda simulates the initial fireball size and temperature, which is shown in Figure 11 for the timestep directly before the Dixie cloud starts rising. Additionally, the time series of the observed top and bottom of the Dixie cloud were recorded in Hawthorne (1979) [30].

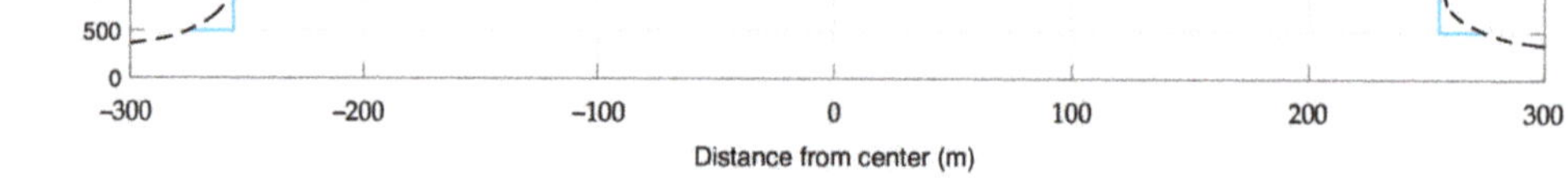

Figure 11. Temperature profile of the initial hot Dixie bubble used in Aeolus (blue solid line) based on a Miranda prediction (black dashed line).

Previous models successfully simulated Dixie cloud rise. Kanarska et al. (2009) compared predictions from a compressible Eulerian model with a low Mach number component [31]. More recently, Arthur et al. (2021) used the Weather Research and Forecasting model to simulate Dixie cloud rise, finding good agreement with observations [32]. However, neither of these prior modeling studies contained detonation gas and debris in the cloud, which is included in the Aeolus simulation.

5.2. Model Setup

In setting up the model grid, there is a tradeoff between having cells that are small enough to resolve the turbulent flow, but not too small because a large domain with many cells is required to simulate cloud rise throughout the troposphere. For this simulation, we selected a model resolution of Δx, Δy, Δz = 30 m in the x-, y-, and z-directions and a domain size of x, y, z = 9000 m, 9000 m, 15,000 m, resulting in 45 million grid cells.

The ambient conditions are specified as vertical profiles of temperature and pressure at the detonation location. These input data are at a resolution of 304.8 m, and Aeolus linearly interpolates the temperature and pressure profiles to its vertical grid (T_k, P_k). The profiles utilized for the Dixie case from radiosonde measurements [30] are shown in Figure 12. The potential temperature of every grid cell $\theta_{i,j,k}$ in Aeolus is set according to the vertical profile:

$$\theta_{i,j,k} = T_k \left(\frac{10^3\ mb}{P_k} \right)^{0.286} \tag{7}$$

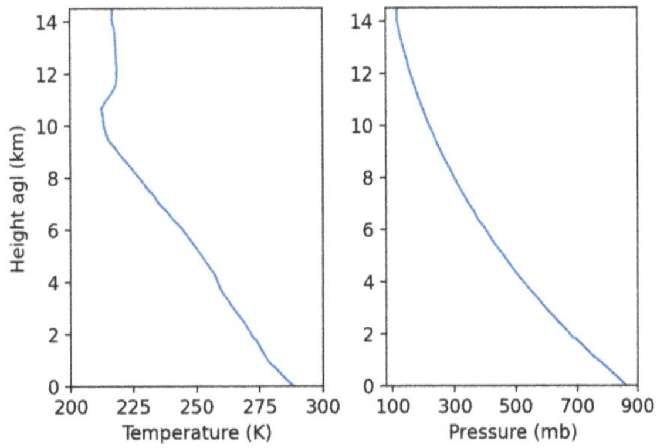

Figure 12. Ambient meteorology utilized to simulate the Dixie test. The ambient temperature and pressure vertical profiles are shown with respect to height above ground level (agl).

The source inputs defining the initial hot bubble include the detonation time t_{det}, bubble diameter D_{src} and temperature T_{src}, the mass M_{src} and number of Lagrangian particles $N_{p,\ src}$ representing materials in the hot bubble, the density ρ_{src} and size $d_{p,src}$ of materials in the hot bubble, and the source position $(x_{src}, y_{src}, z_{src})$. Aeolus replaces the ambient potential tempertaure with the hot bubble potential temperature at the source location based on T_{src} and the grid pressure at t_{det}. Additionally, 1.8 million Lagrangian particles are released at random locations within the bubble volume, representing the hot cloud at t_{det}.

5.3. Entrainment Parameterization

The momentum and energy balances are solved for the velocity and potential temperature fields at each grid cell center. Turbulent viscosity is determined based on the shear rate and Smagorinsky constant C_s, but it is also enhanced by entrainment of ambient air that is not tracked in Aeolus. To account for the induced mixing from entrainment, we add an entrainment term ($E_{i,j,k}$) to eddy viscosity for momentum and potential temperature equation ($v_{i,j,k}$) at the grid cell at index i, j, and k in the x-, y-, and z-direction, respectively.

$$v_{i,j,k} = E_{i,j,k} + \left(C_s \sqrt[3]{\Delta x \Delta y \Delta z} \right)^2 |\overline{S}| \tag{8}$$

where $\overline{S}$ is contraction of the rate-of-strain tensor, $\overline{S}_{ij} = \frac{1}{2}\left(\frac{\partial \overline{u}_i}{\partial x_j} + \frac{\partial \overline{u}_j}{\partial x_i}\right)$.

This enhancement due to entrainment is determined based on the vertical velocity of that cell $w_{i,j,k}$ the potential temperature of that cell $\theta_{i,j,k}$, the ambient potential temperature at that vertical level θ_k^a, and a dimensionless empirical parameter f_{ent} entrainment factor.

$$E_{i,j,k} = max\left\{0,\ f_{ent}\sqrt[3]{\Delta x \Delta y \Delta z}\left|w_{i,j,k}\right|\frac{\theta_{i,j,k} - \theta_k^a}{\theta_k^a}\right\} \tag{9}$$

Using the model inputs and entrainment parameterization described above, we performed seven Aeolus simulations. The model set-up was identical for all seven simulations except that we varied f_{ent} between zero and one. An entrainment factor of zero is equivalent to running a simulation without the entrainment parameterization. The results are shown in Figure 13 where the green profile corresponding to f_{ent} of 0.5 is the closest match to the observations.

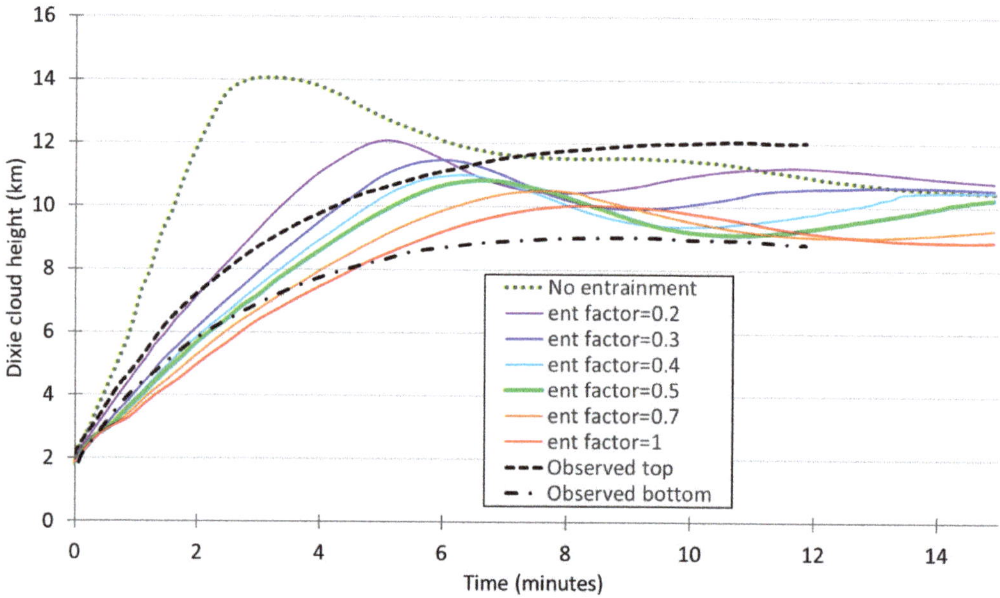

Figure 13. Dixie cloud rise dynamics for different entrainment factors. The black dashed profiles, green dotted profile, and colored solid profiles show the observed cloud top and bottom, the simulated cloud center without entrainment, and the simulated cloud center with varying f_{ent} from 0.2 to 1, respectively.

The entrained ambient air slows the cloud rise velocity, decreasing the maximum cloud height. Additionally, entrainment results in oscillations around the stabilized cloud height since the cloud does not overshoot the tropopause height. The hot cloud rises above its neutral buoyancy height due to its inertia until the inertia cannot sustain the imbalance in density between the cloud and environment. At this point, the cloud is denser than its surroundings and it falls, carried past stabilization by its momentum. The oscillations continue as the cloud height converges to its stabilization height. Theoretically, the frequency of oscillations in a stratified environment can be described with the Brunt–Väisälä frequency N.

$$N^2 = \frac{g}{\theta}\frac{d\theta}{dz} \tag{10}$$

Figure 13 shows the cloud center of mass, which is not the best comparison to the observations of cloud top and bottom. Instead, Figure 14 shows the normalized cloud mass ($f_k(t)$) calculated from the average concentration in the x- and y-directions for each vertical level k ($\overline{C}_k(t)$) where m_k is the number of horizontal grid cells in each vertical level. The mean cloud mass is normalized so the maximum value of $f(t)$ is one, so the average concentration is divided by its maximum value across the vertical levels.

$$\overline{C}_k(t) = \frac{\Sigma_j \Sigma_i C_{i,j,k}(t)}{m_k} \quad (11)$$

$$f_k(t) = \frac{\overline{C}_k(t)}{max\overline{C}(t)} \quad (12)$$

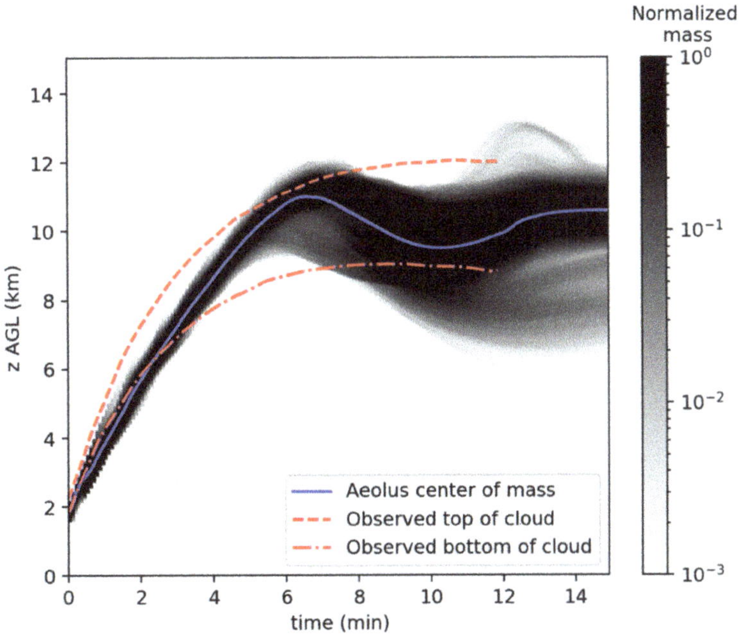

Figure 14. Simulated cloud rise with entrainment factor of 0.5. The average concentration of cloud gas normalized by the maximum value at each time is shown in the shaded grey, the center of mass of simulated cloud is shown in solid blue profile, and the red dashed and dash-dotted profiles show the observed cloud top and bottom.

Similar to simulations by Arthur et al. (2021), the cloud height is underpredicted in the first 2 min post detonation [32]. At later times, the majority of the cloud mass is contained between the observed cloud top and bottom, as shown in Figure 14.

Figure 15 shows the evolution of the cloud temperature and gas and debris concentrations at 1.0, 3.5, 6.5, and 11.5 min after detonation. The potential temperature is shown across the y- and z-grid cells at x = 4.5 km, which corresponds to the cloud center. The gas and debris concentrations are averaged over the x-direction and normalized by their respective source mass to determine their dilution ratios shown in Figure 15. The cloud vertical extent, shown in the grey shaded area, is defined as the altitudes $z^{cloud}(t)$ in which the concentration is more than 10% of the current maximum concentration.

$$z^{cloud}(t) \in \left\{ z_k \; s.t. \; \overline{C}_{j,k}(t) > 0.1 \; max \, \overline{C}_{j,k}(t) \right\} \quad (13)$$

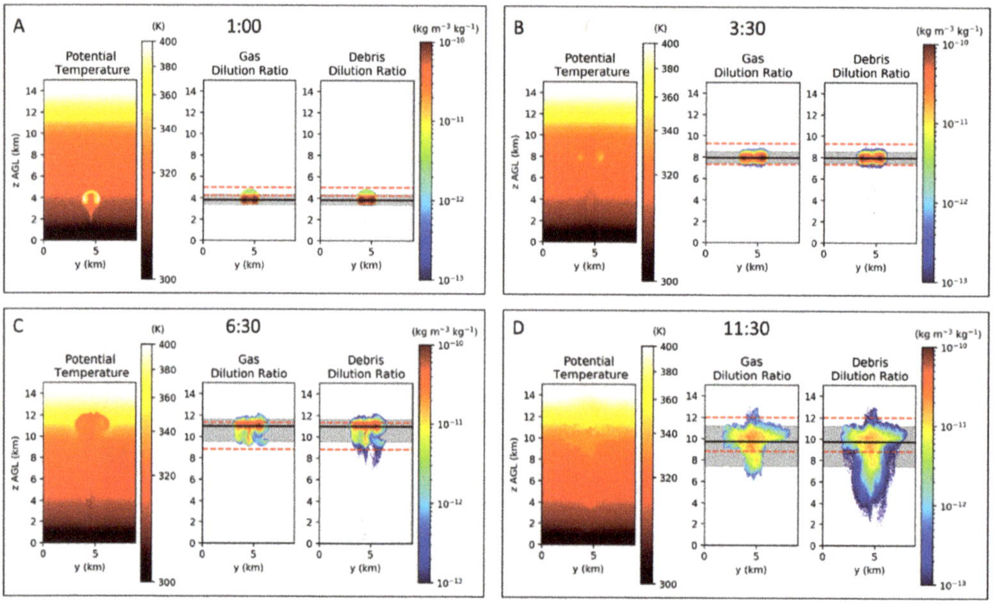

Figure 15. Cloud position and extent at several times using f_{ent} of 0.5. The potential temperature at the cloud center, the average gas dilution ratio, and the average debris dilution ratio are shown at 1.0, 3.5, 6.5, and 11.5 min after detonation in panels (**A–D**), respectively. In the 2D concentration figures, the shaded grey area shows the "cloud vertical extent" described in the text, the black solid line shows the cloud center of mass, and the red dashed lines show the observed cloud top and bottom height.

The cloud center is defined as the cloud center of mass within the cloud extent, which is shown as the black solid lines in Figure 15.

$$z_{center}(t) = \frac{\Sigma_i \Sigma_j \Sigma_k C_{i,j,k}(t)\ z_k^{cloud}(t)}{\Sigma_i \Sigma_j \Sigma_k C_{i,j,k}(t)} \tag{14}$$

The observed cloud top and bottom is also shown in Figure 15 as the red dashed lines. The simulated cloud rise matches observations well on average, with an underestimate of the observed height initially and a slight overestimate at the maximum height near 6.5 min post-detonation. Additionally, even though the simulated cloud bottom extends lower than the observed after about 8 min (as shown in Figure 14), most of the cloud below the observed bottom could be considered the stem instead of the cap.

6. Conclusions

Aeolus is a fast-running computational fluid dynamics urban dispersion model that has been validated using several experimental data sets. Using the Aeolus model, complex dispersal experiments can be completed with simulation run times small enough for use in emergency response, to provide consequence management information. The model is coupled to all the relevant databases required to setup and run the model and produce products which are useful for first responders.

In this work, we have simulated flow and dispersion using the large eddy simulation version of the Aeolus model in three different regimes—complex terrain, urban domain, and high-temperature cloud rising into high altitudes. This showcases the flexibility and adaptability of the model in different scenarios.

Comparing Aeolus predictions to field experiments, the model generally shows good agreement with the measured data. This report details model validation to the Askervein hill field campaign conducted in 1982 and 1983, the Joint Urban field experiments conducted in 2003 for both continuous and instantaneous tracer gas releases, and explosive cloud rise data from the Dixie nuclear test conducted at the Nevada Test Site in 1953. Aeolus results compare well with measured data both qualitatively and quantitatively and were found to compare well with the data.

Expanding the capabilities of a fast-running urban dispersion model and validating its simulation results against field data greatly advances NARAC's ability to make predictions of the fate of material released in an urban environment and complex terrain. The improved and validated Aeolus model represents a significant capability for NARAC, and improved support to the USA's Department of Energy.

In future, we plan to validate the LES model for additional complex terrain regions and other real and mock urban areas. We also plan on validating the model for different release types, including buoyant and dense gas releases in urban areas. Given the simplicity of the model to adapt to complex grids, we intend to extend the model for modeling flow and dispersion pattern in indoor environments. To further increase the efficiency of the large eddy simulation capability of the Aeolus model, we plan to implement the code on a Graphics Processing Unit (GPU) platform which will truly help in operationalizing the model.

In addition, we plan to validate the entrainment parameterization in nuclear cloud rise simulations using other test shots.

Author Contributions: A.A.G., D.L.M. and L.G.G. conceived of the presented ideas. A.A.G. is the main developer of the model. D.D.L., D.L.M. and S.J.N. helped develop and validate parts of the model. D.D.L., D.L.M. and O.A. performed the computations. D.L.M. and L.G.G. encouraged and supervised the findings of this work. All authors discussed the results and contributed to the final manuscript. All authors have read and agreed to the published version of the manuscript.

Funding: This work was performed under the auspices of the U.S. Department of Energy by Lawrence Livermore National Laboratory under Contract DE-AC52-07NA27344. This research was also partially supported by LLNL Strategic Initiative project number 20-SI-006.

Institutional Review Board Statement: This document was reviewed and released by Lawrence Livermore National Laboratory with a release number of LLNL-JRNL-824266.

Informed Consent Statement: Not applicable.

Data Availability Statement: Processed Aeolus model data used in support of the analyses in this paper will be available at ftp://gdo148.ucllnl.org/pub/aeolus-les (accessed on 1 October 2020).

Acknowledgments: We acknowledge Andy Cook and Mindy Cook for their contribution of Miranda simulation results used to initialize the Dixie cloud rise simulations.

Conflicts of Interest: The authors declare no conflict of interest.

References

1. Britter, R.E.; Hanna, S.R. Flow and dispersion in urban areas. *Ann. Rev. Fluid Mech.* **2003**, *35*, 469–496. [CrossRef]
2. Belcher, S.E. Mixing and transport in urban areas. *Phil. Trans. R. Soc. A* **2005**, *363*, 2947–2968. [CrossRef] [PubMed]
3. Theurer, W.; Baechlin, W.; Plate, E. Model study of the development of boundary layers above urban areas. *J. Wind. Eng. Ind. Aerodyn.* **1992**, *41*, 437–448. [CrossRef]
4. Hall, D.J.; Sharples, H.; Walker, S.; Kukadia, V. *Attribution of Pollutant Concentrations on Buildings from Local Traffic—Effects on Ventilation Requirements*; BRE Client Report; 2003; p. 81461. Available online: https://www.aivc.org/resource/attribution-pollutant-concentrations-buildings-local-traffic-effects-ventilation (accessed on 2 April 2021).
5. Röckle, R. Bestimmung der Stomungsverhaltnisse im Bereich Komplexer Bebauugsstrukturen. Ph.D. Thesis, Vom Fachbereich Mechanik, der Technischen Hochschule Darmstadt, Hessen, Germany, 1990.
6. Gowardhan, A.A.; Brown, M.J. *A Study of the Effects of Different Urban Wind Models on Dispersion Patterns Using Joint Urban 2003 Data*; No. LLNL-CONF-531414; Lawrence Livermore National Lab. (LLNL): Livermore, CA, USA, 2012.
7. Wiersema, D.J.; Lundquist, K.A.; Chow, F.K. Mesoscale to microscale simulations over complex terrain with the immersed boundary method in the Weather Research and Forecasting Model. *Mon. Weather. Rev.* **2020**, *148*, 577–595. [CrossRef]
8. Bao, J.; Chow, F.K.; Lundquist, K.A. Large-eddy simulation over complex terrain using an improved immersed boundary method in the Weather Research and Forecasting Model. *Mon. Weather. Rev.* **2018**, *146*, 2781–2797. [CrossRef]

9. Gowardhan, A.A.; Pardyjak, E.R.; Senocak, I.; Brown, M.J. A CFD-based wind solver for an urban fast response transport and dispersion model. *Environ. Fluid Mech.* **2011**, *11*, 439–464. [CrossRef]
10. Neophytou, M.; Gowardhan, A.; Brown, M. An inter-comparison of three urban wind models using Oklahoma City Joint Urban 2003 wind field measurements. *J. Wind. Eng. Ind. Aerodyn.* **2011**, *99*, 357–368. [CrossRef]
11. Gowardhan, A.A. Towards Understanding Complex Flow Phenomenon in Urban Areas Using Numerical Tools. Ph.D. Thesis, Department of Mechanical Engineering, University of Utah, Salt Lake City, UT, USA, 2008.
12. Coirier, W.J.; Kim, S. Summary of CFD urban results in support of the Madison square garden and urban dispersion program field tests. In Proceedings of the 6th Symposium on the Urban Environment, American Meteorological Society, Atlanta, GA, USA, 27 January–3 February 2006.
13. Coirier, W.J.; Kim, S.; Chen, F.; Tuwari, M. Evaluation of urban scale contaminant transport and dispersion modeling using loosely coupled CFD and mesoscale models. In Proceedings of the 6th Symposium on the Urban Environment, American Meteorological Society, Atlanta, GA, USA, 27 January–3 February 2006.
14. Taylor, P.A.; Teunissen, H.W. The Askervein Hill project: Overview and background data. *Bound.-Layer Meteorol.* **1987**, *39*, 15–39. [CrossRef]
15. Allwine, K.J.; Leach, M.J.; Stockham, L.W.; Shinn, J.S.; Hosker, R.P.; Bowers, J.F.; Pace, J.C. Overview of Joint Urban 2003—An atmospheric dispersion study in Oklahoma City. In Proceedings of the Symposium on Planning, Nowcasting and Forecasting in the Urban Zone, AMS, Seattle, WA, USA, 10–15 January 2004.
16. Nasstrom, J.S.; Sugiyama, G.; Baskett, R.L.; Larsen, S.C.; Bradley, M.M. The NARAC modeling and decision support system for radiological and nuclear emergency preparedness and response. *Int. J. Emerg. Manag.* **2007**, *4*, 524–550. [CrossRef]
17. Sugiyama, G.; Nasstrom, J.; Baskett, R.; Simpson, M. National atmospheric release advisory center (NARAC) capabilities for homeland security. In Proceedings of the 5th International Symposium on Computational Wind Engineering, Chapel Hill, NC, USA, 23–27 May 2010; pp. 23–27.
18. Leonard, B.P. A stable and accurate convective modelling procedure based on quadratic upstream interpolation. *Comput. Methods Appl. Mech. Eng.* **1979**, *19*, 59–98. [CrossRef]
19. Durbin, P.A. *NASA Reference Publication*; NASA: Cleveland, OH, USA, 1983.
20. DeLeon, R.; Senocak, I. Turbulent Inflow Generation for Large-eddy Simulation of Incompressible Flows Through Buoyancy Perturbations. In Proceedings of the 23rd AIAA Computational Fluid Dynamics Conference, Denver, CO, USA, 5–9 June 2017; p. 3294.
21. Clawson, K.L.; Carter, R.G.; Lacroix, D.J.; Biltoft, C.A.; Hukari, N.F.; Johnson, R.C.; Rich, J.D. Joint urban 2003 (JU03) SF6 atmospheric tracer field tests. In *NOAA Technical Memorandum OARARL-254*; Air Resources Lab: Richland, WA, USA, 2005.
22. Flaherty, J.; Allwine, K.J.; Allwine, E. Vertical tracer concentration profiles measured during the joint urban 2003 dispersion study. *J. Appl. Meteorol. Clim.* **2007**, *46*, 2019–2037. [CrossRef]
23. Nelson, M.A.; Pardyjak, E.R.; Klewicki, J.C.; Pol, S.U.; Brown, M.J. Properties of the wind field within the Oklahoma City Park Avenue Street canyon. Part I: Mean flow and turbulence statistics. *J. Appl. Meteorol.* **2007**, *46*, 2038–2054. [CrossRef]
24. Brown, M.J.; Boswell, D.; Streit, G.; Nelson, M.; McPherson, T.; Hilton, T.; Pardyjak, E.R.; Pol, S.; Ramamurthy, P.; Hansen, B.; et al. Joint urban 2003 street canyon experiment. In Proceedings of the Symposium on Planning, Nowcasting and Forecasting in the Urban Zone, AMS, Seattle, WA, USA, 10–15 January 2004.
25. Hanna, S.; Chang, J. Acceptance criteria for urban dispersion model evaluation. *Meteorol. Atmos. Phys.* **2012**, *116*, 133–146. [CrossRef]
26. Warner, S.; Platt, N.; Heagy, J.F.; Jordan, J.E.; Bieberbach, G. Comparisons of transport and dispersion model predictions of the mock urban setting test field experiment. *J. Appl. Meteorol.* **2006**, *45*, 1414–1428. [CrossRef]
27. United State, Department of Energy, Nevada Operations Office. *United States Nuclear Tests, July 1945 through September 1992*; No. DOE/NV-209-Rev16; USDOE National Nuclear Security Administration Nevada Field Office: Oak Ridge, TN, USA, 2015.
28. Spriggs, G.; Neuscamman, S.; Nasstrom, J.S.; Knight, K.B. Fallout Cloud Regimes. *Count. WMD J.* **2020**, *21*, 103–113.
29. Cook, A.W. Artificial fluid properties for large-eddy simulation of compressible turbulent mixing. *Phys. Fluids* **2007**, *19*, 055103. [CrossRef]
30. Hawthorne, H.A. *Compilation of Local Fallout Data from Test Detonations 1945–1962 Extracted from DASA 1251. Volume I. Continental US Tests*; General Electric Company: Santa Barbara, CA, USA, 1979.
31. Kanarska, Y.; Lomov, I.; Glenn, L.; Antoun, T. Numerical simulation of cloud rise phenomena associated with nuclear bursts. *Ann. Nucl. Energy* **2009**, *36*, 1475–1483. [CrossRef]
32. Arthur, R.S.; Lundquist, K.A.; Mirocha, J.D.; Neuscamman, S.; Kanarska, Y.; Nasstrom, J.S. Simulating nuclear cloud rise within a realistic atmosphere using the Weather Research and Forecasting model. *Atmos. Environ.* **2021**, *254*, 118363. [CrossRef]

atmosphere

Article

A Graphics Processing Unit (GPU) Approach to Large Eddy Simulation (LES) for Transport and Contaminant Dispersion

Paul E. Bieringer [1,*], Aaron J. Piña [1,2], David M. Lorenzetti [3], Harmen J. J. Jonker [4], Michael D. Sohn [3], Andrew J. Annunzio [1] and Richard N. Fry, Jr. [5]

1 Aeris, 1849 Cherry St., Suite 6, Louisville, CO 80027, USA; aaron.pina@nasa.gov (A.J.P.); aannunzio@aerisllc.com (A.J.A.)
2 National Aeronautics and Space Administration, 300 E St SW, Washington DC 20546-0002, USA
3 Lawrence Berkeley National Laboratory, One Cyclotron Road, Berkeley, CA 94720, USA; dmlorenzetti@lbl.gov (D.M.L.); mdsohn@lbl.gov (M.D.S.)
4 Whiffle, Molengraaffsingel 12, 2629 JD Delft, The Netherlands; harmen.jonker@whiffle.nl
5 Defense Threat Reduction Agency 8725 John J Kingman Rd, Fort Belvoir, VA 22060, USA; richard.n.fry.civ@mail.mil
* Correspondence: paulb@aerisllc.com

Abstract: Recent advances in the development of large eddy simulation (LES) atmospheric models with corresponding atmospheric transport and dispersion (AT&D) modeling capabilities have made it possible to simulate short, time-averaged, single realizations of pollutant dispersion at the spatial and temporal resolution necessary for common atmospheric dispersion needs, such as designing air sampling networks, assessing pollutant sensor system performance, and characterizing the impact of airborne materials on human health. The high computational burden required to form an ensemble of single-realization dispersion solutions using an LES and coupled AT&D model has, until recently, limited its use to a few proof-of-concept studies. An example of an LES model that can meet the temporal and spatial resolution and computational requirements of these applications is the joint outdoor-indoor urban large eddy simulation (JOULES). A key enabling element within JOULES is the computationally efficient graphics processing unit (GPU)-based LES, which is on the order of 150 times faster than if the LES contaminant dispersion simulations were executed on a central processing unit (CPU) computing platform. JOULES is capable of resolving the turbulence components at a suitable scale for both open terrain and urban landscapes, e.g., owing to varying environmental conditions and a diverse building topology. In this paper, we describe the JOULES modeling system, prior efforts to validate the accuracy of its meteorological simulations, and current results from an evaluation that uses ensembles of dispersion solutions for unstable, neutral, and stable static stability conditions in an open terrain environment.

Keywords: large eddy simulation; graphics processing unit computing; atmospheric dispersion modelling; microscale dispersion; model validation

Citation: Bieringer, P.E.; Piña, A.J.; Lorenzetti, D.M.; Jonker, H.J.J.; Sohn, M.D.; Annunzio, A.J.; Fry, R.N., Jr. A Graphics Processing Unit (GPU) Approach to Large Eddy Simulation (LES) for Transport and Contaminant Dispersion. *Atmosphere* **2021**, *12*, 890. https://doi.org/10.3390/atmos12070890

Academic Editor: Patrick Armand

Received: 18 June 2021
Accepted: 1 July 2021
Published: 8 July 2021

1. Introduction

The methods used to simulate outdoor dispersion of airborne materials range from simple, computationally efficient empirical approaches to complex computational fluid dynamics (CFD)-based approaches. The simplest methods use empirical formulations that utilize information on the meteorological conditions to control the corresponding dispersion behavior produced by the model. More complex CFD methods, on the other hand, are capable of resolving (explicitly or implicitly) the time-varying wind, turbulence, and dispersion patterns that drive the downwind transport and dispersion of the material. This enables the development of dispersion simulations that can reconstruct the detailed structures in the contaminant dispersion that qualitatively resembles the "single-realization" visual depictions of smoke dispersion observed in photographs. The challenge inherent with creating "single-realizations" of dispersion with a CFD model is that atmospheric measurements are typically not sufficient to adequately initialize these microscale atmospheric simulations (e.g., at spatial resolutions in the 10s of meters and timescales on the order of 1 s). To produce ensembles of "single-realization" dispersion solutions, the CFD model is initialized with the mean atmospheric conditions that can be measured and then an ensemble of uncorrelated dispersion solutions can be created by moving the release (e.g., in location and/or time) within the turbulent flow. This approach provides a distribution of dispersion solutions that can then be averaged (in time and space) to determine the mean properties of downwind dispersion analogous to the products produced by standard Gaussian plume and puff models. Alternatively, this distribution of dispersion solutions can be sampled/analyzed to understand the variance from the mean and the skewness of the distribution if present [1]. While this approach can be used to provide a wealth of information on the properties of dispersion for a given scenario, the production of ensembles of single-realization dispersion solutions is computationally expensive when compared to traditional Gaussian plume/puff and Lagrangian particle modeling techniques. The computational expense associated with producing these ensembles has, until recently, limited the use of CFD models for dispersion to research and academic applications.

Here, we introduce a large eddy simulation (LES) CFD model that has been implemented to run on a GPU computing platform and discuss the computational performance advantages provided by this GPU-LES approach. The accuracy of the dispersion solutions is critical for providing confidence in the use of this emerging technology. In this paper, we provide an illustration of an approach for validating dispersion simulations at these spatial and temporal scales. Our validation discussion includes a description of the observational data sets used to evaluate the model and the methodology used to simulate these experiments and to compare predictions to the observations. Detailed results across environmental conditions ranging from unstable to stable planetary boundary layer (PBL) conditions (i.e., daytime to nighttime) are provided. We conclude with a summary of the findings, and a brief description of the plans to extend this capability to support modeling urban environments and building interiors.

1.1. Ensemble-Average and Single-Realization Dispersion Solutions

A common simulation method for estimating the dispersion of airborne contaminants are ensemble-average approaches. These are largely empirical approaches in which the model is designed to represent the apparent stochasticity of atmospheric turbulence and its corresponding impact on downwind dispersion. Here, empirical parameters are used to describe dispersion that might have occurred over many atmospheric and dispersion conditions. When formulated for a Eulerian reference frame, these empirical parameters are used to describe airborne material dispersion relative to the mean wind direction where the rate of dispersion changes as a function of downwind distance for a given atmospheric condition. When formulated for a Lagrangian reference frame, airborne material dispersion in these tools is computed from the perspective of a hypothetical air parcel following the mean winds. The empirical parameters used to control the material spread following the flow are typically represented by a series of Gaussian puffs or individual particles. For the Lagrangian puff models, an assumption is made to represent localized concentration patterns using a Gaussian distribution, where the empirical parameters control the localized crosswind spread of the material as a function of downwind distance and atmospheric conditions [2,3]. In Lagrangian particle models, the empirical parameters are typically used to control the magnitude of particle spread within a random-walk method that redistributes the particles at each model time step [4,5]. In both approaches, there is an implicit assumption that the material concentration at each point in space and time is an ensemble average over a large number of realizations of this apparent stochastic process.

Individual realizations of dispersion can differ significantly from the ensemble-average solution, and it is typically not possible to recover specific details of a given dispersion realization from the ensemble-average statistics [6]. While ensemble-average solutions can in general provide reasonable model assessments, they have been shown to provide incorrect and misleading results, particularly for scenarios involving high-frequency data sampling, strong spatial and temporal correlations between variables, and nonlinear effects [1,7]. For example, the consequences of exposure to airborne contaminants may depend nonlinearly on the concentration, so that brief exposure to a high concentration can have greater impact than longer exposure to the same time-averaged concentration. CFD models can be used to create ensembles of what we refer to as "single-realization" dispersion solutions that use a very short time average relative to the eddy turn-over time of the eddies in the flow field. Individual uncorrelated dispersion realization ensemble members can be created in a couple of ways. One approach is to create multiple independent simulations where the model is started from slightly different initial conditions. One way to do this is by modifying the random number seed on the initial heat flux distribution that initiates the turbulence within the model. A second approach is to create uncorrelated releases by spacing a set of releases within a simulated boundary layer with spatially homogenous conditions (e.g., roughness length, heat flux, etc.). A third way this can be done is to create realizations by creating releases within a spatially and homogeneous environment at different times. The use of this approach with a CFD and dispersion model has been demonstrated to produce an ensemble of dispersion solutions that when averaged can closely replicate the dispersion solution from a Gaussian puff model that calculated dispersion based on empirical relations designed to replicate an ensemble averaged dispersion. A more detailed discussion of this can be found in Bieringer et al. [1].

The large eddy simulation (LES) modeling approach has demonstrated the ability to successfully predict unique simulations of many types of atmospheric scenarios relevant for airborne dispersion and defense analyses described above [1,6,8,9]. LES models have been developed to explicitly resolve the motions associated with the largest eddies in the atmospheric boundary layer and use parameterizations or closures to represent the small-scale turbulent eddies. The large eddies are resolved or separated from the small, unresolved eddies by filtering the governing Navier–Stokes equations in the inertial subrange, whereby the eddies smaller than the filter width are treated by a subgrid model and the eddies larger than the filter are explicitly resolved [10]. The filter size used is dependent upon the type of atmospheric conditions being modeled and the corresponding turbulence length scales involved. Air-flow solutions from LES models can be produced at spatial grid increments approaching 1 m and be solved at very short time-steps that when used to drive dispersion models (in both Lagrangian and Eulerian reference frames), can produce single-realization dispersion patterns [1,6,11–14]. When combined with an immersed boundary method (IBM) surface layer parameterization, necessary to keep the model numerically stable at the walls of the buildings, LES models have also been demonstrated to be capable of simulating the detailed winds and dispersion that occur in urban environments [15,16]. When compared to observational data, the results indicate that an LES approach has been shown to accurately represent both the mean properties of dispersion and the variance present in the observations from open terrain convective boundary layer field trials [11,12,17].

1.2. GPU-Enabled Atmospheric Computing

General purpose computing on graphics processing unit (GPGPU) hardware, hereafter referred to as a GPU, has emerged as a high-performance computing (HPC) option for a variety of applications, including scientific computing. GPUs typically have many times the number of computational cores relative to central processing unit (CPU) computers and are purpose built to solve problems in graphics shaders for the calculation of the levels for light, darkness, and color for the rendering of graphics on computer screens. Scientific computing with this technology began in the early 2000s with a demonstration using a matrix multiplication application that was shown to run considerably faster on a GPU than on a CPU [18]. Since then, programming languages and standards, such as OpenCL and Nvidia's CUDA, provide an application programming interface that enables the use of C and C++ software to be executed on GPUs. Modern GPUs, such as the Nvidia A100, have even more densely packed computational cores (nearly 7000), and higher memory bandwidth (2 Terabytes per second) [19], which collectively enable faster computations and reduced data latency (reduced time spent waiting for data load/store processes to complete). Collectively, these characteristics enable the GPUs to significantly outperform CPUs for calculations in data-parallel applications, which require the same instruction (or calculation) to be made concurrently. Many scientific calculations, including CFD and LES calculations, fall into this data parallel paradigm.

The atmospheric science modeling community recognized the computational potential provided by the GPU in the mid to late 2000s. Early examples where the use of a GPU for an atmospheric modeling application explored the porting of computationally expensive elements of the Weather Research and Forecast (WRF) model [20] to run on a GPU. This work, commonly referred to as GPU acceleration, includes work by Michalakes and Vachharajani [21], Mielikainen et al. [22], Silva et al. [23], and Wahib and Maruyama [24]. An alternative approach that has been more recently taken by atmospheric modelers is to port the entire atmospheric model to run resident on the GPU. Examples of models that take this approach include the GPU Resident Atmospheric Simulation Program (GRASP) (previously referred to as the GPU-resident Atmospheric Large-Eddy Simulation (GALES) model) by Schalkwijk et al., 2012 [25]; Schalkwijk et al., 2015 [26]; Schalkwijk et al., 2016 [27]; the Parallelized Large Eddy Simulation Model (PALM) by Maronga et al., 2015 [28]; the MicroHH model by van Heerwaarden et al., 2017 [29] and the FastEddy model (Sauer, J. A., & Muñoz-Esparza, D. (2020) [30]. Using the CPU to handle data input and output (I/O) and moving all of the core atmospheric calculations to the GPU has been shown to provide an increase in the calculation speed of an atmospheric simulation by more than an order of magnitude over comparable calculations on CPU hardware [29].

To characterize these benefits, our team conducted benchmark LES simulations using the Weather Research and Forecast (WRF) model with LES turbulence closure [31,32] and compared the computational performance to the GRASP model. The WRF-LES and GRASP simulations used a 128 $\times$ 128 $\times$ 64 (X, Y, Z) grid with a spatial resolution of 20 m $\times$ 20 m $\times$ 17 m. The simulation used a periodic lateral boundary condition to spin up convective eddies and turbulence over a 1-h period. The WRF-LES simulation was performed on a Dell R640 running Red Hat v7.6 Linux on an Intel Xenon E5 v4, 8-core CPU, and was configured to use the distributed memory, Message Passing Interface (MPI), option. On this hardware, a 1-h WRF-LES simulation required approximately 1 h and 32 min of wall clock time. A comparable GRASP simulation was performed on an NVIDIA Tesla K40 with 2880 cores operating at 745 MHz and 12 GB of onboard fast access memory. The GRASP simulation completed in 36 s of wall clock time. This represents a GPU-LES simulation that is approximately 150 times faster than the comparable CPU-LES simulation.

The GPU-LES has substantially lower equipment costs, power consumption, cooling requirements, and physical space requirements than a CPU platform that can provide comparable simulations. Based on the benchmark described above, and assuming linear performance scaling, we estimate that a cluster of 19 Dell R640 Linux servers and a high-speed network switch would be required to match the performance of a single NVIDIA Tesla K40. Evidence on the hardware performance of the GPU vs. CPU suggests that this type of computational performance difference is a sustained characteristic of these platforms. Figure 1 shows the theoretical computational capacity and memory bandwidth benchmarks for a series of NVIDIA GPUs and Intel CPUs over the past decade [33,34]. This figure illustrates how the GPU continues to maintain a significant advantage over its Intel CPU counterpart, as measured by floating point operations per clock cycle and memory bandwidth speed, and that this advantage has continued to grow as new architecture designs are released. For reference, the NVIDIA K40 used for the performance benchmark cited above is a 2014 graphics card based on the NVIDIA Kepler architecture.

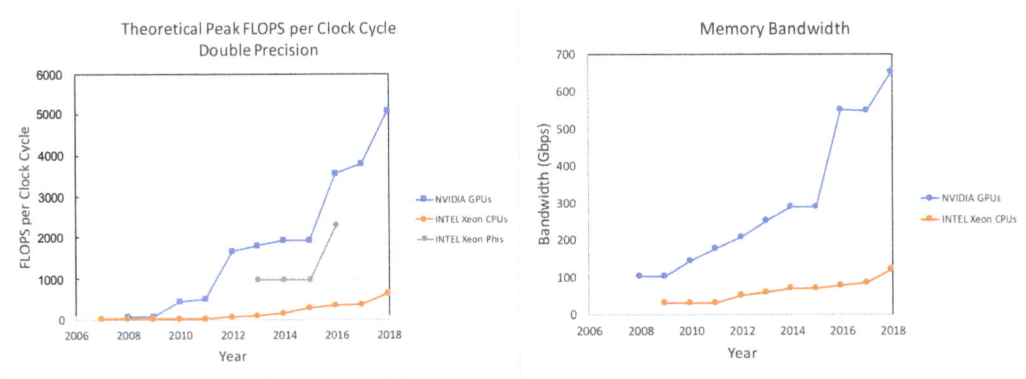

Figure 1. Computational benchmarks for NVIDIA GPU vs. Intel CPU hardware platforms. The left panel illustrates theoretical floating-point operations (FLOPS) per clock cycle and the panel on the right illustrates theoretical memory bandwidth in gigabits per second (Gbps).

1.3. *Atmospheric Dispersion Modeling on a GPU-LES Model*

To form an ensemble of single-realization dispersion solutions using a LES and coupled atmospheric transport and dispersion (AT&D) model comes at a high computational expense. In spite of the computational burden that has historically been associated with LES AT&D, it has been used in a variety of studies ranging from characterizing convective boundary layers [13], the development of more accurate Lagrangian dispersion modeling through the incorporation of subgrid turbulence [11,12], and the use of dispersion ensembles for hazard prediction and sensor placement calculations [1,17]. In recent years, LES models are being increasingly used for atmospheric dispersion studies [25,26] and when implemented on GPU-based computing platforms, showed great promise for enabling a range of atmospheric boundary layer research topics and applications by significantly reducing the computational resource requirements to make the calculations [27]. Here, we present a dispersion modeling system designed to take advantage of the GPU-LES modeling technology called the joint outdoor-indoor urban large eddy simulation (JOULES). JOULES is a collection of modeling capabilities designed to calculate atmospheric conditions and corresponding airborne contaminant transport in open terrain and urban locations both inside and outside of buildings. At the core of JOULES is the GPU Resident Atmospheric Simulation Program (GRASP) described above. GRASP was originally developed to provide high-resolution simulations of clouds, winds, and turbulence and designed to be run on central processing unit (CPU) computing platforms. It was later adapted to run on GPU-based architectures by scientists at Delft University of Technology (TU Delft) and Whiffle B.V. Additional details on the origins of this model as well as information on the turbulence closures and other formulations used within the GRASP model can be found in [25–27].

Since its initial development, GRASP has undergone a number of evaluations to assess its ability to provide high-resolution reconstructions of atmospheric variables and short-term weather predictions. Of particular relevance to its use for atmospheric dispersion applications is the work by Schalkwijk et al. [25,26] to couple GRASP to a regional-scale atmospheric model in order to predict a continuous, year-long, three-dimensional time series of turbulence and clouds. The predictions were compared to detailed boundary layer observations collected at the Cabauw Experimental Site for Atmospheric Research (CESAR). This study included favorable comparisons between the measured and simulated power spectrum of horizontal and vertical wind speed variance across a variety of weather conditions and temporal scales.

2. Materials and Methods

In this section, we describe the implication of coupling JOULES with an AT&D model, which solves for the advection and diffusion of a passive scalar (i.e., a neutrally buoyant airborne tracer). Our study evaluates the accuracy of the GPU-LES dispersion solutions produced by this model over flat open terrain. The evaluation was performed using data from three separate dispersion experiments, with atmospheric conditions ranging from unstable daytime "convective" PBLs, to stable environments that typically occur at night. This section describes the observational data used in the model evaluation, the approach used to develop the ensembles of dispersion solutions, the analysis methodology, and the metrics used to compare the simulations to the observations.

2.1. Observational Data

This study uses data from three dispersion experiments, representing open terrain environments under unstable, neutral, and stable conditions. The observations during unstable or "convective" conditions were taken from the classic convective water tank experiments conducted by Willis and Deardorff [35], and two outdoor atmospheric trials, Project Prairie Grass [36,37], and the COnvective Diffusion Observed by Remote Sensor (CONDORS) experiments [38–40]. Collectively, these three experiments provide data from trials representing both near-surface airborne tracer releases and observations of material concentrations at the surface and aloft. Since vertical dispersion is less significant in stable conditions, near-surface observations from Project Prairie Grass were used to assess our implementation for neutral and stable simulations. The following subsections describe these data.

2.1.1. Willis and Deardorff Water Tank Experiments

Historically, the development of ensembles of dispersion realizations from outdoor dispersion measurements has been difficult to construct because the atmospheric conditions in which the measurements are taken typically do not repeat with sufficient consistency or frequency during the experiments. Laboratory experiments are one way of addressing this challenge and have been demonstrated to provide comprehensive ensembles of dispersion for repeatable conditions. In the 1970s and 1980s, Willis and Deardorff conducted a series of water-tank experiments, to measure the temperature, heat flux, wind velocity, and fluctuations in temperature and velocity in a convectively forced fluid [35]. They later extended these experiments to include dispersion in the convective boundary layer [41]. These observations, and a series of numerical simulations enabled the characterization of dispersion in the convective atmospheric boundary layer, and showed how the large variability in concentration measurements observed in convective boundary layers can be explained by the location of the release relative to the updrafts and downdrafts [42,43].

This work demonstrated that an ensemble of dispersion realizations is typically needed to characterize the statistical properties of dispersion in convective boundary layers. Later experiments measured crosswind-integrated concentration as a function of the downstream distance from the source. This led to a further extension of these earlier convective water tank experiments, where additional observational experiments were designed to produce ensembles of dispersion solutions [41–43] (including measurements of crosswind integrated concentrations as a function of the downstream distance from the source for a series of near-surface and elevated releases. The Willis and Deardorff observations have been used extensively to understand the convective boundary layer and in a variety of model evaluation studies [11,12,42,43] involving LES models. In this study, data from the water tank experiment were used in the assessments of the GPU-LES model simulations of downwind dispersion at the surface and aloft during convective conditions.

2.1.2. Project Prairie Grass Experiment

Project Prairie Grass [36] is a classic outdoor atmospheric dispersion field trial that has been widely used to understand the properties of atmospheric dispersion from near surface releases. Briefly, the experiment featured a series of 10-min continuous sulfur dioxide (SO_2) releases from a point 0.46 m above the ground. Downwind concentration measurements were made along five semi-circular arcs, located 50, 100, 200, 400, and 800 m from the release, at spacings of 2° for the four innermost arcs, and at 1° for the 800 m arc. Surface-based samplers collected 10-min integrated concentration measurements. The mean winds were measured at 8 heights, from 0.125 m to 16 m. The micro-meteorological information (friction velocity (u_*) and Obukhov length (L)) were determined by fitting the friction velocity and temperature measurements to the tower data. Approximately 70 releases were conducted, and data collected across a range of atmospheric stability regimes. In this study, data from the Prairie Grass experiment were used in the assessments of near-surface GPU-LES dispersion model simulations for unstable, neutral, and stable boundary layer conditions. The Project Prairie Grass data were obtained from Arhus University, Denmark [37]. We refer the reader to [36] for a full description of the experiment.

2.1.3. COnvective Diffusion Observed by Remote Sensors (CONDORS) Experiment

The COnvective Diffusion Observed by Remote Sensors (CONDORS) experiments measured both near-surface and vertical dispersion, extending to the full depth of the atmospheric boundary layer [38–40]. Measurements aloft included rawinsondes released near the time of the tracer experiments, acoustic sounders, and observations from a 300-m tower, which enabled estimates of the mixing layer depth, heat, and momentum fluxes. The trials used three tracers (oil fog, chaff, and a passive gas), in daytime releases near Erie, Colorado during August and September. Concentration measurements were collected by samplers at the surface, and with lidar and radar aloft. Twenty-six hours of data were collected across 12 separate mid-day periods. Of these data, over 11 h of data were processed into 29- to 60-min averaging periods during conditions where the convective boundary layer or mixed layer depth normalized by the Obukhov length (z_i/L) ranged from 24 (moderately unstable) to 1125 (extremely unstable) [38]. Obukhov length is defined as:

$$L = -u_*^3 T_a / \left(kg\overline{w'\theta_0'} \right) \tag{1}$$

where u_* is the surface friction velocity, T_a is the ambient temperature, k is the von Kármán constant (k = 0.4), g is the gravitational acceleration, and $\overline{w'\theta_0'}$ is the surface kinematic heat flux. This experiment provides a comprehensive set of measurements of crosswind-integrated concentration, lateral dispersion, plume height, and vertical dispersion. Data from the CONDORS experiment were used in the unstable boundary layer evaluations of the GPU-LES model dispersion at both the surface and aloft.

2.2. *Categorization of the Observations*

For each of the categories of static stability, we combined dispersion measurements from a combination of observations collected at multiple distances downwind from the release location and from a variety of individual trials conducted during similar weather conditions. Combining the observational data in this way enabled us to then compare both the mean and spread of the full distribution of observations to the model simulations. Taken together, the water tank, Prairie Grass, and CONDORS experiments cover a range of weather and stability conditions, near-surface and above-surface data, and distances downwind from the release location. Broadly, Prairie Grass provides near-surface dispersion estimates for unstable, neutral, and stable boundary layer conditions. For unstable boundary layers, these are supplemented by the Willis and Deardorff water tank and CONDORS data for both near-surface and vertical dispersion, including crosswind-integrated concentration, plume height, and lateral and vertical dispersion.

In order to synthesize complete datasets against which to evaluate our GPU-LES model, we organized the data around elements of the Pasquill–Gifford stability categories [44,45]. We chose to represent dispersion in convective (unstable), neutral, weakly stable, moderately stable, and strongly stable conditions, based on the ranges of the Obukhov length that have been correlated to these stability categories [46,47]. Table 1 lists the Obukhov ranges used to select specific trials from the Project Prairie Grass and CONDORS data sets. To the extent possible, care was taken to group measurements taken under similar weather conditions. Specific trials selected by these criteria included:

1. The convective water tank experimental data from Case 1 in Willis and Deardorff [26], comprising data from seven experimental trials;
2. All the surface-based releases of oil and chaff, for eight releases and five locations, in the CONDORS experiment;
3. Seven Project Prairie Grass trials—Trials 7, 8, 10, 16, 25, 44, and 51.

Table 1. Obukhov values used to define stability categories for the GPU-LES evaluation.

Stability	L (m)
Unstable	<−2
Neutral stability	>75
Slightly Stable	35 to 75
Moderately Stable	8 to 35
Extremely Stable	1 to 8

Table 2 gives details on the 81 individual trials distributed across the four categories and the corresponding mean atmospheric properties for each category.

Table 2. The mean environmental conditions of the observational data used in the evaluation.

Stability	Number of Trials	U @ 1m ($m\ s^{-1}$)	u_* ($m\ s^{-1}$)	Heat Flux ($W\ m^{-2}$)	T (°C)	θ (K)	w_* ($m\ s^{-1}$)	L (m)
Unstable	47	4.054	0.331	200.517	28.96	307.10	1.613	−18.33
Neutral stability	12	4.84	0.38	−29.39	22.33	300.36	−0.54	167.25
Slightly Stable	7	2.74	0.21	−25.94	22.15	300.19	−0.50	54.20
Moderately Stable	6	2.21	0.16	−22.07	21.73	299.68	−0.49	18.91
Extremely Stable	9	1.24	0.07	−10.74	19.03	297.13	−0.36	4.29

2.3. Scaling Methodology

In addition to their water tank experiments, Willis and Deardorff [26] also introduced methods for scaling dispersion measurements. Their approach acknowledges that dispersion in the PBL depends primarily on the time over which the fluid disperses through the media. Their initial work considered highly convective PBLs, where non-dimensional scaling factors were derived using the convective boundary layer depth, z_i, and the convective velocity scale, $w_* = \left(g\overline{w'\theta'_0}z_i/T_a\right)^{\frac{1}{3}}$. They further defined the turbulence time scale, or eddy-turnover time, as z_i/w_*. In the along-wind direction, they recognized that, because dispersion in the PBL is a function of time, downwind distance, X, from the source for convective conditions can be scaled as:

$$X = \frac{w_* x}{U z_i} \tag{2}$$

where x is the distance downwind of the release location, and U is the average PBL wind speed. Willis and Deardorff [26] also suggested a dimensionless parameter for the crosswind integrated concentration (CWIC or C^y), defined as $C^y = \int_{-\infty}^{\infty} C(x, y, z)dy$. To do so, they scaled CWIC by the concentration that would be present in a uniformly mixed PBL, well downwind of the release. This uniformly mixed concentration is Q/Uz_i where Q is the mass-release rate. Based on this approach, a normalized dimensionless CWIC can be calculated as:

$$Normalized\ CWIC = \frac{C^y U z_i}{Q} \tag{3}$$

This approach was first used by Willis and Deardorff [35] to visualize the dispersion measurements from their convective water tank experiments. Our work here leverages these methodologies and in particular plots of CWIC like Figures 8 and 9 from Willis and Deardorff [35] that depict measurements with a non-dimensional mean CWIC from the water tank data that was plotted as a function of the scaled downwind distance X and height.

Since its introduction, this scaling approach has been extended and adapted for use in atmospheric dispersion applications that range from interpreting observational data [48] to the development of empirical models [49,50] and the evaluation of dispersion models [11,12]. Here, we used this scaling approach to combine observations taken at different times from a single experiment, and observations taken from different experiments, into a collection of measurements that represent dispersion in each stability class. Similar methods were used by Weil et al. [11,12] for unstable conditions, and Venkatram et al. [51] for neutral and stable conditions. The scaling used in this study and by Weil et al. [11,12] for the convective cases closely follows the approach found in [35]. For evaluating plume height, and vertical and lateral dispersion, we also followed the Weil et al. [11,12] approach of normalizing the results by z_i.

The downwind distance scaling method used in this evaluation for the neutral and stable static stability conditions follows the approach published in [50]. The Venkatram et al. [50] scaling differs slightly from that of Willis and Deardorff [35], normalizing the downwind distance using:

$$X = \frac{x}{|L|} \tag{4}$$

where $|L|$ is the absolute value of the Obukhov length from Equation (1). Because the convective scale velocity of neutral-to-stable cases would be negative, we instead use a scaling factor that utilizes friction velocity and Obukhov length. For neutral and stable scenarios, normalized CWIC is calculated using the following relationship:

$$Normalized\ CWIC = \frac{C^y u_* L}{Q} \tag{5}$$

The scaling parameters, as described above, are applied to all of the point-location observations used in this evaluation. The same relationships are applied to the simulations enabling a comparison with the observations.

2.4. GPU-LES Model Simulations

The development of the GPU-LES dispersion simulations involved a two-step process. The first step was the design and execution of the atmospheric simulation. Five configurations, corresponding to the stability categories defined in Tables 1 and 2, were developed. The unstable configuration was based on the specifications of Weil et al. [11,12]. The neutral and stable configurations required finer spatial resolution, and a different approach for establishing the desired stability conditions. To achieve a simulation with the desired Obukhov length, we set the roughness length, heat flux, temperature, and initial convective boundary layer depth to values representative of the environmental conditions in the ensemble of field trials. Next, we adjusted the geostrophic wind (U_g and V_g) until the Obukhov length matched the observations within the category. Table 3 provides details on the model configurations and initial conditions (e.g., domain size, spatial resolutions, and core meteorological parameters). The wind speeds in Table 3 are the initial values (i.e., at the model start time) for the entire PBL. The turbulent PBL simulation is first spun up for 60 min from a cold start using cyclic (i.e., periodic) lateral boundary conditions. Figure 2 shows sample horizontal and vertical cross-sections of vertical velocity after the spin-up of a convective PBL simulation over a 2 km × 2 km × 1.0 km domain. Following the initial spin-up, we then began the tracer release, allowing the tracer mass to reach steady state. This required another 60 min of simulated time. Hence, we did not use meteorological or dispersion results from the first 120 min of the simulations. Our development here emphasized model selections that aided in separating model calibration to the experiment conditions from model calibration to the resulting data.

Table 3. The JOULES GPU-LES model configuration parameters.

Simulation Category	Grid Points (nx,ny,nz)	Model Resolution (Hor,Vert) (m)	U (m/s)	U_g (m/s)	V_g (m/s)	Heat Flux (W/m^2)	z_i (m)	L (m)
Unstable	(192,192,96)	(52.1,20.8)	2.8	2.8	−1.5	240	1000	−11.5
Neutral stability	(256,256,64)	(6.25,6.25)	8	8	−5.5	−10	190	372.2
Slightly Stable	(256,256,64)	(6.25,6.25)	4	3.5	−4.5	−10	116	47.6
Moderately Stable	(256,256,64)	(6.25,6.25)	4	3.2	−4.2	−10	103	28.9
Extremely Stable	(256,256,64)	(6.25,6.25)	3	2.5	−2.5	−10	78	7.6

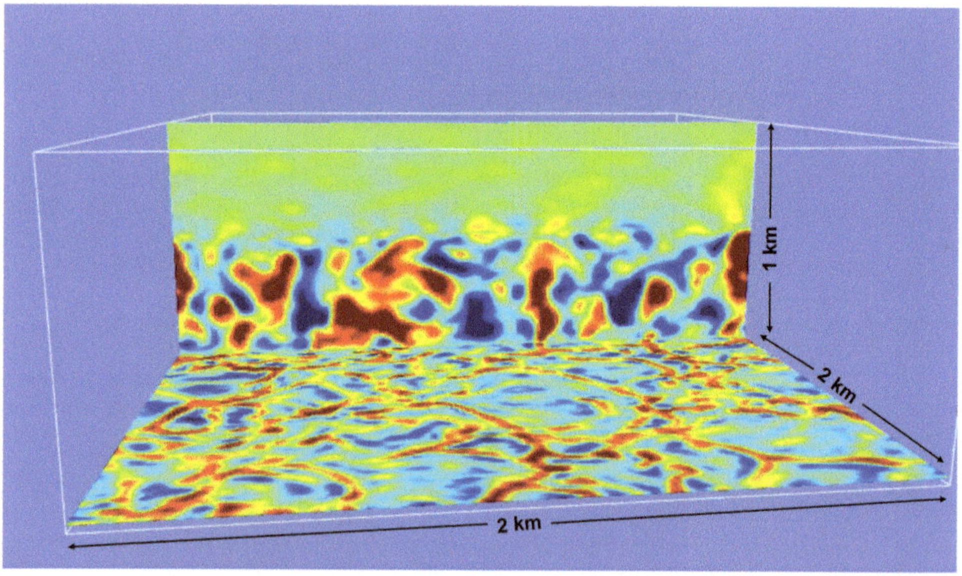

Figure 2. Illustration of a horizontal and vertical cross-section of vertical velocities for a convective planetary boundary layer (PBL) simulation. Red represents updrafts; blue represents downdrafts; and green represents areas where vertical motions are low. In this simulation, the depth of the boundary layer was set at 1000 km, clearly visible as the level where vertical velocities abruptly transition from large to small values.

The second step in developing dispersion simulation data for this evaluation is a process we used to develop an ensemble of uncorrelated dispersion solutions. Once the

GPU-LES model has spun up the turbulence and corresponding dispersion to a quasi-steady state, we calculated the model evaluation metrics. A continuous set of values in the along-wind direction were computed for each of the evaluation metrics. The calculations used a 10-min average or integrated value, depending on the metric in question. Individual uncorrelated realizations of the dispersion patterns, following the methodology used in [12], were created by varying a combination of the source location and the start time of the release. For example, Figure 3 depicts the predicted near-surface concentrations from five uncorrelated simulations. It illustrates the type of variability in the dispersion patterns that can be created within a convective PBL simulation using this methodology. Because the GPU-LES model runs quickly, we were able to generate an ensemble of 130 realizations that we believe described the unstable PBL dispersion comprehensively and do so in approximately two hours of wall-clock time for these experiments. Subsequently, for the neutral and stable simulations, we reverted to using 30 realizations per ensemble, as was done in [12], and we found suitable for our comparisons. The model evaluation metrics were then computed for each ensemble member, using the sampling characteristics of the field program. This enabled us to create a comprehensive distribution of model evaluation metrics that captured the range of dispersion patterns associated with how the airborne material release responds to the winds and turbulence in the PBL.

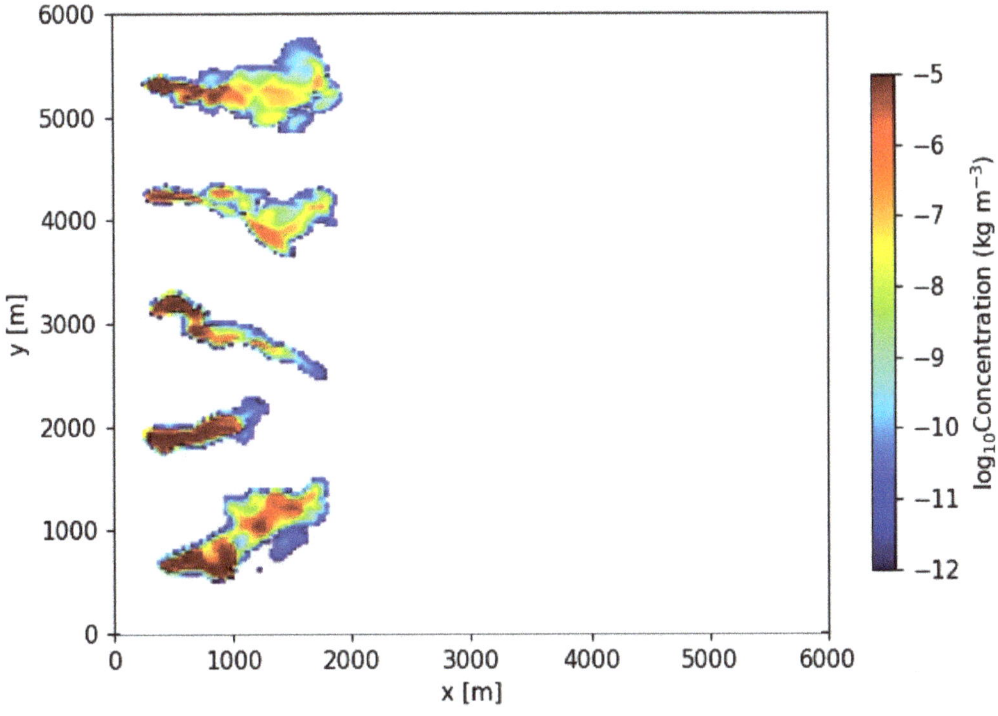

Figure 3. Variability in the dispersion patterns. Five sample realizations of near-surface concentration, 10 min after the start of the tracer release, were generated by varying the release start time or release location in the time-varying turbulent environment produced by the GPU-LES model.

3. Results

We assessed the suitability of the GPU-LES model to simulate dispersion in the PBL by comparing the plume height normalized by PBL height, vertical profiles of CWIC, downwind surface-based CWIC, surface crosswind dispersion, and vertical dispersion to comparable dispersion metrics computed from observations. Calculations of the metrics computed from observations are represented by symbols/markers in the forthcoming figures. Calculations of the metrics for a single dispersion realization from the GPU-LES model simulations are represented by either grey lines or blue dots and are plotted on the figure to illustrate the distribution of solutions. The following sections present results comparing the JOULES GPU-LES model predictions to observational data and are organized by stability category as described in Table 1.

3.1. Unstable PBL Comparison

The observational data sets used for the unstable PBL evaluation were from the Willis and Deardorff [35] water-tank, the Project Prairie Grass, and CONDORS experiments. The crosswind integrated concentration was computed from the release location to a downwind distance of 9750 m, for the full vertical depth of the simulation (2000 m). Figure 4 shows the CWIC calculation from the unstable GPU-LES dispersion solution. Qualitatively, the results closely match the measurements illustrated in Figure 8 from the Willis and Deardorff [35] experiment for scaled downwind distances less than three. Overall, the GPU-LES dispersion solution shows strong qualitative agreement with the Willis and Deardorff [35] observations, notably the near surface concentration minimum that occurs between a downwind distance of ~1 to 3. Because the GPU-LES simulation extends beyond the downwind distance measured by Willis and Deardorff, we are able to predict an increase in average CWIC near surface extending out to a scaled downwind distance of X = ~5.

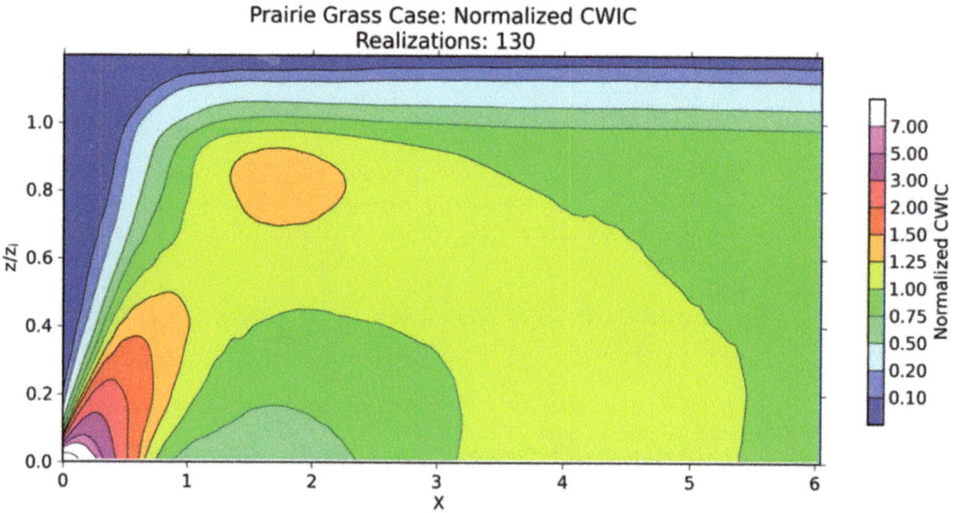

Figure 4. Normalized crosswind integrated concentration as a function of the downwind distance from the release location. This figure and corresponding calculation are frequently used in the atmospheric dispersion community to compare dispersion simulations to observational data. The plot represents the average CWIC from 130 individual GPU-LES realizations.

Figure 5 shows a quantitative comparison of the GPU-LES CWIC solution with the Willis and Deardorff [35] data. The figure illustrates how CWIC varies with height, out to a scaled downwind distance of approximately 3 from the release location. Past X = 3, the CWIC value shows unity up to the height of the boundary layer. It shows promising agreement between the ensemble mean of CWIC from the GPU-LES model (black line) and the measured vertical profiles of CWIC from the water tank experiments (black dots).

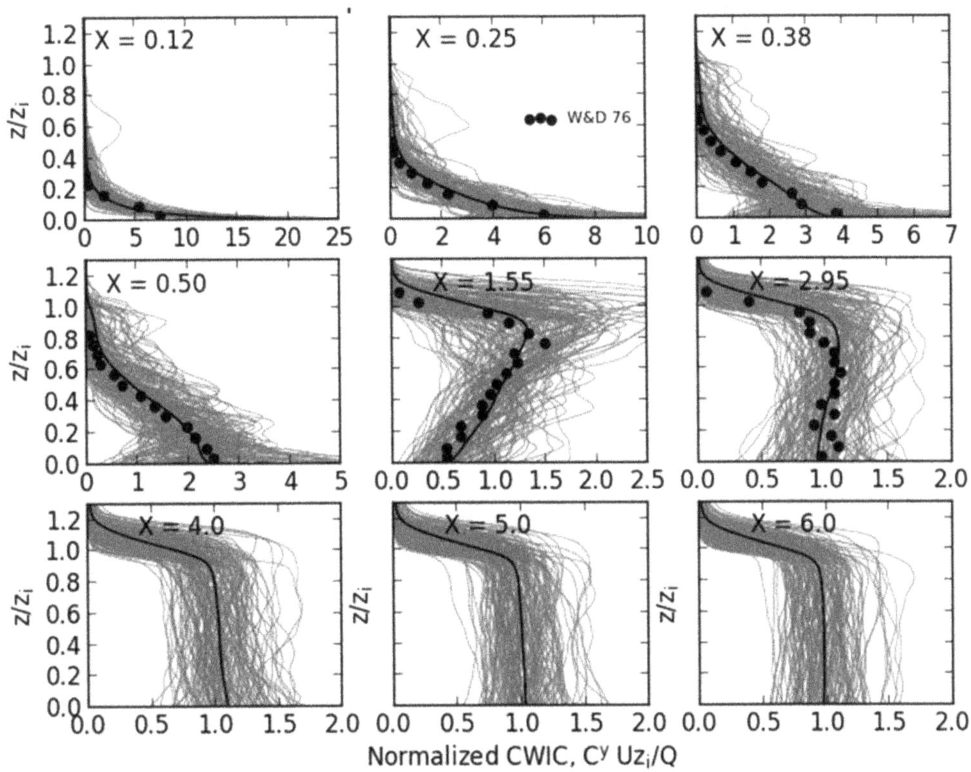

Figure 5. Crosswind integrated vertical dispersion as a function of the downwind distance from the release location, for the Willis and Deardorff [35] water tank experiments. The gray lines depict 130 realizations from the JOLES model, and the black lines show the average of these realizations. The circles represent data measured by Willis and Deardorff.

Surface measurements for calculating CWIC were available from the Project Prairie Grass and CONDORS experiments. Figure 6 compared surface CWIC measurements (represented by the black circles, dots, squares, and stars) with calculations from 130 GPU-LES realizations (represented by the small blue dots). The green line denotes a calculation of CWIC using surface layer similarity (SLS) theory [51]. At locations nearest the source, the LES model under predicts CWIC compared to SLS theory and Prairie Grass data. This is due to the resolution of the computational grid. Near the source, the tracer is immediately dispersed uniformly throughout each grid cell, making the model over-dispersive at this location. The underprediction near the source can be addressed (when required) by decreasing the volume of the grid cells (i.e., increasing the spatial resolution), or through the use of a Lagrangian particle dispersion model (LPDM). Figure 6 shows that, except very near to the source, the simulations of surface-based normalized CWIC show promising agreement with the measurements from the field trials. The figure also illustrates how the spread in the observations and model simulations vary as a function of the scaled downwind distance from the source. Notably, both show the pattern of smaller spread near the source location, an increase in spread near a scaled downwind distance of one, and then a collapse of the spread to a CWIC value near one further downwind as the material becomes well-mixed in the PBL.

Figure 7 compared lateral or horizontal crosswind dispersion, computed from surface observations from both the Project Prairie Grass and CONDORS experiments, and corresponding calculations of lateral dispersion from the GPU-LES model. Here, again, there is fair agreement between the observations and the mean lateral dispersion from the ensemble of simulations. A majority of observations fall within the ensemble distribution, and the slope of the increase in plume spread with distance is well aligned with the observations. However, the mean of the ensemble appears to be slightly below the observations, including the formulation fit to the data by Briggs [48], for scaled downwind distances greater than approximately 0.1.

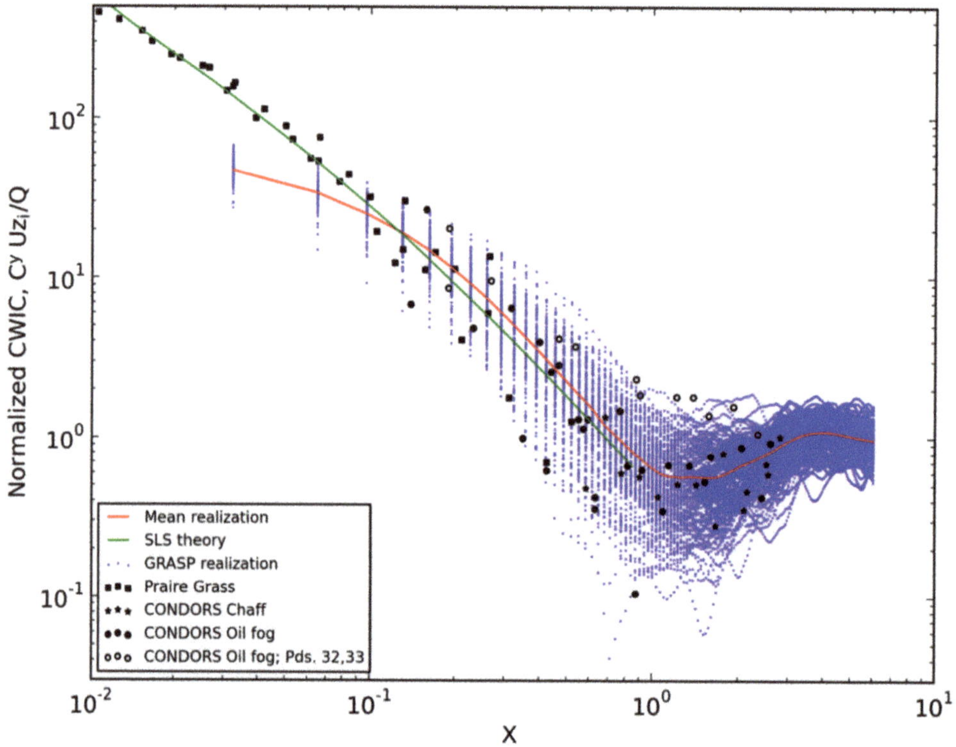

Figure 6. Crosswind integrated concentration as a function of the downwind distance from the release location. The green line represents the dispersion based on surface layer similarity (SLS) theory. The blue dots depict 130 individual realizations from the LES model, and the red line represents the average of these realizations. The markers (closed circles, open circles, squares, and stars) represent observations measured during convective trials in the Prairie Grass and CONDORS experiments.

In Figure 8 we illustrate the vertical dispersion results. For this metric, the mean of the ensemble aligns well with the observations, and the spread of the ensemble corresponds nicely to the spread observed in the CONDORS data. This metric also demonstrates that the GPU-LES dispersion model exhibits the expected increase in vertical dispersion as the scaled downwind distance increases. Finally, it appears to correctly capture the properties in the vertical dispersion observations, where normalized vertical dispersion stabilizes at scaled downwind distances greater than one.

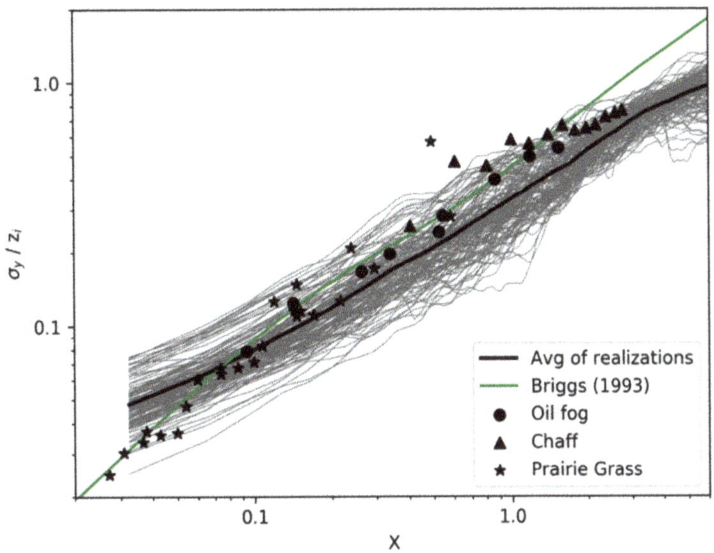

Figure 7. Surface crosswind dispersion from the GPU-LES model as a function of the downwind distance, compared to observations from the CONDORS field program. The gray lines depict a series of 30 individual realizations from the LES model. The black line represents the average of those realizations. The circles and triangles represent observations of normalized concentrations measured during the CONDORS and Prairie Grass experiments.

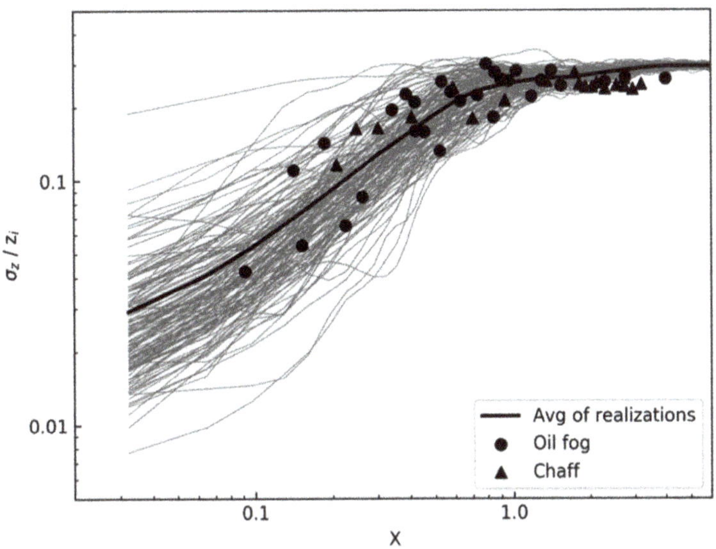

Figure 8. Normalized vertical dispersion as a function of the downwind distance. The gray lines depict 30 individual realizations from the LES model, while the black line represents the average of those realizations. The circles and triangles represent observations of normalized concentrations measured during the CONDORS experiments.

Figure 9 shows the normalized average plume height, $\overline{z_p}$, as a function of the scaled downwind distance. Excluding outlier observations during sampling periods 32 and 33 (pds 32,33) in the CONDORS field trials, noted by Briggs [49], there is good agreement between the CONDORS observations of $\overline{z_p}$ and the dispersion realizations. The average value of $\overline{z_p}$ from the ensemble of simulations (black line) is near the center of the scatter of observations. The ensemble of the plume height calculations presents little variability near the release location and far downwind of the release, but a greater range of results at scaled downwind distances between X = 0.75 and 2.5. This pattern in the ensemble spread agrees closely with the scatter observed in the CONDORS observations. It also matches expectations based on the characteristics of dispersion in a strongly convective PBL environment.

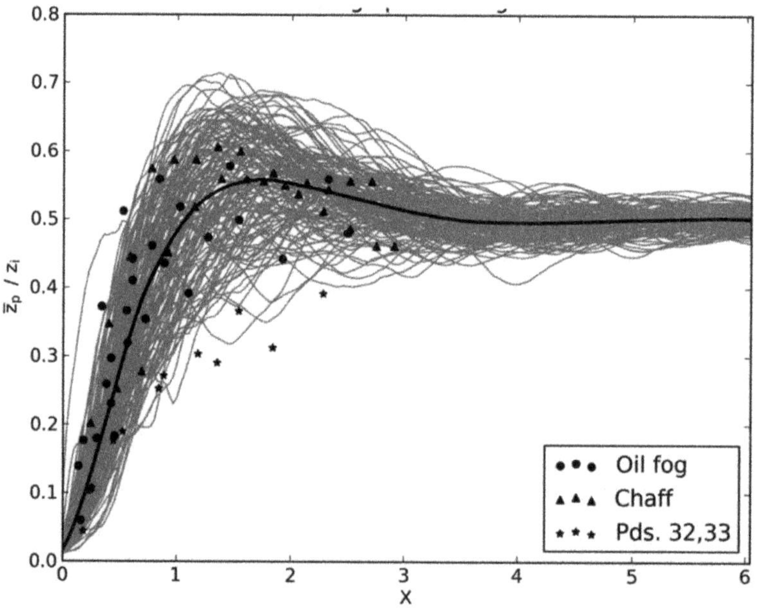

Figure 9. Normalized plume height as a function of the downwind distance. The gray lines depict 100 individual realizations from the JOULES and the black line represents the average of those realizations. The circles and triangles represent observations measured during the CONDORS experiments.

3.2. Neutral and Stable PBL Comparison

Data from the Project Prairie Grass field experiment were also used to evaluate the dispersion solutions from the GPU-LES dispersion simulations for neutral and stable PBL conditions. Our selection criteria identified surface concentration data from 34 trials for use in the model evaluation. As described above, in order to account for variability of dispersion conditions within this range of stability categories, we categorized these data into four distinct subsets; see Table 1. Surface CWIC and lateral dispersion results are provided for each category.

3.2.1. Neutral PBL Comparison

Figure 10 shows a vertical cross-section of normalized CWIC as a function of the downwind distance calculated from the ensemble of 30 neutral GPU-LES dispersion simulations with an Obukhov length of ~372 m. It illustrates that for the neutral condition simulations, the airborne materials from a surface release mix up to about half the depth of the PBL (~95m). In Figure 11, the calculation of CWIC from the GPU-LES model was compared to calculations of CWIC from the surface observations from the Prairie Grass experiment, where the Obukhov length during the experiment was greater than 75 m. This figure shows good agreement between the CWIC observations (black dots) and the values computed from the GPU-LES simulations (blue dots). The red line depicts the ensemble average of the CWIC simulations. Note that the slope of this line matches the slope of the observations. The figure also indicates that the width of the distribution from the ensemble of simulations is very consistent with the scatter suggested by the observations.

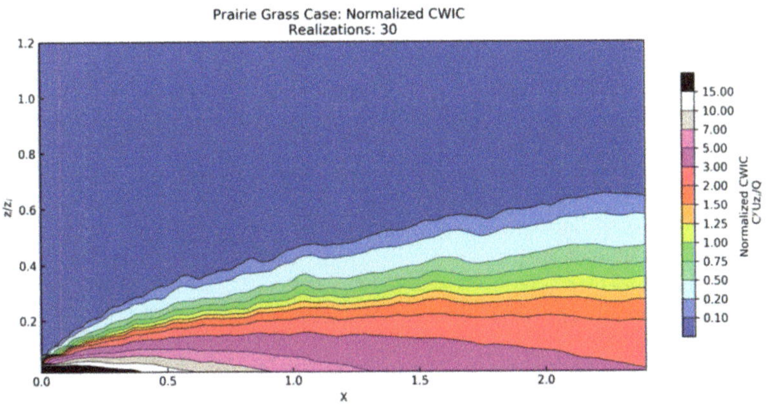

Figure 10. Normalized crosswind integrated concentration as a function of the downwind distance from the release location, for the neutral stability. The plot shows the average CWIC from 30 individual GPU-LES realizations.

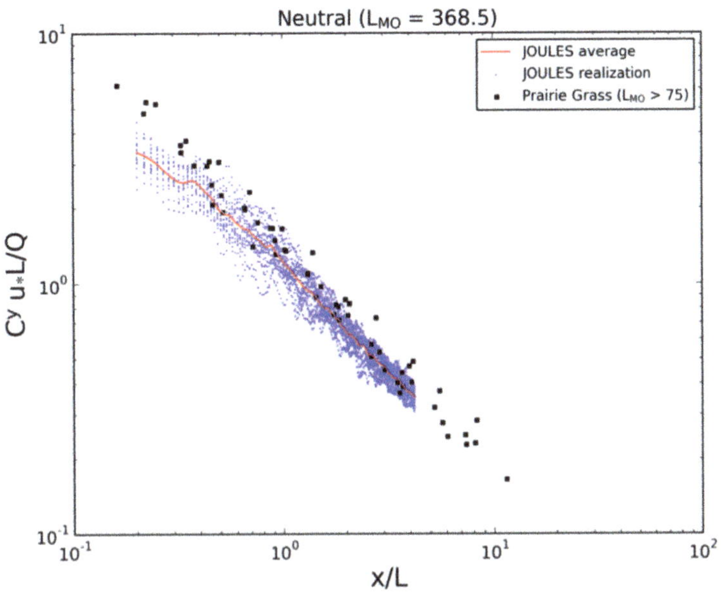

Figure 11. Normalized crosswind integrated concentration as a function of the downwind distance from the release location, for neutral stability conditions. The blue dots represent individual realizations of dispersion from the GPU-LES model, where the simulation was designed to produce neutral conditions with L = ~368. The red line represents the average of this ensemble. The black squares represent observations from the Prairie Grass experiment during neutral conditions.

3.2.2. Stable PBL Comparison

Similar calculations were made for the simulations produced for the slightly, moderately, and extremely stable PBL conditions. Figure 12 depicts the vertical cross-sections of CWIC for each and illustrates how the simulated CWIC values vary as the environmental conditions become more stable. As expected, the depth to which the airborne materials from a surface release are mixed decreases as the static stability increases (approximately 62, 41, and 31 m for the weakly, moderately, and extremely stable conditions, respectively). This results in higher concentrations being seen further downwind of the release location as the atmosphere becomes more stable. The data from these three stability categories were also directly compared to normalized CWIC calculations made using Project Prairie Grass observations. The results, summarized in Figure 13, show good agreement between the model and observations for the weakly and moderately stable conditions. There was considerably more scatter in the CWIC data computed from the ensemble of extremely stable observations and the ensemble average of the extremely stable GPU-LES CWIC simulations was on the high end of the pattern of observations (though still within the range of the scatter). This suggests that the turbulence is much lower in our GPU-LES simulation than what was present in the experimental data and that we do not have a sufficient understanding of the sources of turbulence in these very stable cases to incorporate it into the simulation.

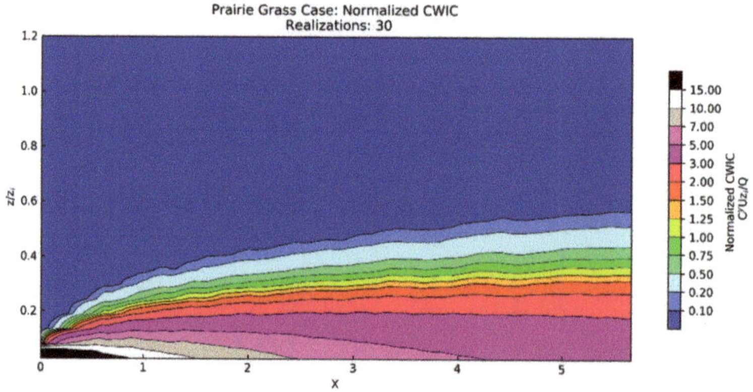

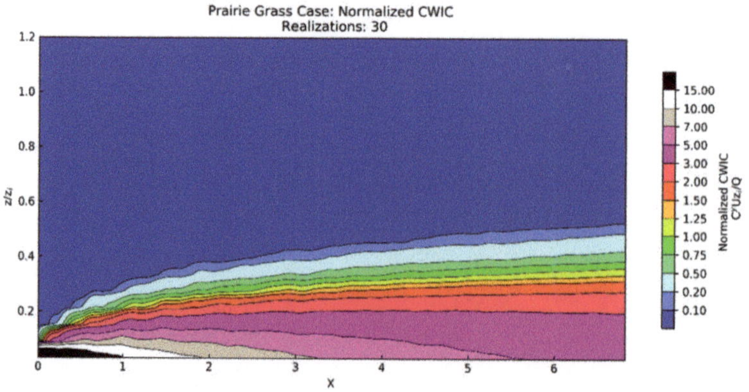

Figure 12. *Cont.*

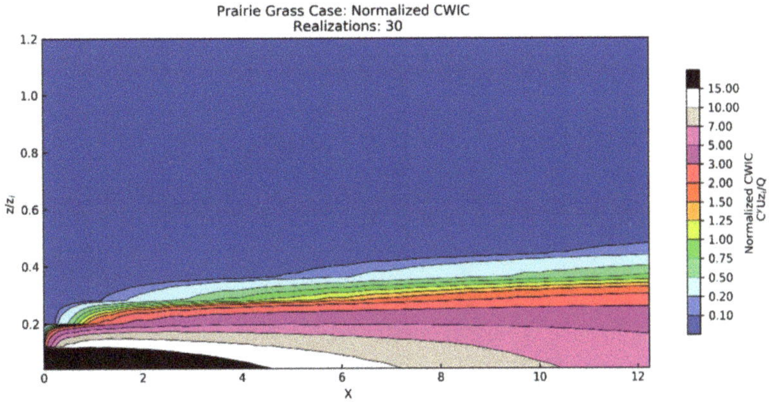

Figure 12. Normalized crosswind integrated concentration as a function of the downwind distance from the release location, for the slightly (**top**), moderately (**middle**), and extremely (**bottom**) stable conditions. Each plot represents the average CWIC from 30 individual GPU-LES dispersion simulation realizations.

In Figure 14, results from the neutral through extremely stable cases are plotted to depict the collective information on model accuracy for this range of stabilities. These results, and the results shown earlier for the convective PBL environments, indicate that the GPU-LES modeling system may be able to be configured from first principal parameters to provide ensembles of single-realization dispersion solutions that are representative of this range of environmental conditions.

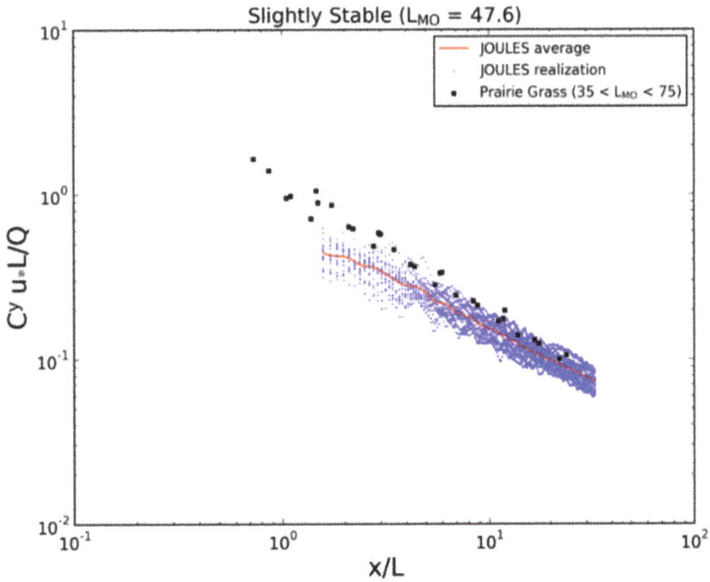

Figure 13. *Cont.*

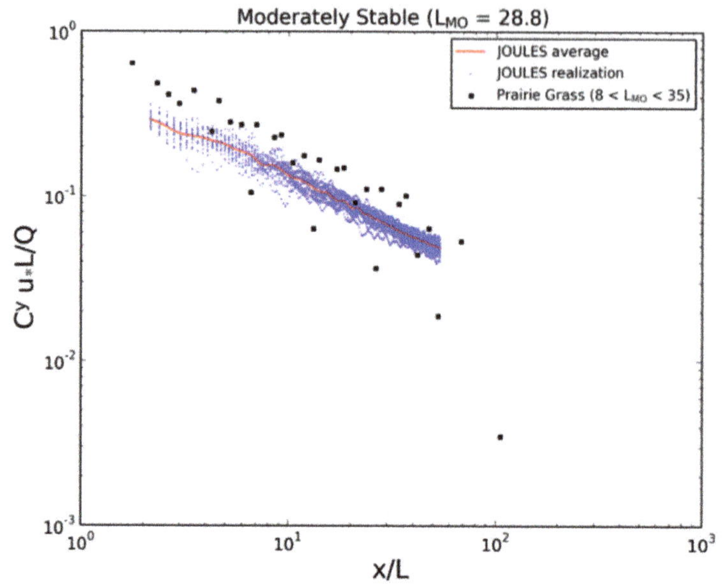

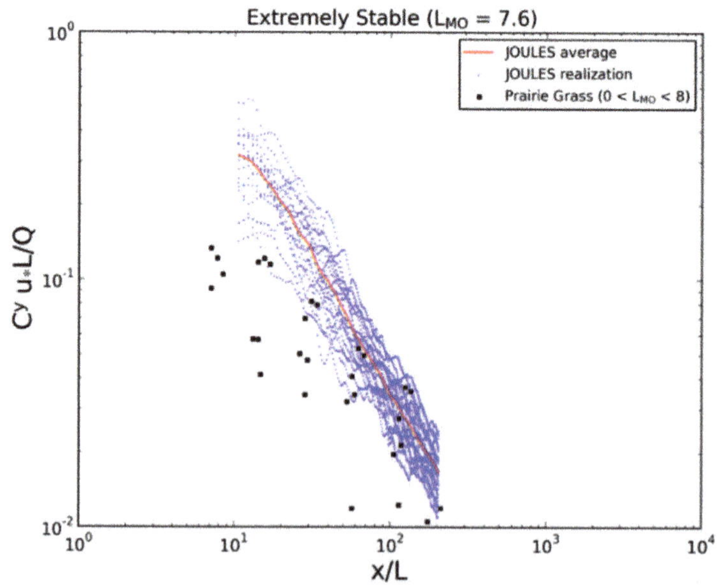

Figure 13. Normalized crosswind integrated concentration as a function of downwind distance from the release location, for weakly (**top**), moderately (**middle**), and extremely (**bottom**) stable conditions. The black square markers are derived from the Prairie Grass experiments. Blue dots represent individual realizations of dispersion from the GPU-LES model. The red line represents the average of this ensemble of simulated results.

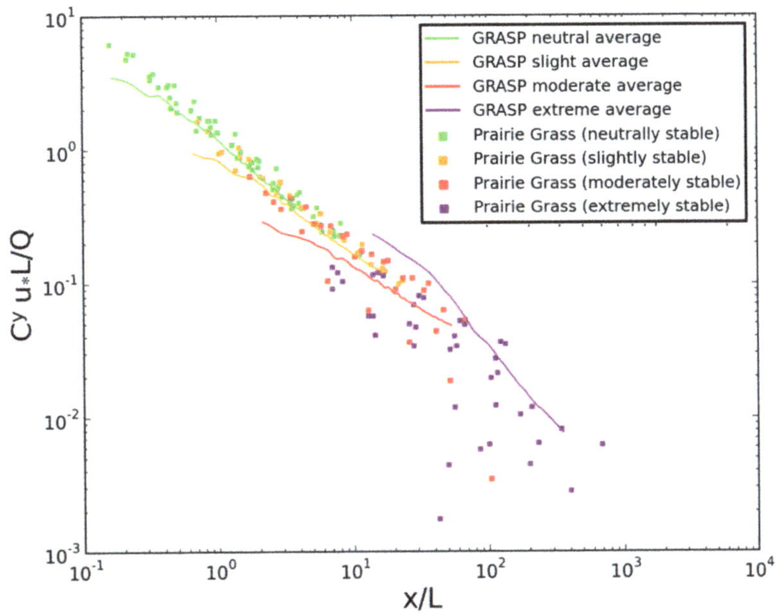

Figure 14. A summary of the GPU-LES dispersion simulation results for the neutral and stable boundary layer simulations. Each of the lines represents an ensemble average of 18 simulations. The dots represent observations from the Prairie Grass experiment during neutral and stable conditions. The results demonstrate that the GPU-LES can accurately represent atmospheric dispersion for conditions ranging from neutral to stable conditions.

4. Discussion and Conclusions

A research goal for developing our GPU-LES approach, JOULES, is to design a system for computing single-realizations of detailed, coupled urban (outdoor-indoor) contaminant dispersion. To describe the variability inherent in the atmospheric and urban conditions, our design required that that each simulation is completed quickly so that we could generate many realizations, all equally probable. A resulting capability would allow us to compute mass-conserving transport in a complete urban setting for various applications, many which cannot be analyzed using ensemble-averaging methods. The capability also allows us to derive synthetic data to test the suitability of existing operational tools.

Here, we present the evaluation of the JOULES dispersion solutions for open terrain environments. These tests are critical for many applications. It is also essential before testing it for more complex urban settings. This study used observational data from three field trials, following peer-reviewed methods and evaluation metrics that have been extended to evaluate the GPU-LES dispersion model's suitability and promise across a range of environmental conditions, including convective (daytime) and extremely stable (nighttime) conditions. The open terrain convective comparisons showed very close agreement, both at the surface and aloft, for surface-based releases across this range of stability regimes. The simulations and performance metrics also closely match the performance of the Lagrangian particle dispersion model (LPDM) and National Center for Atmospheric Research (NCAR) LES model published in [12] for convective conditions.

This study moved beyond the work presented by Weil et al. [12] and examined the accuracy of dispersion simulations for neutral and stable conditions. JOULES also performed well for the neutral, weakly, and moderately stable cases when compared to surface observations from Project Prairie Grass. While there was some agreement between the dispersion simulations and observation for the extremely stable cases, the normalized CWIC calculations from JOULES were on the high end of the scatter in the observations.

The atmospheric conditions and corresponding dispersion solutions can be produced by configuring first principle PBL parameters in the model to produce a simulated environment across static stability scenarios that range from unstable convective to moderately stable conditions. The results of this model evaluation study suggest that JOULES can produce very promising atmospheric dispersion solutions for open-terrain homogeneous environments. Furthermore, the GPU implementation has been demonstrated to enable simulations to run over 150 times faster than comparable CPU-based LES implementations. This advancement significantly reduces the computational costs associated with developing microscale atmospheric and dispersion simulations and now makes it feasible to produce ensembles of single-realization dispersion solutions that are necessary in a variety of airborne dispersion and defense analyses [1,8,9].

In future work, we plan to implement a simulation capability for urban interiors. Such efforts require deciding on appropriate physics-based models and improving computational performance to allow for even larger simulation domains, terrains, and non-homogeneous land covers.

Author Contributions: Conceptualization, methodology, project administration, writing—original draft preparation was performed by: P.E.B. Formal analysis, investigation and visualization were performed by: A.J.P. and P.E.B., and A.J.A. Software was performed by: H.J.J.J., and A.J.P. Writing—review and editing was performed by: P.E.B., A.J.P., D.M.L., M.D.S., and R.N.F.J. All authors have read and agreed to the published version of the manuscript.

Funding: This research was funded in part by the Defense Threat Reduction Agency and performed under U.S. Department of Energy Contract number DE-AC02- 05CH11231. Aeris research was funded as a subcontractor to Lawrence Berkeley National Laboratory. The views expressed in this paper are solely those of the authors.

Institutional Review Board Statement: Not applicable.

Informed Consent Statement: Not applicable.

Data Availability Statement: Observational data used in this study are available in the peer reviewed literature, through references supplied in the references section of the paper, and courtesy of personal communications with Steven Hanna and Joe Chang who maintain an archive of these historical data sets.

Acknowledgments: The authors acknowledge constructive discussions with George Bieberbach and Jonathan Hurst throughout this effort.

Conflicts of Interest: The authors declare no conflict of interest.

References

1. Bieringer, P.E.; Annunzio, A.J.; Platt, N.; Bieberbach, G.; Hannan, J. Contrasting the use of single-realization versus ensemble-average atmospheric dispersion solutions for chemical and biological defense analyses. *J. Appl. Meteor. Climatol.* **2014**, *53*, 1399–1415. [CrossRef]
2. Bass, A. Modelling long range transport and diffusion. In Proceedings of the 2nd Joint Conference on Applications of Air Pollution Meteorology, New Orleans, LA, USA, 24–27 March 1980; pp. 193–215.
3. Sykes, R.I.; Parker, S.; Henn, D.; Chowdhury, B. *SCIPUFF Version 3.0 Technical Documentation*; Sage Management: Princeton, NJ, USA, 2016; 403p.
4. Thomson, D.J. Criteria for the selection of stochastic models of particle trajectories in turbulent flows. *J. Fluid Mech.* **1987**, *180*, 529–556. [CrossRef]
5. Wilson, J.D.; Sawford, B.L. Review of Lagrangian stochastic models for trajectories in the turbulent atmosphere. *Bound.-Layer Meteor.* **1996**, *78*, 191–220. [CrossRef]
6. Haupt, S.E.; Annunzio, A.J.; Schmehl, K.J. Evolving turbulence realizations of atmospheric flow. *Bound.-Layer Meteorol.* **2013**, *149*, 197–217. [CrossRef]
7. Gunatilaka, A.; Skvortsov, A.; Gailis, R. A review of toxicity models for realistic atmospheric applications. *Atmos. Environ.* **2014**, *84*, 230–243. [CrossRef]

8. Bieberbach, G.; Bieringer, P.E.; Cabell, R.; Hurst, J.; Weil, J.; Wyszogrodzki, A.; Hannan, J. A Framework for Developing Synthetic Chemical and Biological Agent Release Data Sets. In Proceedings of the 13th International Conference on Harmonisation within Atmospheric Dispersion Modeling for Regulatory Purposes, Paris, France, 1–4 June 2010.
9. Platt, N.; DeRiggi, D.; Warner, S.; Bieringer, P.E.; Bieberbach, G.; Wyszogrodzki, A.; Weil, J. Method for comparison of large eddy simulation generated wind fluctuations with short-range observations. *Int. J. Environ. Pollut.* **2012**, *48*, 22–30. [CrossRef]
10. Wyngaard, J.C. *Turbulence in the Atmosphere;* Cambridge University Press: Cambridge, UK, 2010; 393p.
11. Weil, J.C.; Sullivan, P.P.; Moeng, C.H. The use of large-eddy simulations in Lagrangian particle dispersion models. *J. Atmos. Sci.* **2004**, *61*, 2877–2887. [CrossRef]
12. Weil, J.C.; Sullivan, P.P.; Moeng, C.H. Statistical variability of dispersion in the convective boundary layer: Ensembles of simulations and observations. *Bound.-Layer Meteorol.* **2012**, *145*, 185–210. [CrossRef]
13. Henn, D.S.; Sykes, R.I. Large-eddy simulation of dispersion in the convective boundary layer. *Atmos. Environ.* **1992**, *26a*, 3145–3159. [CrossRef]
14. Caulton, D.R.; Li, Q.; Bou-Zeid, E.; Fitts, J.P.; Golston, L.M.; Pan, D.; Lu, J.; Lane, H.M.; Buchholz, B.; Guo, X.; et al. Quantifying uncertainties from mobile-laboratory-derived emissions of well pads using inverse Gaussian methods. *Atmos. Chem. Phys.* **2018**, *18*, 15145–15168. [CrossRef]
15. Lundquist, K.A.; Chow, F.K.; Lundquist, J.K. An immersed boundary method enabling large-eddy simulations of flow over complex terrain in the WRF model. *Mon. Weather. Rev.* **2012**, *140*, 3936–3955. [CrossRef]
16. Tomas, J.M.; Eisma, H.E.; Pourquie, M.J.B.M.; Eisinga, G.E.; Jonker, H.J.J.; Westerweel, J. Pollutant Dispersion in Boundary Layers Exposed to Rural-to-Urban Transitions: Varying the Spanwise Length Scale of the Roughness. *Bound. Layer Meteorol.* **2017**, *163*, 225–251. [CrossRef]
17. Bieringer, P.E.; Piña, A.J.; Sohn, M.D.; Jonker, H.J.J.; Bieberbach, G., Jr.; Lorenzetti, D.M.; Fry, R.N., Jr. Large Eddy Simulation (LES) Based System for Producing Coupled Urban and Indoor Airborne Contaminant Transport and Dispersion Solutions. In Proceedings of the 18th International Conference on Harmonisation within Atmospheric Dispersion Modeling for Regulatory Purposes, Bologna, Italy, 9–12 October 2017.
18. Du, P.; Weber, R.; Luszczek, P.; Tomov, S.; Peterson, G.; Dongarra, J. From CUDA to OpenCL: Towards a performance-portable solution for multi-platform GPU programming. *Parallel Comput.* **2012**, *38*, 391–407. [CrossRef]
19. NVIDIA Corporation. Design and Visualization. Available online: https://www.nvidia.com/en-us/data-center/a100/ (accessed on 13 June 2021).
20. Skamarock, W.C.; Klemp, J.B. A time-split nonhydrostatic atmospheric model for weather research and forecasting applications. *J. Comput. Phys.* **2008**, *227*, 3465–3485. [CrossRef]
21. Michalakes, J.; Vachharajani, M. GPU acceleration of numerical weather prediction. *Parallel Process. Lett.* **2008**, *18*, 531–548. [CrossRef]
22. Mielikainen, J.; Huang, B.; Huang, H.L.A.; Goldberg, M.D. Improved GPU/CUDA based parallel weather and research forecast (WRF) single moment 5-class (WSM5) cloud microphysics. *IEEE J. Sel. Top. Appl. Earth Obs. Remote. Sens.* **2012**, *5*, 1256–1265. [CrossRef]
23. Silva, J.P.; Hagopian, J.; Burdiat, M.; Dufrechou, E.; Pedemonte, M.; Gutiérrez, A.; Ezzatti, P. Another step to the full GPU implementation of the weather research and forecasting model. *J. Supercomput.* **2014**, *70*, 746–755. [CrossRef]
24. Wahib, M.; Maruyama, N. Highly optimized full GPU-acceleration of non-hydrostatic weather model SCALE-LES. In Proceedings of the 2013 IEEE International Conference on Cluster Computing (CLUSTER), Indianapolis, IN, USA, 23–24 September 2013; pp. 1–8.
25. Schalkwijk, J.; Griffith, E.J.; Post, F.H.; Jonker, H.J.J. High-performance simulations of turbulent clouds on a desktop PC: Exploiting the GPU. *Bull. Am. Meteorol.* **2012**, *93*, 307–314. [CrossRef]
26. Schalkwijk, J.; Jonker, H.J.J.; Siebesma, A.P.; Bosveld, F.C. A year-long large-eddy simulation of the weather over Cabauw: An overview. *Mon. Weather Rev.* **2015**, *143*, 828–844. [CrossRef]
27. Schalkwijk, J.; Jonker, H.J.J.; Siebesma, A.P. An Investigation of the Eddy-Covariance Flux Imbalance in a Year-Long Large-Eddy Simulation of the Weather at Cabauw. *Bound.-Layer Meteorol.* **2016**, *160*, 17–39. [CrossRef]
28. Maronga, B.; Gryschka, M.; Heinze, R.; Hoffmann, F.; Kanani-Suhring, F.; Keck, M.; Raasch, S. The Parallelized Large-Eddy Simulation Model (PALM) version 4.0 for atmospheric and oceanic flows: Model formulation, recent developments, and future perspectives. *Geosci. Model Dev. Discuss.* **2015**, *8*, 1539–1637.
29. Van Heerwaarden, C.; Van Stratum, B.J.; Heus, T.; Gibbs, J.A.; Fedorovich, E.; Mellado, J.P. MicroHH 1.0: A computational fluid dynamics code for direct numerical simulation and large-eddy simulation of atmospheric boundary layer flows. *Geosci. Model Dev.* **2017**, 3145–3165. [CrossRef]
30. Sauer, J.A.; Muñoz-Esparza, D. The FastEddy®resident-GPU accelerated large-eddy simulation framework: Model formulation, dynamical-core validation and performance benchmarks. *J. Adv. Model. Earth Syst.* **2020**, *12*, e2020MS002100. [CrossRef]
31. Moeng, C.H.; Dudhia, J.; Klemp, J.; Sullivan, P. Examining two-way grid nesting for large eddy simulation of PBL using the WRF model. *Mon. Weather. Rev.* **2007**, *135*, 2295–2311. [CrossRef]
32. Rybchuk, A.; Alden, C.B.; Lundquist, J.K.; Rieker, G.B. A Statistical Evaluation of WRF-LES Trace Gas Dispersion Using Project Prairie Grass Measurements. *Mon. Weather. Rev.* **2021**, *149*, 1619–1633.
33. Intel Corporation. Product Specifications. Available online: https://ark.intel.com (accessed on 10 November 2018).

34. NVIDIA Corporation. Design and Visualization. Available online: https://www.nvidia.com/en-us/design-visualization/ (accessed on 10 November 2018).
35. Willis, G.E.; Deardorff, J.W. A laboratory model of diffusion into the convective planetary boundary layer. *Q. J. R. Meteorol. Soc.* **1976**, *102*, 427–445. [CrossRef]
36. Barad, M.L. Project Prairie Grass. A field program in diffusion. In *Geophysical Research Paper No. 59, Vols I and II AFCRF-TR-235*; Air Force Cambridge Research Center: Bedford, MA, USA, 1958.
37. Arhus. The Project Prairie Grass Data. Available online: http://envs.au.dk/en/knowledge/air/models/background/omlprairie/excelprairie/ (accessed on 5 May 2017).
38. Eberhard, W.L.; Moninger, W.R.; Briggs, G.A. Plume dispersion in the convective boundary layer. Part I: CONDORS field experiment and example measurements. *J. Appl. Meteorol.* **1988**, *27*, 599–616. [CrossRef]
39. Briggs, G.A. Plume dispersion in the convective boundary layer. Part II: Analyses of CONDORS field experiment data. *J. Appl. Meteorol.* **1993**, *32*, 1388–1425. [CrossRef]
40. Kaimal, J.C.; Eberhard, W.L.; Moninger, W.R.; Gaynor, J.E.; Troxel, S.W.; Uttal, T.; Briggs, G.A.; Start, G.E. *PROJECT CONDORS Convective Diffusion Observed by Remote Sensors*; NOAA/ERL Wave Propagation Laboratory, U.S. Department of Commerce: Boulder, CO, USA, 1986.
41. Willis, G.E.; Deardorff, J.W. A laboratory model of the unstable planetary boundary layer. *J. Atmos. Sci.* **1974**, *31*, 1297–1307. [CrossRef]
42. Lamb, R.G. A numerical simulation of dispersion from an elevated point source in the convective planetary boundary layer. *Atmos. Environ.* **1978**, *12*, 1297–1304. [CrossRef]
43. Deardorf, J.W.; Willis, G.E. Ground level concentration fluctuations from a buoyant and a non-buoyant source within a laboratory convectively mixed layer. *Atmos. Environ.* **1984**, *18*, 1297. [CrossRef]
44. Pasquill, F. The Estimation of the Dispersion of Windborne Material. *Meteorol. Mag.* **1961**, *90*, 33–49.
45. Gifford, F.A. Use of Routine Meteorological Observations for Estimating Atmospheric Dispersion. *Nucl. Saf.* **1961**, *2*, 47–57.
46. Pasquill, F.; Smith, F.B. The physical and meteorological basis for the estimation of the dispersion of windborne material. In Proceedings of the Second International Clean Air Congress, Washington, DC, USA, 6–11 December 1970; Englund, H.M., Beery, W.T., Eds.; Academic Press: New York, NY, USA, 1971; pp. 1067–1072.
47. Woodward, J.L. Appendix A: Atmospheric stability classification schemes. In *Estimating the Flammable Mass of a Vapor Cloud: A CCPS Concept Book*; Center for Chemical Process Safety of the American Institute of Chemical Engineers: New York, NY, USA, 1998; pp. 209–212.
48. Briggs, G.A. Analysis of diffusion field experiments. In *Lectures on Air Pollution Modeling*; Venkatram, A., Wyngaard, J.C., Eds.; American Meteorological Society: Boston, MA, USA, 1988.
49. Khaled, S.M.E.; Etman, S.M.; Embaby, M. New analytical solution of the dispersion equation. *Atmos. Res.* **2007**, *84*, 337–344.
50. Venkatram, A.; Snyder, M.G.; Heist, D.K.; Perry, S.G.; Petersen, W.B.; Isakov, V. Re-formulation of plume spread for near-surface dispersion. *Atmos. Environ.* **2013**, *77*, 846–855. [CrossRef]
51. Stull, R. *An Introduction to Boundary Layer Meteorology*; Springer: Amsterdam, The Netherlands, 1988; 670p.

atmosphere

Article

Lattice Boltzmann Method-Based Simulations of Pollutant Dispersion and Urban Physics

Jérôme Jacob [1,*], Lucie Merlier [2], Felix Marlow [1] and Pierre Sagaut [1]

1 Aix Marseille Univ, CNRS, Centrale Marseille, M2P2 UMR 7340, 13451 Marseille, France; jan-felix.MARLOW@univ-amu.fr (F.M.); pierre.sagaut@univ-amu.fr (P.S.)
2 Univ Lyon, CNRS, INSA-Lyon, Université Claude Bernard Lyon 1, CETHIL UMR5008, F-69621 Villeurbanne, France; lucie.merlier@insa-lyon.fr
* Correspondence: jerome.jacob@univ-amu.fr

Abstract: Mesocale atmospheric flows that develop in the boundary layer or microscale flows that develop in urban areas are challenging to predict, especially due to multiscale interactions, multiphysical couplings, land and urban surface thermal and geometrical properties and turbulence. However, these different flows can indirectly and directly affect the exposure of people to deteriorated air quality or thermal environment, as well as the structural and energy loads of buildings. Therefore, the ability to accurately predict the different interacting physical processes determining these flows is of primary importance. To this end, alternative approaches based on the lattice Boltzmann method (LBM) wall model large eddy simulations (WMLESs) appear particularly interesting as they provide a suitable framework to develop efficient numerical methods for the prediction of complex large or smaller scale atmospheric flows. In particular, this article summarizes recent developments and studies performed using the hybrid recursive regularized collision model for the simulation of complex or/and coupled turbulent flows. Different applications to the prediction of meteorological humid flows, urban pollutant dispersion, pedestrian wind comfort and pressure distribution on urban buildings including uncertainty quantification are especially reviewed. For these different applications, the accuracy of the developed approach was assessed by comparison with experimental and/or numerical reference data, showing a state of the art performance. Ongoing developments focus now on the validation and prediction of indoor environmental conditions including thermal mixing and pollutant dispersion in different types of rooms equipped with heat, ventilation and air conditioning systems.

Keywords: lattice Boltzmann method; large eddy simulation; pollutant dispersion; urban physics

Citation: Jacob, J.; Merlier, L.; Marlow, F.; Sagaut, P. Lattice Boltzmann Method-Based Simulations of Pollutant Dispersion and Urban Physics. *Atmosphere* **2021**, *12*, 833. https://doi.org/10.3390/atmos12070833

Academic Editor: Patrick Armand

Received: 30 May 2021
Accepted: 22 June 2021
Published: 28 June 2021

1. Introduction

The capability to accurately predict urban physics and environmental quality for citizens via numerical simulation is nowadays a critical challenge, since it is a key tool for designing future optimized and sustainable urban areas. Such predictive models should account for a very broad range of physical mechanisms, ranging from large-scale meteorological effects to very small scale unsteady fluctuations of physical quantities such as temperature, air velocity, pressure, humidity, etc.

Among the different interacting physical phenomena occurring in the atmosphere at the meso- and microscales are orographic effects, land sea breeze, urban heat islands, deep convection, thunderstorms, convection, thermals, building wakes and turbulence (see Figure 1 in Schlünzen et al. [1]). Mesoscale atmospheric flows that develop in the boundary layer or microscale flows that develop in urban areas are thus very complex, especially due to multiscale interactions, multiphysical couplings, land and urban surface thermal and geometrical properties and turbulence.

Accurately predicting these different atmospheric phenomena and their underlying physical mechanisms is challenging. It is even more the case in the urban roughness sublayer due to the intricate patterns of cities, generally composed of heterogeneous and dense layouts of numerous buildings and trees, as well as various types of surfaces, heat and moisture sources. However, accurately predicting these flows in cities is of the utmost importance, especially as urban air flows are determining pollutant dispersion, pedestrian wind comfort, and structural and thermal loads on buildings.

Therefore, a key feature of predictive numerical models is the capability to handle realistic full scale configurations (including geometrical details and physical mechanisms at play) via high-fidelity unsteady simulation techniques well suited for turbulent flows. Large eddy simulations (LESs [2]) have appeared as a promising numerical approach for that purpose, whose main limitation is related to the simulation complexity and numerical cost [3,4]. To alleviate this problem, Lattice Boltzmann Methods (LBMs [5,6]) have recently been identified as one of the most efficient approaches for "revolutionnary computational fluid dynamics (CFD)" [7] since they allow for a drastic reduction in (i) the computational time compared to classical CFD approaches based on the Navier–Stokes equations and (ii) the preprocessing step including the volumic mesh generation thanks to the coupled use of embedded Cartesian grids and immersed boundary conditions techniques.

The efficiency of LBM-based solvers to handle full scale urban flow simulations has been demonstrated by several authors. As an example, they have been used recently to simulate the flow in a $19.2 \times 4.8 \times 1$ km^3 area of Tokyo to evaluate the gust index at pedestrian level [8] and the turbulence statistics [9]. These simulations were carried out in a complex geometry thanks to the use of Cartesian grids and the immersed boundary conditions strategy that permit drastically reducing the preprocessing cost. Thanks to its high efficiency in terms of parallel computing, the LBM is an attractive method for implementation on Graphic Processing Units (GPUs), which reduce the computational cost. The use of the LBM for implementing GPUs allows real time simulations of the flow over a build area [10] and the dispersion of a pollutant inside Oklahoma City [11] to be achieved with quite good accuracy. The efficiency of the LBM compared to a Navier–Stokes (NS) approach has been analysed in the case of the cross flow through an open window in an isolated cubical building. The comparison of the velocity field inside the building shows a good accuracy of both NS-LES and LBM-LES simulations compared to experimental data and the LBM simulation on GPU was up to 700 times faster than the NS simulations [12,13].

The goal of the present paper is to illustrate and summarize an innovative CFD LES approach for the simulation of atmospheric and urban flows, developed in the framework of the ProLB software [14,15] and based on the LBM. Already used to study the aerodynamics of vehicles and airfoils [16–18], the approach was adapted to deal with atmospheric and urban physics problems. Thanks to its formulation and the treatment of boundary conditions, the approach is computationally effective. It is thus possible to perform efficient but detailed and high-fidelity simulations of complex built environments with multiscale interactions and multiphysical couplings, in order to study a large scope of atmospheric and urban physics problems.

To illustrate the ability of the developed LBM-LES approach to study mesoscale and urban microscale atmospheric flows, including transport and dispersion processes, the paper is organized as follows. First, Section 2 synthesizes the main numerical aspects of the developed method. Then, Sections 3–5 summarize recent developments and validated studies performed using the hybrid recursive regularized (HRR) collision model for the simulation of complex or/and coupled turbulent atmospheric flows, including a micrometeorological model, with different applications. In particular:

- Section 3 reviews four applications of the developed method to the prediction of convective humid air flows with condensation with increasing meteorological likeness: a double diffusive Rayleigh Bénard convection (Section 3.1), rising two- and three-dimensional moist thermal bubbles (Section 3.2) and a shallow cumulus convection (Section 3.3).
- Section 4 reviews three use cases focusing on pollutant dispersion, considering different geometrical or physical complexity levels: neutral gas dispersion in a street canyon

without or with tree planting effects (Section 4.1), neutral gas dispersion behind an isolated building without or with unstable thermal stratification (Section 4.2) and neutral or dense gas dispersion in a complex realistic urban environment (Section 4.3).
- Lastly, Section 5 reviews the validation and study of velocities (Sections 5.1 and 5.3) and pressure distribution on building facades (Sections 5.2 and 5.4) in a complex urban environment including high-rise buildings, with uncertainty quantification towards a more relevant assessment of pedestrian wind comfort and wind loads on urban buildings.

To conclude, Section 6 discusses the ongoing developments, especially regarding indoor dispersion problems with thermal and moving people effects, and Section 7 closes the present paper by summarizing the main findings of the different studies performed, highlighting the benefit of the current approach to support the development of fast response models and decision making.

2. The HRR-LBM-WMLES Numerical Model for Atmospheric Flows

This section summarizes the main elements of the developed LB-based numerical model. More details about the general LBM can be found in Krüger et al.'s work [5].

2.1. Generalities about LBM

The LBM mainly aims at simulating the macroscopic behaviour of fluids. However, as compared to the macroscopic description of the flow underlying common Navier–Stokes-based approaches, the LBM is based on a mesoscopic description of the flow. The fluid dynamics is simulated through streaming and collision steps based on the lattice Boltzmann equation:

$$f_i(\mathbf{x} + \mathbf{c_i}\Delta t, t + \Delta t) - f_i(\mathbf{x}, t) = \Omega_i(\mathbf{x}, t) \tag{1}$$

where $\mathbf{c}_i$ is a set of discrete velocities, usually a D3Q19 for three-dimensional problems (3 dimensions, 19 discrete velocities), $f_i(\mathbf{x}, t)$ is the discrete distribution function, and $\Omega_i(\mathbf{x}, t)$ is the collision operator, i.e., the source term representing the redistribution of f_i induced by collision. From this lattice Boltzmann equation, we can see that in practice only the first order neighbours are used in the algorithm, which permits increasing the parallel computation efficiency and allows the LBM to be faster than classical Navier–Stokes approaches.

Multiscale expansions show that the three-dimensional weakly compressible Navier–Stokes equations can be recovered to the second order by the LBM.

2.2. Key Features of the Present Lattice Boltzmann Method

Based on the general LBM framework, different developments were made in the ProLB solver to deal more efficiently with atmospheric and urban problems. In particular, the collision term was estimated using a regularized BGK model with a hybrid recursive procedure [19] as follows:

$$\Omega_i(x,t) = \sigma f_i^{neq,LBM}(x,t) + (1-\sigma) f_i^{neq,FD}(x,t) \tag{2}$$

where $0 \leq \sigma \leq 1$, $f_i^{neq,LBM}(x,t)$ is the nonequilibrium part of the distribution function computed from the projection of distribution function on Hermite polynomials and $f_i^{neq,FD}(x,t)$ is the nonequilibrium part of distribution functions approximated by finite differences. This procedure allows for both improved robustness as compared to usual BGK or MRT collision models and accuracy for large scale simulations, including urban and atmospheric flows.

Different forcing mechanisms (e.g., buoyancy, Earth rotation, etc.) can also be taken into account in the model following Guo et al. [20]. In particular, buoyancy forces (F_g), due to differences in gas composition or temperature, can be modeled using a usual Boussinesq approach, as follows:

$$\mathbf{F}_{g,c} = \rho_0 \, \beta_c \, (C - C_0) \, \mathbf{g} \tag{3}$$

$$\mathbf{F}_{g,T} = \rho_0 \, \beta_T \, (T - T_0) \, \mathbf{g} \tag{4}$$

with:

$$\beta_c = -\frac{1}{\rho_{air}} \left(\frac{C \, \rho_{poll} - \rho_{air}}{C} \right) \tag{5}$$

$$\beta_T = \frac{1}{T_0} \tag{6}$$

where C is the concentration, C_0 is the reference concentration that is equal to 0, β_c is the expansion coefficient of concentration, T is the temperature, T_0 is the reference temperature and β_T is the expansion coefficient of temperature. Note that thermal buoyant flows can also be modeled thanks to the perfect gas law. Several other external forcings such as mesoscale or Coriolis effects can also be taken into account.

The domain boundaries are integrated using the cut cell method, which enables complex geometries to be modeled while substantially reducing the preprocessing costs (see Feng et al.'s work [21] for details), and a synthetic eddy method (SEM, [22]) was developed to generate inlet turbulence [23].

Since the LBM exhibits the same turbulent closure problems as the Navier–Stokes-based methods, classical subgrid models and wall models (WMs) for atmospheric flow simulations (including neutral, convective and stable cases) are used in the present solver, leading to the definition of the present HRR-LBM-WMLES tool for atmospheric flow simulation. Details are omitted here for the sake of brevity and can be found in Feng et al.'s work [24].

The developed approach also relies on a hybrid strategy to address more complex problems than purely aerodynamic problems, such as dispersion or thermal problems. This approach does not consider multidistributions, but the mass and momentum conservation equations are solved using the LBM, while the scalar transport equations are solved using a usual finite volume/finite difference method for the sake of efficiency. The corresponding temporal and spatial discrete coordinates are the same as the ones used for the LBM. Thus, it is possible to consider only one additional unknown per additional equation and optimize the numerical efficiency of the model.

3. Application of the Present HRR-LBM-WMLES to Convective Humid Boundary Layers with Cloud Formation

This section reviews different use cases studied to finally address large scale buoyant meteorological flows accounting for cloud dynamics thanks to a condensation scheme. More details can be found in Feng et al.'s work [25,26].

For these applications, the potential temperature (θ) is defined as:

$$\theta = T \left(\frac{p_0(z)}{p_0} \right)^{-\frac{R_d}{c_p}} \tag{7}$$

where R_q is the air gas constant per mass unit, C_p is the averaged mass heat capacity, p_0 is the reference pressure at ground and $p_0(z)$ is the height-dependent reference state pressure.

The air was assumed as a mixture of liquid water (mass fraction q_l), water vapour (mass fraction q_v) and dry air (mass fraction q_d), the rate of phase change was assumed to be infinitely fast, and the two phases were assumed to be in thermo-chemical equilibrium. The vapour and liquid water were modeled following Equations (8) and (9).

$$\frac{\partial q_v}{\partial t} + u_\alpha \frac{\partial q_v}{\partial x_\alpha} = \frac{\partial}{\partial x_\alpha} \left(D_q \frac{\partial q_v}{\partial x_\alpha} \right) - \dot{Q} \tag{8}$$

$$\frac{\partial q_l}{\partial t} + u_\alpha \frac{\partial q_l}{\partial x_\alpha} = \frac{\partial}{\partial x_\alpha} \left(D_q \frac{\partial q_l}{\partial x_\alpha} \right) + \dot{Q} \tag{9}$$

where $\dot{Q}$ is a source term (typically the condensation) and D_q is the water fraction diffusivity coefficient.

3.1. Double Convective Rayleigh-Bénard with Humid Air

The first validation case aims at validating the capability of the method to capture buoyancy effects, including both temperature and humidity gradients. It addresses the flow in a 2D square domain with temperature and humidity differences between its bottom and top boundaries as shown in Figure 1. Two Rayleigh numbers of 10^4 and 10^5 as well as two grids with $\delta x = 0.02$ m or half and $\delta t = 0.0115$ s or half were considered. Condensation was neglected and details of the model settings can be found in Feng et al.'s work [25].

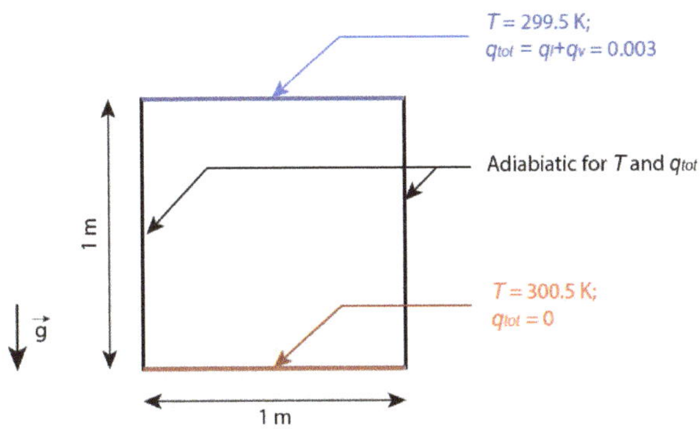

Figure 1. Configuration of the double convective Rayleigh-Bénard.

HRR-LBM simulation results were compared to the reference solution of Ouertatani et al. [27], which used a finite volume method discretized using the QUICK scheme in the momentum equation, and a second order central difference scheme in the energy equation, with a nonuniform grid of 256^2 points.

The comparison between the simulated and reference velocity profiles along the domain midlines (Figure 2), as well as local Nusselt number through the hot wall, potential temperature and total water humidity contours when the steady state had been achieved (not shown here), highlights a good performance of the developed HRR-LBM.

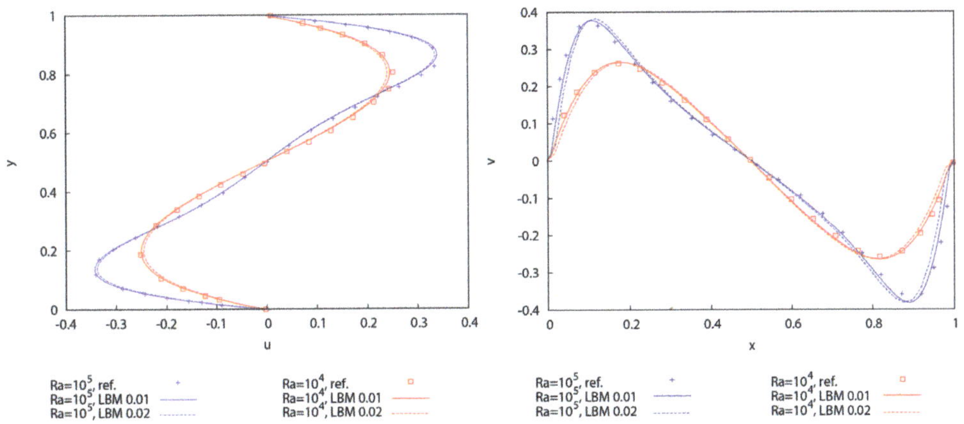

(**a**) u along y (**b**) v along x

Figure 2. Reference and simulated velocity profiles along the midlines for the double Rayleigh-Bénard convection.

3.2. 2D and 3D Rising Moist Bubbles

The second and third validation cases focus on a 2D and a 3D moist bubbles rising in a 2D or a 3D rectangular domain as shown in Figure 3. In both cases, the domain height was 2400 m and the domain width was 3600 m.

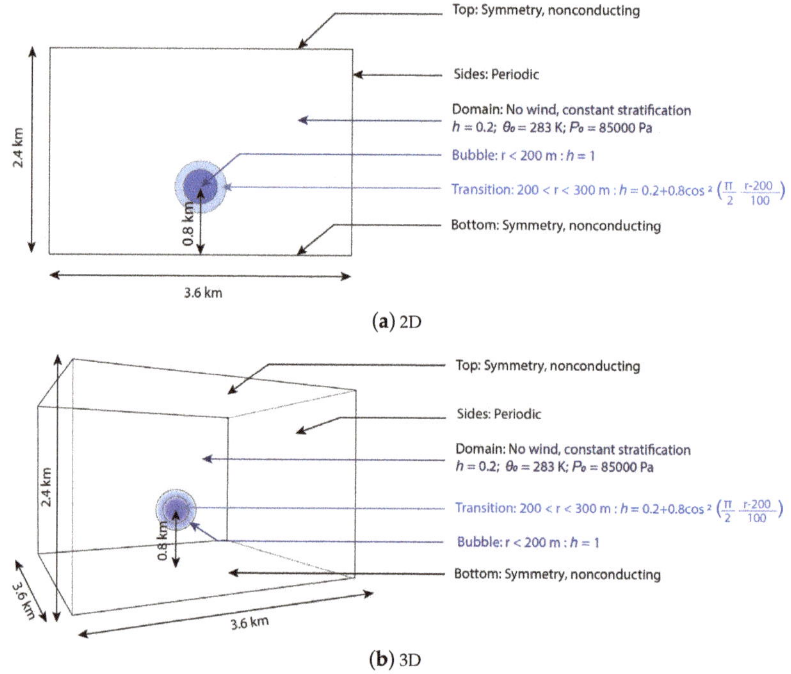

Figure 3. Configurations of the rising moist thermal bubbles.

In the 2D configuration, a uniform grid with $\delta x = 5\,\text{m}$ and $\delta t = 0.034\,\text{s}$ was used, while in the 3D configuration a nonuniform grid with finest $\delta x = 6.25\,\text{m}$ and $\delta t = 0.18\,\text{s}$ was considered. Other model settings can be found in Feng et al.'s work [25]. 2D simulations were performed for 10 min physical time and the results were compared to reference data at 3, 5 and 7 min. 3D simulations were run for 6 min physical time and the results were compared to reference data at 2, 4 and 6 min.

HRR-LBM simulation results were compared to benchmark solutions, which used a multidimensional positive definitive advection transport algorithm and an anelastic solver with a 2.5 m grid in 2D configuration and on a grid with a 6.25 m finest grid spacing in 3D configuration [28,29].

Regarding the 2D bubble, HRR-LBM and reference results were compared in terms of highest vertical position of the 20% maximum q_l contours and vertical fluid velocity at the top central position of the interface at different times (Table 1). As in the reference study, the HRR-LBM results show the rising and expansion of the moist bubble as the phase changes. HRR-LBM results are thus generally in good agreement with the reference data except for the highest vertical position of the 20% maximum q_l contours and the vertical fluid velocity at the top central position of the interface, especially at 7 min, which could be explained by the Boussinesq and not the aneleastic approximation used in the current study.

Table 1. Highest vertical location of the 20% maximum q_l contour (H_{20}) and vertical fluid velocity at the top central position of the interface (W_f) at 3, 5 and 7 min.

Physical Time	Variables	HRR-LBM (2 eq.)	[28]
3 min	H_{20} (m)	1128	1194
	W_f (m s^{-1})	1.60	1.59
5 min	H_{20} (m)	1278	1363
	W_f (m s^{-1})	1.21	1.21
7 min	H_{20} (m)	1374	1468
	W_f (m s^{-1})	0.56	0.72

Regarding the 3D case, the good performance of the HRR-LBM model was highlighted by the comparison of the vertical velocity profiles along the midline of the domain (Figure 4) as well as liquid humidity profiles and the liquid and vapour contours (not shown).

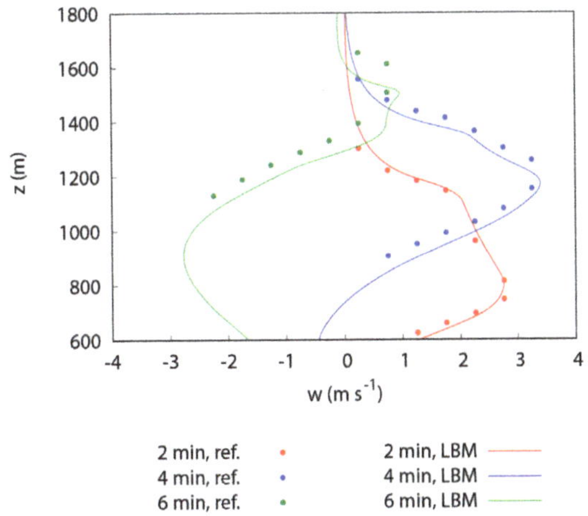

Figure 4. Reference and simulated vertical velocity profiles along the midline for the 3D moist bubble at 3 different times.

3.3. *Atmospheric Cloud Formation*

Finally, the fourth validation case addresses a convective atmospheric boundary layer case with a shallow cumulus formation to assess the capability of the HRR-LBM-WMLES model to predict the moist thermodynamics and its interactions with meteorological flows. The configuration refers to the model intercomparison case of the Barbados Oceanographic and meteorological Experiment (BOMEX). The considered domain is a 5 km long and 3 km high square-based domain with temperature and humidity surface fluxes and underwent an altitude-dependent geostrophic wind. Table 2 gives the related initial conditions.

Table 2. Initial conditions set for the shallow cumulus convection case.

Height (m)	q_l (g · kg^{-1})	θ_l (K)	u (m · s^{-1})	v (m · s^{-1})
0	17.0	298.7	−8.75	0
520	16.3	298.7		
700			−8.75	
1480	10.7	302.4		
2000	4.2	308.2		
3000	3.0	311.85	−4.61	0

For this configuration, HRR-LBM-WMLES simulations considered a $\delta x = 40$ m grid with $\delta t = 0.27$ s. The Monin Obukhov similarity theory was used as the surface model of the horizontal momentum components, temperature and humidity, and additional sources terms were added to the model to represent the large scale effects that cannot be included in LES. Details of the model settings can be found in Feng et al.'s work [24,25].

Simulations were run for 6 h physical time and statistics were computed on 1 h.

The comparison of HRR-LBM and reference results of Siebesma et al. [30] in terms of profiles of mean velocity (Figure 5), potential temperature, vapour and liquid water (not shown) highlights a good performance of the developed HRR-LBM-WMLES model, as the different atmospheric layers (mixed, conditionally unstable and inversion, from the ground upwards) and the instantaneous formation of the cloud are well captured.

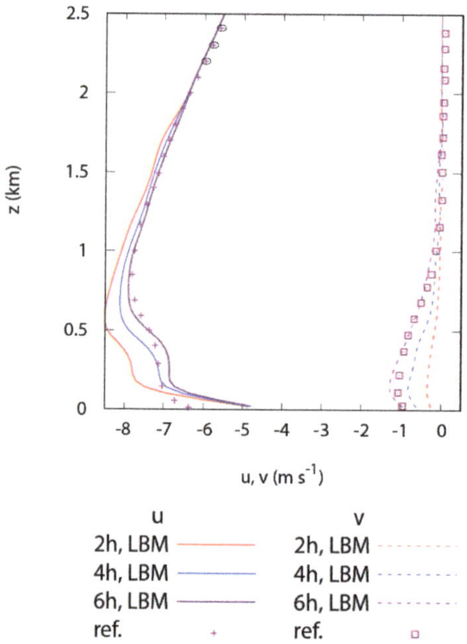

Figure 5. Reference and simulated profiles of mean velocity for the shallow cumulus convection case.

Hence, the hybrid LBM-based atmospheric numerical model, including the HRR collision model with forcing terms, and a finite volume method for temperature and water transport equation appear well suited to predict atmospheric humid convection with cloud formation, even considering large scale sources.

4. HRR-LBM-WMLES Application to Urban Pollutant Dispersion

This section reviews different case studies performed to assess the performance and the benefits of the proposed approach regarding the prediction of outdoor urban pollutant dispersion. Geometric complexity (trees, urban patterns) and buoyancy effects (pollutant or thermals) were especially studied.

4.1. Dispersion in a Street Canyon Including Tree Planting

A first validation study deals with dispersion of neutral traffic such as pollutant emissions in an ideal street canyon planted or not along a centred row of more or less dense tree crowns ($\lambda = 0$, 80 and 200 m^{-1}, respectively) and undergoing a perpendicular wind (Figure 6).

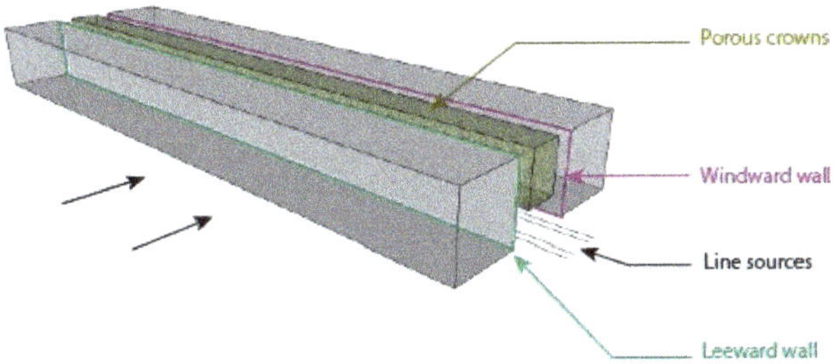

Figure 6. Configuration of street canyon with tree planting case.

The drag force (F_{por}) induced by porous media is modeled via a Forchheimmer formulation, as follows:

$$F_{por} = -\rho \times R \times |\mathbf{u}| \times \mathbf{u} \times \Phi \tag{10}$$

where $R = \lambda/2$ is the drag force coefficient and Φ is the ratio of porous media immersed in the volume cell.

Simulations were performed at reduced scale (1:150, $H = 0.12$ m) as in the reference experiment [31], with a domain of $3 \times 2 \times 1$ m^3 and the street canyon located 0.84 m from the inlet. Table 3 gives the different boundary conditions. Accounting for δx = 0.00125 m, $\delta t = 1.44 \times 10^{-5}$ s and five refinement levels, the total number of grid points was 41×10^6. Simulations were run on 240 cores for 25 s physical time, with the last 10 s being kept for postprocessing. Details can be found in Merlier et al.'s work [32].

Table 3. Model settings for the street canyon dispersion case.

Domain dimensions (m^3)	$3 \times 2 \times 1$
Reference height (m)	$H = 0.12$ m
Reference velocity (m s^{-1})	4.65
Inlet	Imposed velocity and turbulent intensity values
Outlet	Imposed pressure value
Top boundary condition	Slip wall
Lateral boundary condition	Slip wall
Ground boundary condition	Wall law
Building boundary condition	Wall law

Quantitative (Table 4) and qualitative (not shown) comparisons of experimental and numerical mean concentrations on the leeward and windward walls of the street canyon highlight a state of the art performance of the HRR-LB model as compared to other reference studies.

Table 4. Quality metrics for the street canyon. References according to Chang and Hanna [33].

Case	**FAC2**	**FB**	**RNMSE**
Target	1	0	0
Tolerance	>0.5	$]-0.3; 0.3[$	<1.2
$\lambda = 0$ m^{-1}	0.9	−0.1	0.3
$\lambda = 80$ m^{-1}	0.7	0.1	0.4
$\lambda = 200$ m^{-1}	0.4	0.2	0.5

Table 4 especially shows that the global statistical performance indicators generally match the acceptance criteria suggested by Chang and Hanna [33]. Nonetheless, the results show a better agreement between numerical and experimental concentrations on the leeward wall, where the concentration is higher than on the windward wall, and for less dense tree crowns ($\lambda = 0$ and $80\ \mathrm{m}^{-1}$). The analysis of concentration distributions on walls also highlights higher concentrations on the streamwise symmetry plane as in the reference experiment.

In addition, thanks to high-fidelity unsteady simulations, the analysis of concentration statistics at different locations in the street canyon highlighted that high peaks of concentration can occur notably in the presence of dense tree crowns, which could be particularly harmful for short time exposure problems.

4.2. Dispersion behind a Building under Unstable Stratification

A second validation study deals with the dispersion of a gas released from a ground source just downwind of an isolated building located in an unstable boundary layer (Figure 7).

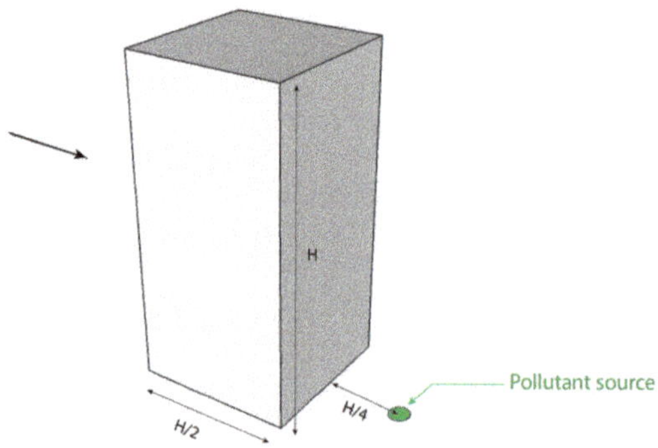

Figure 7. Configuration of the isolated building located in an unstable boundary layer case.

Simulations were performed at a reduced scale ($H = 0.16\,\mathrm{m}$) as in the reference study [34], with a domain of $2 \times 1.2 \times 1\ \mathrm{m}^3$ according to AIJ guidelines [35], the building being located 0.32, 0.6 and 0.84 m from the inlet, lateral and top domain boundaries, respectively. Table 5 gives the corresponding boundary conditions. Accounting for $\delta x = 0.002\,\mathrm{m}$, $\delta t = 12.8 \times 10^{-5}$ s and five refinement levels for the coarse grid case and $\delta x = 0.001\,\mathrm{m}$, $\delta t = 6.4 \times 10^{-5}$ s and six refinement levels for the medium grid case, the total number of grid points was 4.7×10^6 for coarse grid and 11.2×10^6 for the medium grid. Simulations were carried out on 28 or 56 cores for 16.25 s physical time, the last 6.4 s being kept to compute the statistics.

Table 5. Model settings for the isolated building located in an unstable boundary layer case.

Domain dimensions (m^3)	$2 \times 1.2 \times 1$
Reference height (m)	0.16
Reference velocity ($\mathrm{m\,s^{-1}}$)	1.37
Inlet	Imposed velocity, temperature and turbulent kinetic energy (Figure 8)
Outlet	Imposed pressure
Top boundary condition	Slip Wall
Lateral boundary condition	Slip Wall
Ground boundary condition	Wall model with imposed temperature ($T_f = 45.3\,^\circ\mathrm{C}$)
Building boundary condition	Wall model with imposed temperature ($T_{build} = 41.7\,^\circ\mathrm{C}$)

The pollutant source with a diameter of 0.005 m is located 0.04 m downstream the building. A gas flux of $q = 9.17 \times 10^{-6}$ m^3 s^{-1} with a temperature of 30.4 °C was imposed. The inflow velocity, temperature and turbulent kinetic energy were interpolated from experimental data provided in the TPU database [36] and are plotted in Figure 8.

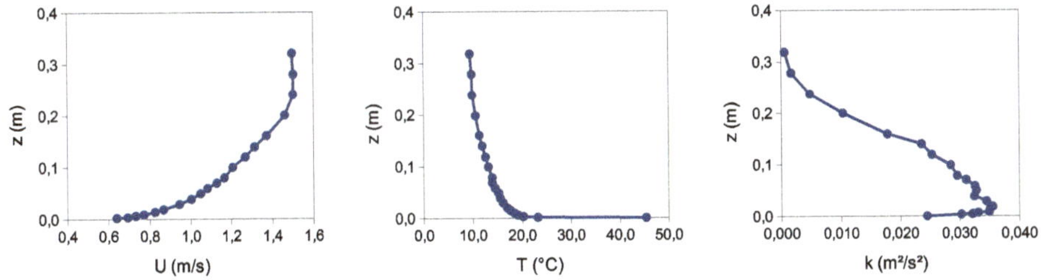

(**a**) velocity (**b**) temperature (**c**) turbulent kinetic energy

Figure 8. Inlet profiles for the isolated building located in an unstable boundary layer.

The velocity, temperature and concentration of tracer gas were compared to experimental data at four different locations downstream of the building. Velocity was normalized with the reference velocity at building height ($U* = U/U_H$), the temperature was normalized using floor temperature (T_f and the difference between floor temperature and temperature at building height ($T* = (T - T_f)/(T_f - T_H)$) and the concentration was normalized by release gas flux, building height and reference velocity ($c* = (cU_H H^2)/q$).

Figures 9 and 10 present the comparison of the HRR-LBM results obtained with a coarse and medium grid with the experimental data of Yoshie et al. [37]. A good agreement was obtained for concentration field downstream of the building and the present results are more accurate than other literature LES simulations related to the same case [34,38]. A fairly good agreement was found for streamwise velocity profiles downstream of the building, although velocity is generally underestimated compared to experimental data.

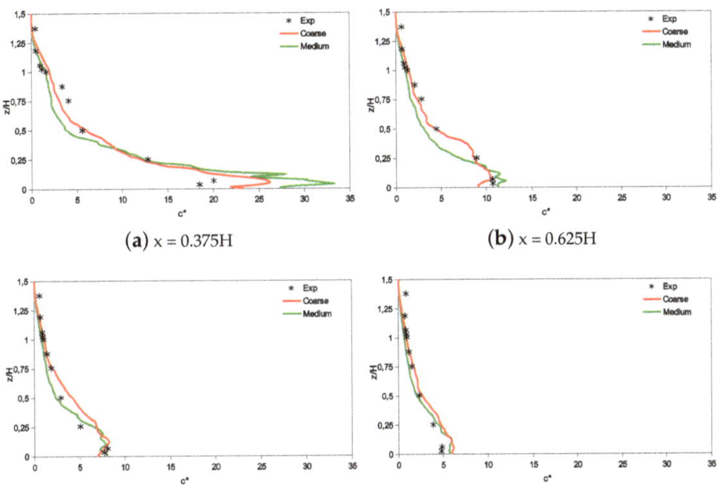

(**a**) x = 0.375H (**b**) x = 0.625H (**c**) x = H (**d**) x = 1.5H

Figure 9. Comparisons of normalized pollutant concentrations at different positions downstream of the building.

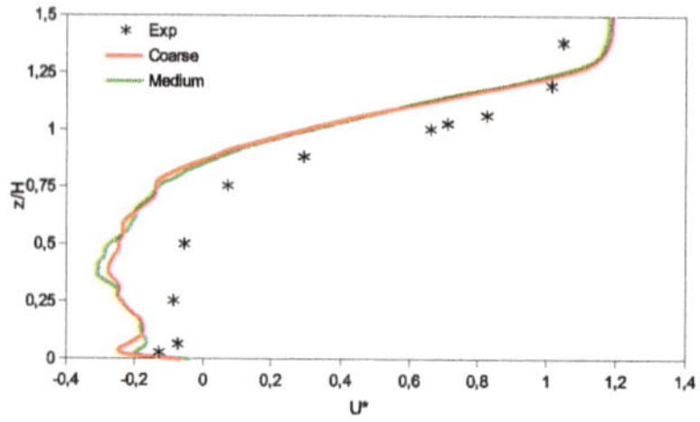

(a) x = 0.375H

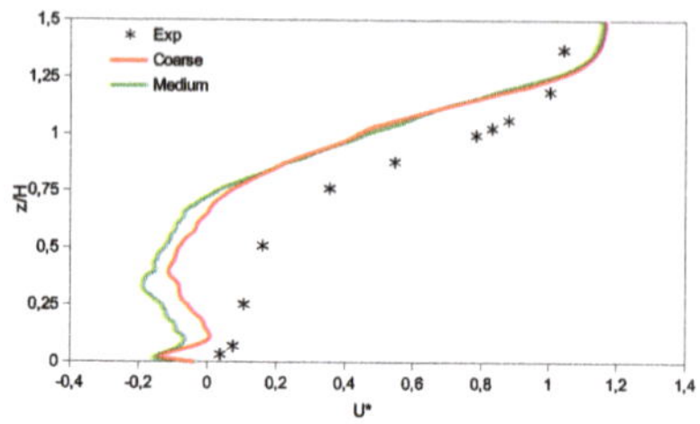

(b) x = 0.625H

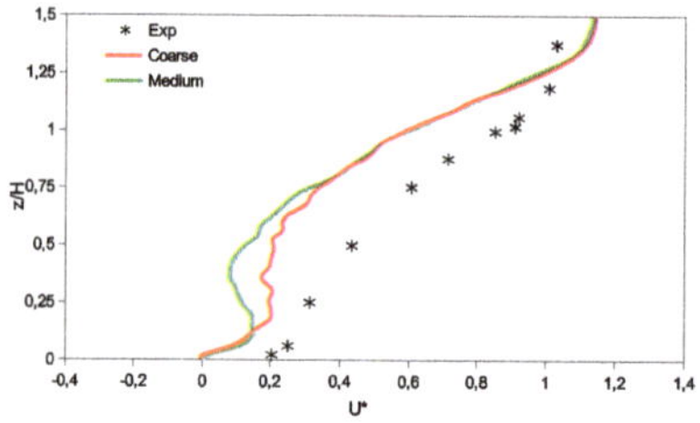

(c) x = H

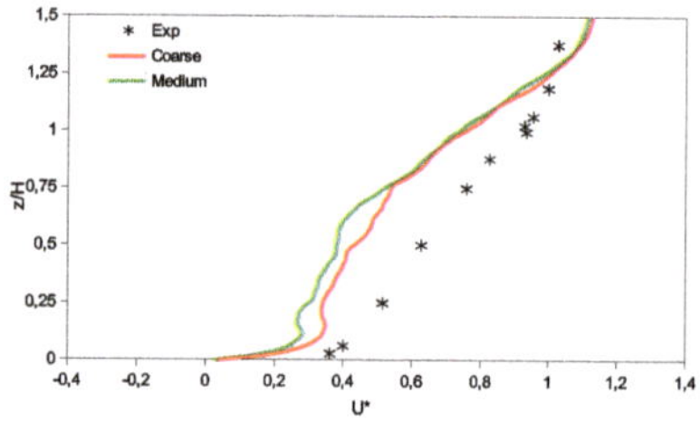

(d) x = 1.5H

Figure 10. Comparisons of normalized streamwise velocity at different positions downstream of the building.

4.3. *Dispersion of Neutral and Dense Gas in an Urban Area*

A third validation study deals with the dispersion of neutral and dense gas from a ground source in a realistic urban environment. Two different source locations and wind incidences (C1: neutral and dense gas emitted in a rather channeled flow in a large avenue, C2: neutral gas emitted in a surround built crossroad, see Figure 11) are considered.

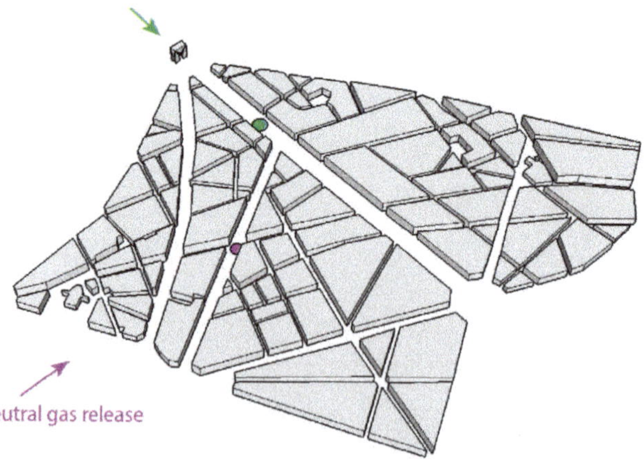

Figure 11. Configuration of the urban area dispersion case.

Simulations were performed at reduced scale (1:350, $H_{moy} = 0.078\,\text{m}$) as in the reference experiment [39], with a domain of 8.75 (C1) or 9.5 (C2) $\times 3.5 \times 1.5\,\text{m}^3$, with the model located 1.15 m from the inlet of the domain. Table 6 gives the different boundary conditions.

Table 6. Model settings for the urban area dispersion case.

Domain Dimensions (m^3)	8.75 or 9.8 × 3.5 × 1.5
Reference height (m)	0.078
Reference velocity ($m\,s^{-1}$)	1
Inlet	Imposed velocity
Outlet	Imposed pressure value
Top boundary condition	Slip wall
Lateral boundary condition	Wall law
Ground boundary condition	Wall law
Building boundary condition	Wall law

Accounting for $\delta x_{min} = 1.75 \times 10^{-3}$ m and $\delta t = 1.5 \times 10^{-4}$ s with six refinement levels, the total number of grid points was 175×10^6 (C1) or 220×10^6 (C2). Simulations were run on about 10^3 cores for 27.5 s physical time and statistics were computed over 5 s. More details can be found in the work of Merlier et al. [40].

The performance indicators given in Table 7 for concentration at street level highlight a good performance of the model, especially for configuration 2.

Table 7. Quality metrics for the urban area dispersion case. References according to Hanna and Chang [41].

Case	**FAC2**	**FB**	**NMSE**
Target	1	0	0
Tolerance	≥0.3	[−0.67; 0.67]	≤6
C.1, neutral gas	0.46	−0.14	0.54
C.1, dense gas	0.38	0.47	0.62
C.2, neutral gas	0.55	−0.01	0.23

Results match the different acceptance criteria for urban dispersion models suggested in the work of Hanna and Chang [41], although the dense gas configuration shows a worse, but still acceptable, accuracy. Indeed, the analysis of the spatial distribution of concentration (not shown) highlights that:

- The neutral gas emitted in the channeled flow is carried mainly in the main street by the prevailing flow and progressively mixes upwards;
- The dense gas emitted in the channeled flow follows a rather circular and flat dispersion all around the source;
- The dispersion of the neutral gas emitted in the crossroad is notably vertical due to the presence of the obstacles built upstream and downstream.

The comparison of the HRR-LBM and reference vertical as well as horizontal concentration profiles above the canopy (not shown) are generally also satisfactory regarding both the mean and the standard deviation of concentration levels. These results suggest that the dynamics of dispersion processes are well reproduced by the model. Being capable of providing the temporal statistics of velocity and concentration, the developed approach appears well suited to support the design of fast response models.

5. HRR-LBM-WMLES-Based Urban Wind Prediction with Application to Pedestrian Wind Comfort and Building Wind Loads under Uncertainty

A next step toward an improved reliability of numerical simulations for realistic full scale urban applications is the capability to account for uncertainties that appear in the prescription of the atmospheric conditions, instantaneous wind conditions, surface roughness, pollutant source features, etc. The variability of the numerical solution with respect to uncertain parameters must be quantified in order to provide users with useful information, since a single fully deterministic solution is meaningless in such problems. To this end, efficient techniques for Uncertainty Quantification (UQ) must be used. A challenging issue is that most existing UQ techniques require the use of a significant number of samples, each sample being currently an unsteady high-fidelity LES of the case under consideration. Therefore, adequate UQ methods must be defined that minimize the number of required simulations while preserving the accuracy of the uncertainty propagation in the results. Such a method, the c-APK method, was developed by Margheri and Sagaut [42] with application to urban flow simulations in complex areas.

The c-APK method will not be detailed here for the sake of brevity, and the reader is referred to Margheri and Sagaut [42] for details. Nonetheless, to highlight two additional applications of the c-APK method, this section reviews two validation studies and two projected applications related to the prediction of urban air flows with UQ: pedestrian wind comfort and wind loads on buildings. The configuration studied for the projected applications is a new hypothetical tower located in a complex and dense urban area.

5.1. *Prediction of Mean Flow Field in a Complex Urban Environment*

The first validation study focuses on a full scale urban area of Tokyo, which includes an area of low-rise buildings upstream a cluster of towers. Simulations were performed at full scale, considering $H_{max} = 225$ m, to evaluate the performance of the model at a realistic Reynolds number, as both full scale and reduced scale data are provided in the reference experimental data [43]. More precisely, the configuration studied corresponds to the case F of the open source Architectural Institute of Japan's database, which gathers full scale and wind tunnel measurements, wind tunnel tests corresponding to the 1977 full scale measurements campaign.

Figure 12 details the different model dimensions and boundary conditions used for the simulations.

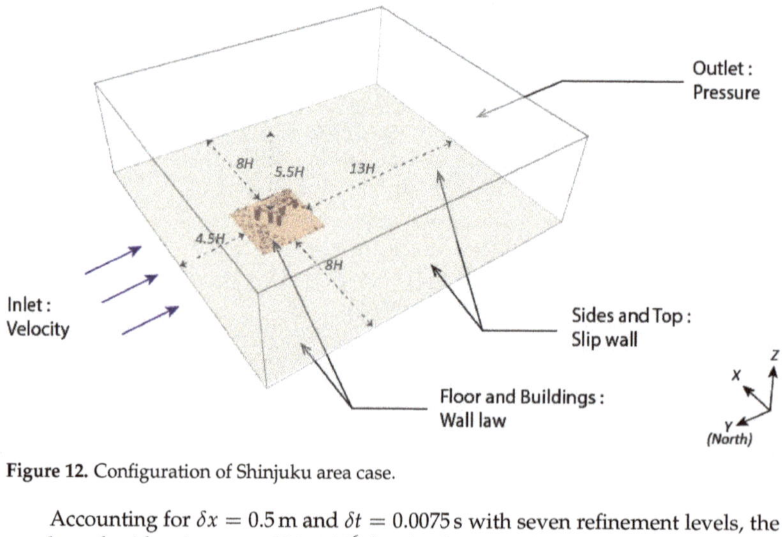

Figure 12. Configuration of Shinjuku area case.

Accounting for $\delta x = 0.5$ m and $\delta t = 0.0075$ s with seven refinement levels, the total number of grid points was 136×10^6 for the finest grid (respectively, $\delta x = 1$ m, $\delta t = 0.015$ s, six refinement levels and 54×10^6 grid points for medium grid). Simulations were run on 504 cores (240 for medium grid) for 2 h physical time, the last 1 h being kept for postprocessing. More details can be found in the work of Jacob and Sagaut [44].

The HRR-LBM and reference mean velocities reported in Figure 13 at 10 m high and for different locations in the area of interest evidence a good performance of the developed model and also for mean velocity.

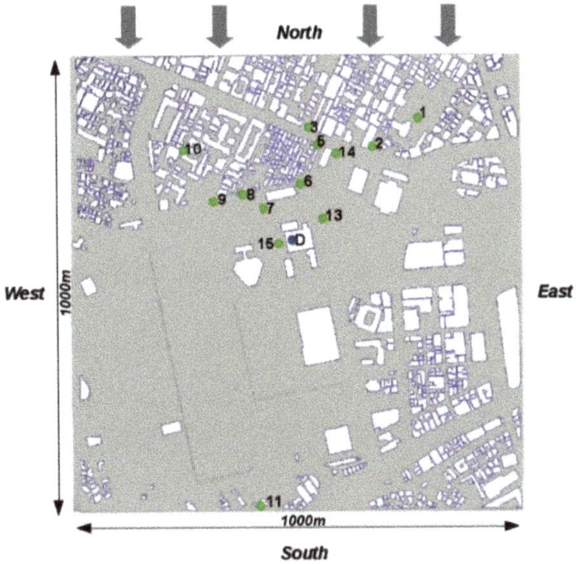

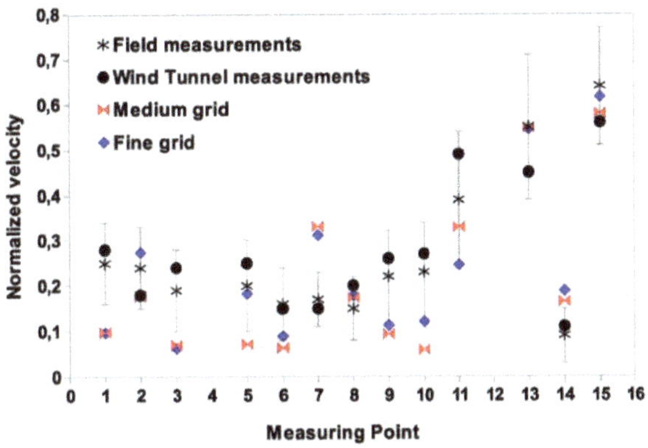

Figure 13. Reference and simulated mean velocities for the Shinjuku area case. Medium grid results are in red; fine grid results are in blue.

5.2. Prediction of Surface Pressure on a High-Rise Building

To evaluate the accuracy of the developed approach also on building surface quantities before coupling the HRR-LBM-WMLES tool with a UQ technique, a second validation study was extensively carried out at a reduced scale isolated high-rise building (1:300, $H = 0.49$ m), for which detailed measurement data of velocity and pressure on the model surface are available [45]. Figure 14 details the different model dimensions and boundary conditions, which included an extension of the original incompressible SEM to reconstruct the inlet turbulence in the current LBM framework, given the importance of the turbulent inflow properties dynamically studying wind loading on isolated structures.

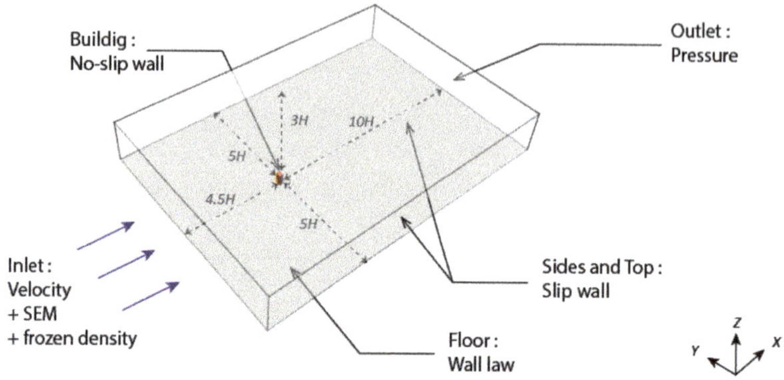

Figure 14. Configuration of the isolated high-rise building case.

With $\delta x = 1.56 \times 10^{-3}$ m and $\delta t = 1.5 \times 10^{-5}$ s and six refinement levels, the total number of grid points was 5.8×10^6. More details can be found in the work of Buffa et al. [23].

The comparison of mean and standard deviation values of pressure coefficients at the building surface for the different faces of the building and at different heights (Figure 15) shows a very satisfactory match for all the tested data, especially for the mean Cp value. Experimental and simulated data were also extensively compared in terms of spectral analysis and local instantaneous pressure maxima (not shown), showing the reliability of the developed approach for wind load prediction.

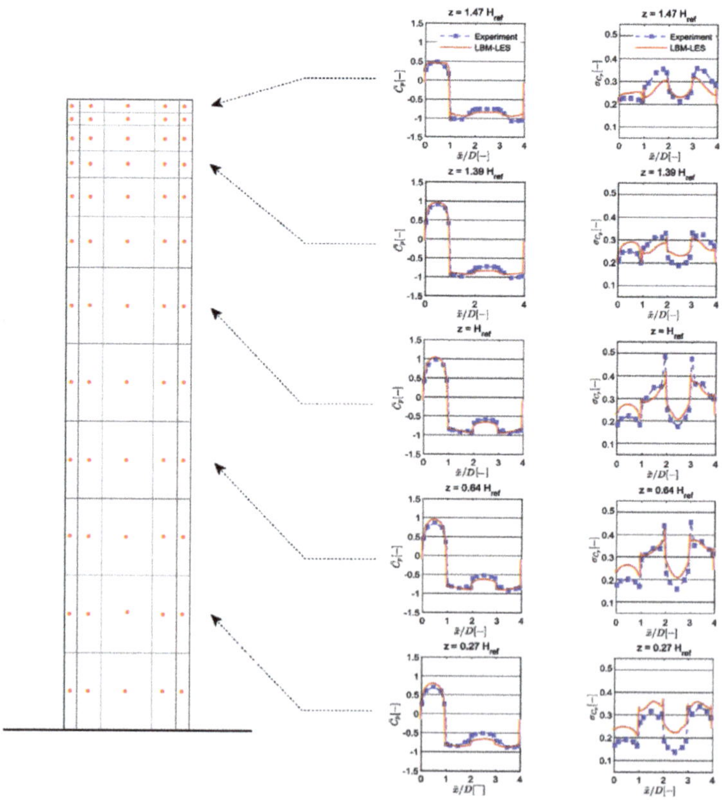

Figure 15. Reference and simulated pressure coefficients for the isolated high-rise building case.

5.3. Wind Comfort Assessment with Uncertainty Quantification

Thanks to the possibilities offered by the developed dynamic model, pedestrian wind comfort at street level ($H = 2\,\text{m}$) was studied considering Melbourne criteria [46] in an area of $400 \times 400\,\text{m}^2$ inside Shinjuku area as shown in red in Figure 16.

Figure 16. Integration of the new hypothetical buildings in the Shinjuku urban area.

The domain used in this study is the same as in Section 5.1. Simulations were run on 1.5 h physical time and the statistics were computed over the last hour. As described by Jacob and Sagaut [44], two buildings were added in Shinjuku area: one in the middle of the area (in blue in Figure 16) for wind comfort assessment and one (in green in Figure 16) for the study of Section 5.4. The case presented by Jacob and Sagaut [44] with the wind coming from the north is here considered as the reference sample for the UQ analysis, and several other samples have been considered by changing the velocity magnitude by a factor α ($0.6 \leq \alpha \leq 1.4$) and wind direction with an angle θ around the north direction ($-30 \leq \theta \leq 30°$). A set of 57 simulations have been performed for this study permitting to estimate first the sensitivity of the mean velocity field to α and θ and to compute the mean value of mean velocity at pedestrian level using the c-APK method.

Figure 17 shows the Sobol indices obtained for the mean velocity field at the pedestrian level, which permits quantification the influence of each parameter on the global variance of the system. The results highlight that the inflow velocity magnitude is generally the main influential parameter in most locations in the area of interest. In some locations, wind direction has a large contribution on the variance, especially in areas that can be located in wake for some inflow directions.

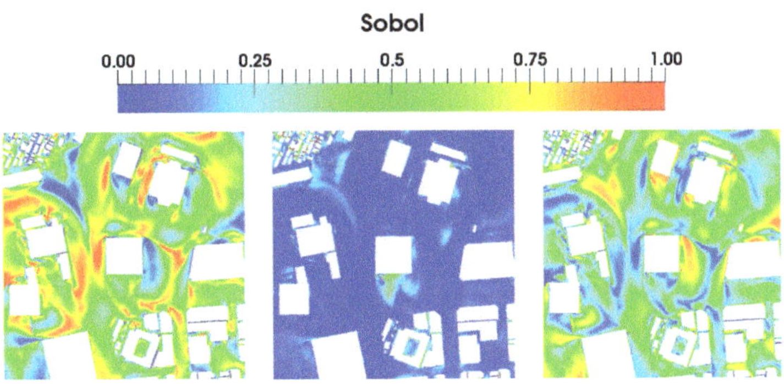

Figure 17. Sobol indices of the mean velocity computed from c-APK model for (**a**) α index, (**b**) $\alpha - \theta$ index and (**c**) θ index.

Figure 18 shows the classification obtained following the Melbourne criteria using the data of the reference sample ($\alpha = 1, \theta = 0$) and the mean data obtained from the 57 samples. The results highlight that by using the c-APK output, a large part of the area is located in Zones A, B and C where pedestrian comfort can be considered as good, whereas a part of the area is unclassified (here the pedestrian comfort is not so good) using only the reference sample.

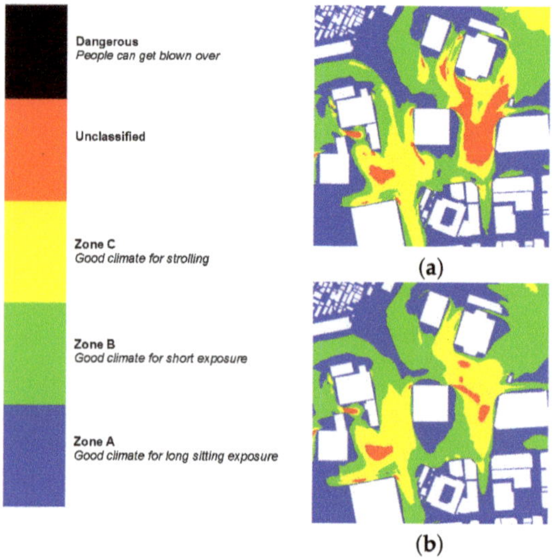

Figure 18. Visualization of the different wind comfort zones following (**a**) Melbourne criteria based on the reference data ($\alpha = 1, \theta = 0$) and (**b**) Melbourne criteria based on the UQ data.

5.4. Prediction of Structural Wind Loads with Uncertainty Quantification

The same configuration as in Section 5.3 was considered to study pressure distribution on the facades of the hypothetical building (green building in Figure 16) for further applications to the prediction of structural wind loads. The analysis is based on the same set of simulations changing the inflow velocity magnitude and direction.

The Sobol indices plotted on the different building faces in Figure 19 highlight that the inflow velocity magnitude is the most influential parameter on the pressure distribution on the building faces, except on a part of the north face, where the wind direction appears to influence pressure values more than the wind velocity magnitude.

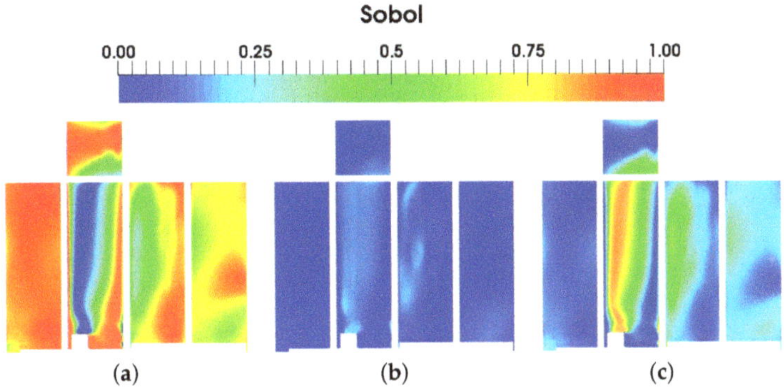

Figure 19. Sobol indices of the mean pressure on building walls computed from c-APK model for (**a**) α index, (**b**) $\alpha - \theta$ index and (**c**) θ index. The building faces are presented here, from left to right, in the order east, north, west and south, with the top face on top of the figure.

This is explained by the fact that a large part of the north face is located in a wake area when the flow is coming from the northeast direction, whereas it is directly exposed to inflow wind when it comes from the northwest direction. The second order term is very low compared to the others and does not significantly contribute to the global variance of the pressure on building faces. Figure 20 presents the pressure coefficients obtained in the case of the reference sample ($\alpha = 1, \theta = 0$) on the different building faces and the pressure coefficients computed from the average pressure given by the c-APK analysis. Few differences were observed on top, east, west and south faces, whereas the estimated pressure coefficient was lowered by the c-APK analysis on the north face compared to the reference sample.

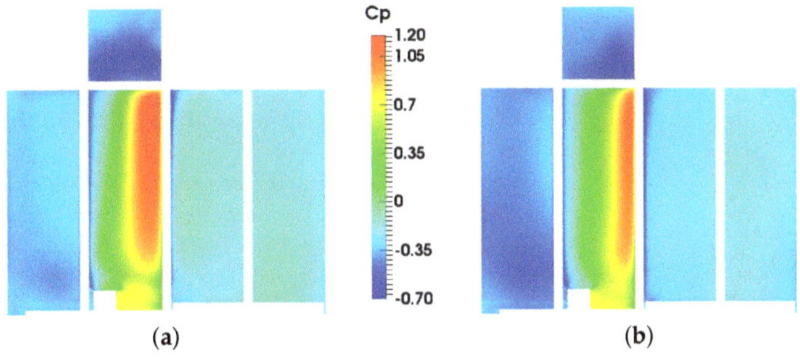

Figure 20. Pressure coefficients obtained on the building faces for (**a**) the reference sample ($\alpha = 1$, $\theta = 0$) and (**b**) the c-APK average over the 57 samples. The building faces are presented here, from left to right, in the order east, north, west and south, with the top face on top of the figure.

This result suggests that uncertainties on the inflow wind should be accounted for in urban simulations since it has an impact on the dynamics of the flow around high-rise buildings and on the related surface quantities.

6. Indoor Pollutant Dispersion with Thermal and Moving Body Effects

Another step toward the numerical simulation of realistic full scale configuration is the use of high-fidelity CFD for evacuation problems. In such cases, human agents present in the domain under consideration may have an effect on the pollutant dispersion because they trigger some physical mechanisms (mixing induced by the wakes of moving persons, natural convection due to the heat release by persons, breathing effects, etc.), leading to the definition of a two-way coupling problem. In such a problem, the behaviour of human agents has a direct influence on the flow, but it is governed by their responses to external parameters, which result from the conjunction of physical (e.g., heat) and psychological parameters. Therefore, a more complex level of modeling is required that couples classical physical CFD to psychological and behavioural models.

The present HRR-LBM-WMLES model has been coupled with a Social Force Model (SFM) [47–50] that allows us to evaluate individual human displacement, taking into account some psychological effects, and therefore to account for the effect of crowd evacuation on pollutant dispersion. The trajectories of people leaving a room are evaluated using SFM, then the drag force of each person is added into LBM equation using the actuator line method (ALM) [51] to account for the effect of human motion on the fluid, leading to the occurrence of wakes.

An application of this coupled method is shown here. It is related to an evacuation scenario inside a concert hall [52]. This hall (with dimensions $83.2 \times 68.4 \times 15.1\ \mathrm{m}^3$) contains 12 exits, indicated in green in Figure 21, and is initially occupied by 6026 persons. The concert hall is equipped with ventilation systems located on the room ceiling that allow for the balancing of the heat release of the persons in the concert hall.

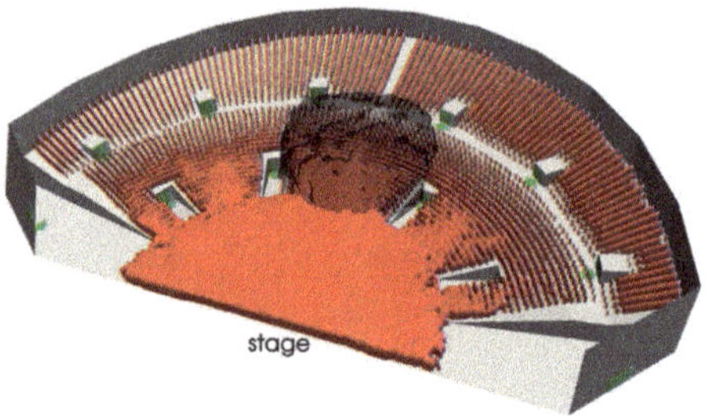

Figure 21. Configuration of the concert hall simulation. The exits are marked in green and the person initial positions of people are indicated in red.

We consider an instantaneous pollutant release from a cloud with a 20 m diameter in the centre of the concert hall, as shown in Figure 21. Evacuation starts at the same time as pollutant release and 299.2 s are necessary to allow all the people inside the concert hall to exit.

The pollutant field at human head level obtained at several times is displayed in Figure 22. We can see here that the pollutant is advected inside persons wake from the initial cloud to the two central upper exits. Pollutant dispersion towards the other exits is not significant since people leaving the concert all through these exits were not initially located inside the cloud.

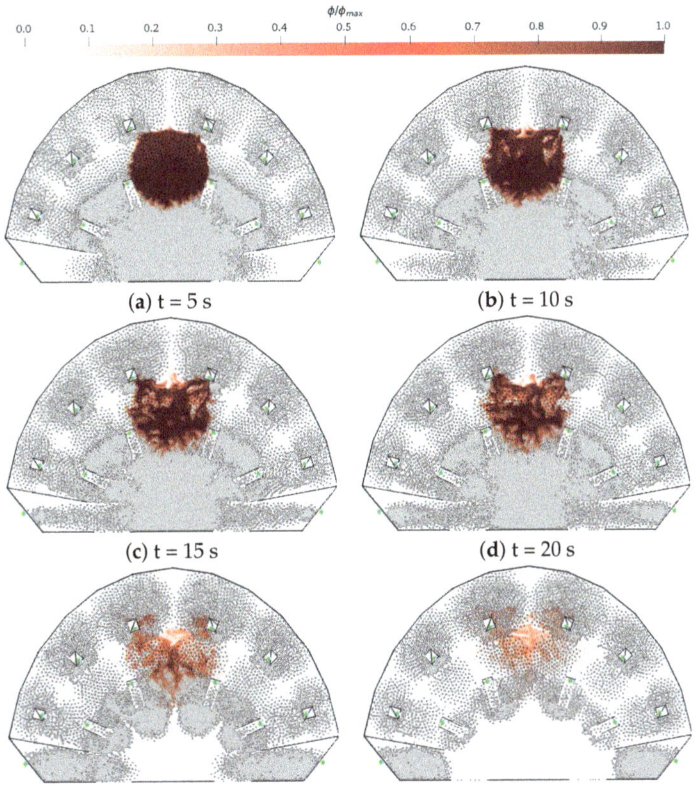

(a) t = 5 s (b) t = 10 s

(c) t = 15 s (d) t = 20 s

(e) t = 40 s (f) t = 60 s

Figure 22. Visualisation of pollutant field at head level at different time step in case of an evacuation of a concert hall (person locations are indicated in grey).

7. Concluding Remarks

The versatility and efficiency of the HRR-LBM-WMLES approach for atmospheric flow simulations, urban physics and pollutant transport prediction have been illustrated by a broad range of applications including humidity effects with phase changes and evacuation prediction with coupling to a social force model.

The main advantages of the present approach are its efficiency in terms of computational time, which is due to the explicit nature of the lattice Boltzmann method, the compactness of the underlying stencil, and the preprocessing time which is drastically reduced thanks to the use of embedded uniform grids along with an immersed boundary approach to handle complex fully arbitrary geometries.

In all cases, a very satisfactory agreement with reference data (if they exist) is reported, demonstrating the accuracy of the simulation tool. The coupling with the c-APK method for uncertainty quantification was also illustrated, showing that the HRR-LBM-WMLES tool is fast enough to allow for the use of UQ tools in complex full scale configurations.

Hence, thanks to its reliability—highlighted through the different validation studies—relevance—highlighted by the different physical analyses carried out—and its computational efficiency—induced by its numerical properties—the developed method appears very to be promising to support the design of fast response models and urban decision making.

Author Contributions: Conceptualization, J.J., L.M., F.M. and P.S.; methodology, J.J., L.M., F.M. and P.S.; software, J.J.; validation, J.J., L.M. and F.M.; formal analysis, J.J., L.M. and F.M.; writing—original draft preparation, J.J. and L.M.; writing—review and editing, F.M. and P.S.; supervision, P.S.; project administration, P.S. All authors have read and agreed to the published version of the manuscript.

Funding: This research received no external funding.

Institutional Review Board Statement: Not applicable.

Informed Consent Statement: Not applicable.

Data Availability Statement: Not applicable.

Acknowledgments: Centre de Calcul Intensif d'Aix-Marseille is acknowledged for granting access to its high-performance computing resources. This work was granted access to the HPC resources of TGCC/CINES under the allocation 2021-A0092A07679 made by GENCI.

Conflicts of Interest: The authors declare no conflict of interest.

Abbreviations

The following abbreviations are used in this manuscript:

CFD	Computational Fluid Dynamics
LES	Large Eddy Simulation
LBM	Lattice Boltzmann Method
SEM	Synthetic Eddy Method
BGK	Bhatnagar Gross Krook
MRT	Multiple Relaxation Time
HRR	Hybrid Recursive Regularized
WMLES	Wall Model Large Eddy Simulation
QUICK	Quadratic Upstream Interpolated Scheme for Convective Kinematic
c-APK	centered Anova POD Kriging
UQ	Uncertainty Quantification
NS	Navier Stokes
ALM	Actuator Line Model

References

1. Schlünzen, K.H.; Grawe, D.; Bohnenstengel, S.I.; Schlüter, I.; Koppmann, R. Joint modelling of obstacle induced and mesoscale changes—Current limits and challenges. *J. Wind Eng. Ind. Aerodyn.* **2011**, *99*, 217–225. [CrossRef]
2. Sagaut, P. *Large Eddy Simulation for Incompressible Flows: An Introduction*; Springer Science & Business Media: Berlin/Heidelberg, Germany, 2006.
3. Blocken, B. 50 years of Computational Wind Engineering: Past, present and future. *J. Wind Eng. Ind. Aerodyn.* **2014**, *129*, 69–102. [CrossRef]
4. Blocken, B. LES over RANS in building simulation for outdoor and indoor applications: A foregone conclusion? In *Building Simulation*; Springer: Berlin/Heidelberg, Germany, 2018.
5. Krüger, T.; Kusumaatmaja, H.; Kuzmin, A.; Shardt, O.; Silva, G.; Viggen, E. *The Lattice Boltzmann Method. Principles and Practice*; Springer: Berlin/Heidelberg, Germany, 2017.
6. Chen, S.; Doolen, G.D. Lattice Boltzmann method for fluid flows. *Annu. Rev. Fluid Mech.* **1998**, *30*, 329–364. [CrossRef]
7. Slotnick, J.; Khodadoust, A.; Alonse, J.; Darmofal, D.; Gropp, W.; Lurie, E.; Mavriplis, D. CFD Vision 2030 study: A path revolutionnary computational aerosciences. *NASA-CR* **2014**, *2014*, 218178.
8. Ahmad, N.H.; Inagaki, A.; Kanda, M.; Onodera, N.; Aoki, T. Large-eddy simulation of the gust index in an urban area using the lattice Boltzmann method. *Bound.-Layer Meteorol.* **2017**, *163*, 447–467. [CrossRef]
9. Inagaki, A.; Kanda, M.; Ahmad, N.H.; Yagi, A.; Onodera, N.; Aoki, T. A numerical study of turbulence statistics and the structure of a spatially-developing boundary layer over a realistic urban geometry. *Bound.-Layer Meteorol.* **2017**, *164*, 161–181. [CrossRef]
10. Lenz, S.; Schönherr, M.; Geier, M.; Krafczyk, M.; Pasquali, A.; Christen, A.; Giometto, M. Towards real-time simulation of turbulent air flow over a resolved urban canopy using the cumulant lattice Boltzmann method on a GPGPU. *J. Wind Eng. Ind. Aerodyn.* **2019**, *189*, 151–162. [CrossRef]
11. Onodera, N.; Idomura, Y.; Hasegawa, Y.; Nakayama, H.; Shimokawabe, T.; Aoki, T. Real-Time Tracer Dispersion Simulations in Oklahoma City Using the Locally Mesh-Refined Lattice Boltzmann Method. *Bound.-Layer Meteorol.* **2021**, *179*, 187–208. [CrossRef]
12. King, M.F.; Khan, A.; Noakes, C. Coupled indoor/outdoor airflow simulation comparing ANSYS Fluent with a GPU-based lattice Boltzmann model for urban environments. In Proceedings of the Indoor Air 2015, Philadelphia, PA, USA, 22 July 2018.

13. King, M.F.; Khan, A.; Delbosc, N.; Gough, H.L.; Halios, C.; Barlow, J.F.; Noakes, C.J. Modelling urban airflow and natural ventilation using a GPU-based lattice-Boltzmann method. *Build. Environ.* **2017**, *125*, 273–284. [CrossRef]
14. CS. ProLB. 2021. Available online: http://www.prolb-cfd.com/ (accessed on 28 May 2021).
15. M2P2. LaBS. 2021. Available online: http://www.m2p2.fr/valorisation-6/transfert-technologique-labs-3202.htm (accessed on 28 May 2021).
16. Wilhelm, S.; Jacob, J.; Sagaut, P. An explicit power-law-based wall model for lattice Boltzmann method—Reynolds-averaged numerical simulations of the flow around airfoils. *Phys. Fluids* **2018**, *30*, 065111. [CrossRef]
17. Wilhelm, S.; Jacob, J.; Sagaut, P. A new explicit algebraic wall model for LES of turbulent flows under adverse pressure gradient. *Flow Turbul. Combust.* **2021**, *106*, 1–35. [CrossRef]
18. Cai, S.G.; Degrigny, J.; Boussuge, J.F.; Sagaut, P. Coupling of turbulence wall models and immersed boundaries on Cartesian grids. *J. Comput. Phys.* **2021**, *429*, 109995. [CrossRef]
19. Jacob, J.; Malaspinas, O.; Sagaut, P. A new hybrid recursive regularised Bhatnagar–Gross–Krook collision model for Lattice Boltzmann method-based large eddy simulation. *J. Turbul.* **2018**, *19*, 1051–1076. [CrossRef]
20. Guo, Z.; Zheng, C.; Shi, B. Discrete lattice effects on the forcing term in the lattice Boltzmann method. *Phys. Rev. E* **2002**, *65*, 046308. [CrossRef]
21. Feng, Y.; Guo, S.; Jacob, J.; Sagaut, P. Solid wall and open boundary conditions in hybrid recursive regularized lattice Boltzmann method for compressible flows. *Phys. Fluids* **2019**, *31*, 126103.
22. Jarrin, N.; Benhamadouche, S.; Laurence, D.; Prosser, R. A synthetic-eddy-method for generating inflow conditions for large-eddy simulations. *Int. J. Heat Fluid Flow* **2006**, *27*, 585–593. [CrossRef]
23. Buffa, E.; Jacob, J.; Sagaut, P. Lattice-Boltzmann-based large-eddy simulation of high-rise building aerodynamics with inlet turbulence reconstruction. *J. Wind Eng. Ind. Aerodyn.* **2021**, *212*, 104560. [CrossRef]
24. Feng, Y.; Miranda-Fuentes, J.; Guo, S.; Jacob, J.; Sagaut, P. ProLB: A lattice Boltzmann solver of large-eddy simulation for atmospheric boundary layer flows. *J. Adv. Model. Earth Syst.* **2021**, *13*, e2020MS002107. [CrossRef]
25. Feng, Y.; Boivin, P.; Jacob, J.; Sagaut, P. Regularized lattice Boltzmann simulation of humid air with application to meteorological flows. *Phys. Rev. E* **2019**, *100*, 023304. [CrossRef] [PubMed]
26. Feng, Y.; Mirand-Fuentes, J.; Jacob, J.; Sagaut, P. Hybrid Lattice Boltzmann model for atmospheric flows under anelastic approximation. *Phys. Fluids* **2021**, *33*, 036607. [CrossRef]
27. Ouertatani, N.; Ben Cheikh, N.; Ben Beya, B.; Lili, T. Numerical simulation of two-dimensional Rayleigh–Bénard convection in an enclosure. *Comptes Rendus Mécanique* **2008**, *336*, 464–470. [CrossRef]
28. Grabowski, W.W.; Clark, T.L. Cloud–Environment Interface Instability: Rising Thermal Calculations in Two Spatial Dimensions. *J. Atmos. Sci.* **1991**, *48*, 527–546. [CrossRef]
29. Grabowski, W.W.; Clark, T.L. Cloud-Environment Interface Instability: Part II: Extension to Three Spatial Dimensions. *J. Atmos. Sci.* **1993**, *50*, 555–573. [CrossRef]
30. Siebesma, A.P.; Bretherton, C.S.; Brown, A.; Chlond, A.; Cuxart, J.; Duynkerke, P.G.; Jiang, H.; Khairoutdinov, M.; Lewellen, D.; Moeng, C.H.; et al. A Large Eddy Simulation Intercomparison Study of Shallow Cumulus Convection. *J. Atmos. Sci.* **2003**, *60*, 1201–1219. [CrossRef]
31. KIT. CODASC: Concentration Data of Street Canyons—Karlsruhe Institute of Technology, Laboratory of Building & Environmental Aerodynamics. Available online: http://www.windforschung.de/CODASC.htm (accessed on 28 August 2017).
32. Merlier, L.; Jacob, J.; Sagaut, P. Lattice-Boltzmann Large-Eddy Simulation of pollutant dispersion in street canyons including tree planting effects. *Atmos. Environ.* **2018**, *195*, 89–103. [CrossRef]
33. Chang, J.C.; Hanna, S.R. Air quality model performance evaluation. *Meteorol. Atmos. Phys.* **2004**, *87*, 167–196. [CrossRef]
34. Zhou, X.; Ying, A.; Cong, B.; Kikumoto, H.; Ooka, R.; Kang, L.; Hu, H. Large eddy simulation of the effect of unstable thermal stratification on airflow and pollutant dispersion around a rectangular building. *J. Wind Eng. Ind. Aerodyn.* **2021**, *211*, 104526. [CrossRef]
35. Tominaga, Y.; Mochida, A.; Yoshie, R.; Kataoka, H.; Nozu, T.; Yoshikawa, M.; Shirasawa, T. AIJ guidelines for practical applications of CFD to pedestrian wind environment around buildings. *J. Wind Eng. Ind. Aerodyn.* **2008**, *96*, 1749–1761. [CrossRef]
36. TPU Database. Flow and Concentrations around an Isolated Building (Wind Tunnel). Available online: http://www.wind.arch.t-kougei.ac.jp/info_center/pollution/Non-Isothermal_Flow.html (accessed on 28 April 2021)
37. Yoshie, R.; Jiang, G.; Shirasawa, T.; Chung, J. CFD simulations of gas dispersion around high-rise building in non-isothermal boundary layer. *J. Wind Eng. Ind. Aerodyn.* **2011**, *99*, 279–288. [CrossRef]
38. Bazdidi-Tehrani, F.; Gholamalipour, P.; Kiamansouri, M.; Jadidi, M. Large eddy simulation of thermal stratification effect on convective and turbulent diffusion fluxes concerning gaseous pollutant dispersion around a high-rise model building. *J. Build. Perform. Simul.* **2019**, *12*, 97–116. [CrossRef]
39. FFI. MODITIC Project. 2016. Available online: https://www.ffi.no/no/Forskningen/totalforsvar/MODITIC/Sider/moditic.aspx (accessed on 28 June 2018).
40. Merlier, L.; Jacob, J.; Sagaut, P. Lattice-Boltzmann large-eddy simulation of pollutant dispersion in complex urban environment with dense gas effect: Model evaluation and flow analysis. *Build. Environ.* **2019**, *148*, 634–652. [CrossRef]
41. Hanna, S.; Chang, J. Acceptance criteria for urban dispersion model evaluation. *Meteorol. Atmos. Phys.* **2012**, *116*, 133–146. [CrossRef]

42. Margheri, L.; Sagaut, P. A hybrid anchored-ANOVA-POD/Kriging method for uncertainty quantification in unsteady high-fidelity CFD simulations. *J. Comput. Phys.* **2016**, *324*, 137–173. [CrossRef]
43. Tominaga, Y.; Yoshie, R.; Mochida, A.; Kataoka, H.; Harimoto, K.; Nozu, T. Cross Comparisons of CFD Prediction for Wind Environment at Pedestrian Level around Buildings. In Proceedings of the Sixth Asia-Pacific Conference on Wind Engineering, APCWE-VI, Seoul, Korea, 12–14 September 2005.
44. Jacob, J.; Sagaut, P. Wind comfort assessment by means of large eddy simulation with lattice Boltzmann method in full scale city area. *Build. Environ.* **2018**, *139*, 110–124. [CrossRef]
45. Sheng, R.; Perret, L.; Calmet, I.; Demouge, F.; Guilhot, J. Wind tunnel study of wind effects on a high-rise building at a scale of 1:300. *J. Wind Eng. Ind. Aerodyn.* **2018**, *174*, 391–403. [CrossRef]
46. Melbourne, W. Criteria for environmental wind conditions. *J. Ind. Aerodyn.* **1978**, *3*, 241–249. [CrossRef]
47. Helbing, D. A mathematical model for the behavior of pedestrians. *Behav. Sci.* **1991**, *36*, 298–310. [CrossRef]
48. Helbing, D.; Molnar, P. Social force model for pedestrian dynamics. *Phys. Rev. E* **1995**, *51*, 4282. [CrossRef] [PubMed]
49. Helbing, D.; Farkas, I.J.; Molnar, P.; Vicsek, T. Simulation of pedestrian crowds in normal and evacuation situations. *Pedestr. Evacuation Dyn.* **2002**, *21*, 21–58.
50. Yu, W.; Johansson, A. Modeling crowd turbulence by many-particle simulations. *Phys. Rev. E* **2007**, *76*, 046105. [CrossRef] [PubMed]
51. Martínez-Tossas, L.A.; Churchfield, M.J.; Meneveau, C. Optimal smoothing length scale for actuator line models of wind turbine blades based on Gaussian body force distribution. *Wind Energy* **2017**, *20*, 1083–1096. [CrossRef]
52. Marlow, F.; Jacob, J.; Sagaut, P. A multidisciplinary model coupling Lattice-Boltzmann-based CFD and a Social Force Model for the simulation of pollutant dispersion in evacuation situations. Submitted.

atmosphere

MDPI

Article

Use and Scalability of OpenFOAM for Wind Fields and Pollution Dispersion with Building- and Ground-Resolving Topography

Daniel Elfverson * and Christian Lejon *

CBRN Defence and Security, Swedish Defence Research Agency, Cementvägen 20, SE-901 82 Umeå, Sweden
* Correspondence: daniel.elfverson@foi.se (D.E.); christian.lejon@foi.se (C.L.)

Abstract: Complex flow and pollutant dispersion simulations in real urban settings were investigated by using computational fluid dynamics (CFD) simulations with the SST $k-\omega$ Reynolds-averaged Navier–Stokes (RANS) equation with OpenFOAM. The model was validated with a wind-tunnel experiment using two surface-mounted cubes in tandem, and the flow features were reproduced with the correct qualitative behaviour. The real urban geometry of the Parade Square in Warsaw, Poland was represented with both laser-scanning data for the ground geometry and the CityGML standard to describe the buildings as an example. The Eulerian dispersion of a passive scalar and the flow behaviour could be resolved within minutes over a computational domain with a size of 958×758 m^2 and a height of 300 m with over 2 M cells due to the good and strong parallel scalability in OpenFOAM. This implies that RANS modelling with parallel computing in OpenFOAM can potentially be used as a tool for situational awareness on a local urban scale; however, entire cities would be too large.

Keywords: RANS; urban dispersion modelling; Reynolds-averaged Navier–Stokes; situational awareness; CityGML

Citation: Elfverson, D.; Lejon, C. Use and Scalability of OpenFOAM for Wind Fields and Pollution Dispersion with Building- and Ground-Resolving Topography. *Atmosphere* **2021**, *12*, 1124. https://doi.org/10.3390/atmos12091124

Academic Editor: Patrick Armand

Received: 10 August 2021
Accepted: 27 August 2021
Published: 31 August 2021

Publisher's Note: MDPI stays neutral with regard to jurisdictional claims in published maps and institutional affiliations.

1. Introduction

The simulation of intended or unintended atmospheric dispersion in a complex geometry—e.g., in an urban area or industrial site—is necessary for the assessment of hazards. In addition to the characteristics of sources, which are not considered here, the governing factors in such events are the wind speed and atmospheric stability, as well as the interaction with 3D structures, such as buildings and rough terrain. For the purpose of situational awareness, a reliable, fast, and fairly accurate tool is needed in the mitigation of risk, in decision making, and among first responders.

Wind-field modelling is imperative for such capabilities, and computational fluid dynamics (CFD) models have gained increased popularity for phenomena occurring at the street level, such as wind wakes and re-circulation patterns, as people who are at risk can be modelled. Among these models, the Reynolds-averaged Navier–Stokes (RANS) model is the fastest and is, thus, likely a suitable choice for moderately large areas [1]. It involves solving the general equations of fluid dynamics (i.e., the continuity, momentum, and energy equations) together with a turbulence closure in order to approximate the turbulent behaviour of the flow. Large eddy simulations (LESs) offer turbulence modelling that is closer to reality, but at the expense of computational effort. The first-responder tool CT-analyst [2] relies on LES-precomputed wind fields. Back in 2010, some authors stated that it would take about two years to get good coverage of an area [2]. Other authors, on the other hand, claimed that pre-computations on the full parameter set of, e.g., wind speeds, wind directions, and turbulences, would not be realistic, even with RANS [3]. Today, the question of the size of the domain that can be modelled with RANS within the timeframe of a first response remains.

Although CFD and pre-computations may be possible, there are also other non-CFD or hybrid approaches that have been developed for risk mitigation and fast response. For an early review, the reader is referred to [4]. Setting a basis for the field, in 1990, Röckle developed [5] a wind-field model based on the forcing mass consistency for the mean flow together with empirical local wind observations around buildings. This renders a Poisson equation that is computationally feasible to solve. See [6] for an evaluation of the Quick Urban and Industrial Complex (QUIC) model. The downside of this diagnostic approach is the lack of generality for handling arbitrary and complex 3D geometries. Gaussian puff models have also been popular; because they do not explicitly resolve urban topography, efforts have been made to combine them with a mechanistic street network model and Lagrangian stochastic model for dispersion [7], or with SIRANERISK [8], which is an operational model. More recently, an interesting approach was presented in which megacities could be modelled based on a forecasted meteorology by combining CFD (RANS) near the source with a diagnostic model for the entire domain; this was referred to as the PMSS modelling system [3].

To investigate the speed achieved by a RANS model in an urban geometry, the popular and open-source CFD environment OpenFOAM [9] was used to solve the system of equations with the shear stress transport (SST) $k - \omega$ [10–12] turbulence closure in parallel on a local server with 80 cores. This closure was validated for urban geometries in [13,14], and it is an adequate choice because of its relatively low computational cost. OpenFOAM has support for both compressible and incompressible turbulent flow simulations in 3D, as well as an extensive set of turbulence closures for RANS modelling. Recently, a new suite of atmospheric boundary conditions was added to OpenFOAM (v2006), which made it a good choice for this application.

Herein, OpenFOAM was firstly validated against wind-tunnel data from two surface-mounted cubes in tandem [15], which showed qualitative agreement. The parallel scalability and computational time from the perspective of a situational awareness tool were then investigated with the atmospheric inflow profiles. By using the Parade Square in Warsaw, Poland as an example, a method was devised so that dispersion events could be simulated in complex and real geometries.

2. Materials and Methods

The Navier–Stokes (NS) equations describe the motion of viscous fluids. Consider the NS equations for incompressible flow together with the continuity equation and an equation for turbulent pollutant transport:

$$\partial_t \boldsymbol{u} + \nabla \cdot (\boldsymbol{u} \otimes \boldsymbol{u}) = -\frac{1}{\rho_0} \nabla p + \nabla \cdot \nu (\nabla \boldsymbol{u} + \nabla \boldsymbol{u}^T) + \frac{1}{\rho_0} \boldsymbol{g}, \tag{1}$$

$$\nabla \cdot \boldsymbol{u} = 0, \tag{2}$$

$$\partial_t c + \nabla (c \boldsymbol{u}) = \nabla \cdot D_c \nabla c + S_c, \tag{3}$$

where $\boldsymbol{u}$ is the velocity, p is the pressure, ν is the kinematic viscosity, ρ_0 is the mass density, g represents body forces, c is the pollutant concentration, D_c is the molecular diffusion of the pollutant, and S_c is a source term. In RANS models, the time averages of velocity ($\bar{\boldsymbol{u}}$), pressure ($\bar{p}$), and concentration ($\bar{c}$) are simulated. Thus, instead of resolving in time, the Reynolds averaging decomposes these quantities into fluctuating and static components or primed and barred notation ($\boldsymbol{u} = \boldsymbol{u}' + \bar{\boldsymbol{u}}$, $p = p' + \bar{p}$, and $c = c' + \bar{c}$, respectively); in the NS equations, the time-averaged flow description is given as follows:

$$\nabla \cdot (\bar{\boldsymbol{u}} \otimes \bar{\boldsymbol{u}}) = -\frac{1}{\rho_0} \nabla \bar{p} + \nabla \cdot \nu (\nabla \bar{\boldsymbol{u}} + \nabla \bar{\boldsymbol{u}}^T) - \nabla \cdot \overline{(\boldsymbol{u}' \otimes \boldsymbol{u}')} + \frac{1}{\rho_0} \boldsymbol{g}, \tag{4}$$

$$\nabla \cdot \bar{\boldsymbol{u}} = 0. \tag{5}$$

The pollutant transport is modelled with time dependence in order to to have a framework that accommodates a time-varying source:

$$\partial_t \bar{c} + \nabla(\bar{c}\bar{\boldsymbol{u}}) = \nabla \cdot D_c \nabla \bar{c} - \nabla \cdot \overline{(c'\boldsymbol{u}')} + S_c. \tag{6}$$

The fluctuating components remain as cross-terms; for ease of the simulations, the Reynolds stress and turbulent pollutant flux are parameterised with the Boussinesq hypothesis [16]:

$$-\overline{(\boldsymbol{u}' \otimes \boldsymbol{u}')} = \nu_t(\bar{\boldsymbol{u}} \otimes \bar{\boldsymbol{u}}) - \frac{2}{3}k\mathbf{I}, \tag{7}$$

$$-\overline{(c'\boldsymbol{u}')} = \frac{\nu_t}{\mathrm{Sc}_t}\nabla \bar{c}, \tag{8}$$

respectively, where ν_t is the eddy/turbulent viscosity and ν_t/Sc_t is the eddy pollutant flux. The turbulent Schmidt number, Sc_t, is case dependent and can vary in the range from 0.2 to 1.3, but here $\mathrm{Sc}_t = 0.7$ according to others [17,18].

To close the system of equations, k and ν_t are modelled with the SST $k - \omega$ model [10–12]. This model is a blend of the standard $k - \epsilon$ model in the free shear flow where it is accurate and the $k - \omega$ in the boundary layer where the standard $k - \epsilon$ model otherwise over-predicts the turbulent kinetic energy. The kinetic turbulent energy k and ω are modelled with the following equations:

$$\nabla \cdot (\bar{\boldsymbol{u}}k) = \nabla \cdot (\nu + \sigma_k \nu_t)\nabla k + G - \beta^* \omega k + S_k, \tag{9}$$

$$\nabla \cdot (u\omega) = \nabla \cdot (\nu + \sigma_\omega \nu_t)\nabla \omega + \frac{\Upsilon}{\nu}G - \beta\omega^2 - (F_1 - 1)CF_{k\omega} + S_\omega, \tag{10}$$

where the terms and default constants for OpenFOAM-v2012 [19] are used. We use a built-in steady-state solver *simpleFoam* [19] and the simple consistent algorithm [20] for the RANS equations and for the turbulent transport ((6) and (8)); a custom solver was also implemented.

2.1. Boundary Conditions

The turbulent inflow parameters for the simulation of the double-bluff wind-tunnel experiment were set to:

$$k = \frac{3}{2}(IU)^2, \quad \omega = \frac{k^{1/2}}{(\beta^*)^{1/4}L}, \tag{11}$$

with the intensity $I = 0.1$ and characteristic length scale $L = 0.04$.

For the ABL flow, a logarithmic inflow boundary [21] based on Monin–Obukhov similarity theory was used:

$$u = \frac{u_*}{\kappa}\log\left(\frac{z + z_0}{z_0}\right), \quad k = \frac{u_*^2}{(\beta^*)^{1/2}}, \quad \omega = \frac{u_*}{(\beta^*)^{1/2}}\frac{1}{z + z_0}, \quad u_* = \frac{u_{\mathrm{ref}}\kappa}{\ln\left(\frac{z_{\mathrm{ref}} + z_0}{z_0}\right)}, \tag{12}$$

where $z_0 = 0.04$ is a roughness parameter.

For the wall boundary conditions, the velocity field $\bar{u}$ uses a no-slip boundary condition on the ground and buildings. For k using a zero gradient and for ω and ν_t, we use wall functions [19]. Symmetric boundary conditions are imposed for the sides and top boundaries.

2.2. Three-Dimensional Geometry and Computational Mesh

The virtual topography of the Parade Square in Warsaw, Poland was built based on a combination of two data sources: one for the buildings [22] in the CityGML format and a numerical terrain model [23] (NTM) to represent ground roughness. CityGML is the international standard of the Open Geospatial Consortium (OGC) and is based on the Geography Markup Language (GML) [24]. The standard defines a three-dimensional description of cities and regional areas at three different levels of detail (LODs). In this work, LOD = 2 was chosen, since it contains sufficient details for accurate wind-field calculations and dispersion events, but omits details that would make the amount of data very large and too complicated for the generation of a computational mesh. NTM data are based on laser scanning. This dataset was filtered using the ground classification (=2) to exclude, e.g., buildings for which CityGML was a better source. The average height error was not greater than 0.20 m, and the data were at a 1 m spatial resolution in the ground plane in the source. However, they were downsampled to a 5 m resolution to aid in having faster simulations. All geometric operations were performed with the Visualization Toolkit [25] to produce a triangulation of the surfaces that was compatible with the native mesh generator *snappyHexMesh* in *OpenFOAM*.

3. Results

3.1. Double Bluff

The comparison of the wind-tunnel data of two surface-mounted cubes in tandem [15,26] showed a good agreement between the simulations and experiments. The main features of the mean velocity field (see Figure 1a,b), the front wake at the first cube, the re-circulation zone between the cubes, and the leeward wake of the second cube were present. The lift at the front wall of the first cube eventually dropped down and was re-attached at the roof of the second cube. This model successfully captured the complex flow pattern of a 3D geometry that was similar to that of an urban environment with sharp-cornered buildings. The Reynolds number was Re = 22,000 based on the free-stream velocity of $u = 8.8$ m/s, cube dimension of $h = 0.04$ m, and viscosity of air of $\nu = 1.6 \times 10^{-5}$. Although small geometric differences were reported, this was well above the threshold for the independence of the Reynolds number [27].

3.2. Wind-Field Simulation of a Scaled-Up Double-Bluff Geometry with Atmospheric Inflow Profiles

Simulations of the scaled-up version of the wind-tunnel experiment with cubes (h = 20 m) that were 500 times larger than those in the original experiment are shown in Figure 1c. The Reynolds number (Re = 44) was still above the threshold for the independence of the Reynolds number (=25) [27]. Here, a logarithmic inflow profile was used in comparison with the flat profile in order to simulate the wind tunnel's geometry; clearly, this had profound effects on the flow characteristics. The re-circulations were still there, but were slightly shifted in their positions. The main difference was that the wind speed was, for the height of the buildings, distinctly lower than in the free stream above the buildings. With a logarithmic inflow, the urban topography effectively acted to shield against the free-blowing wind.

The logarithmic boundary conditions in *OpenFOAM* were developed for neutral atmospheric stability [28]. Inflow boundary conditions based on Monin–Obukhov similarity theory for the case of stable stratification were employed with RANS modelling and validated in atmospheric dispersion field trials called the Mock Urban Setting Test (MUST) [29]. Much better agreement was found when the inflow turbulent kinetic energy (TKE) was fitted to the upstream values measured [29]. *OpenFOAM* offers the possibility of fitting the levels of the TKE of the inflow to the measured data and still satisfying the solution to the RANS equations [30]. Nevertheless, it is perhaps more challenging to accurately assess the present atmospheric turbulence (as well as the wind speed and direction). The merging of atmospheric stratification with numerical dispersion modelling is likely a topic for continued future research; a review is provided in [31].

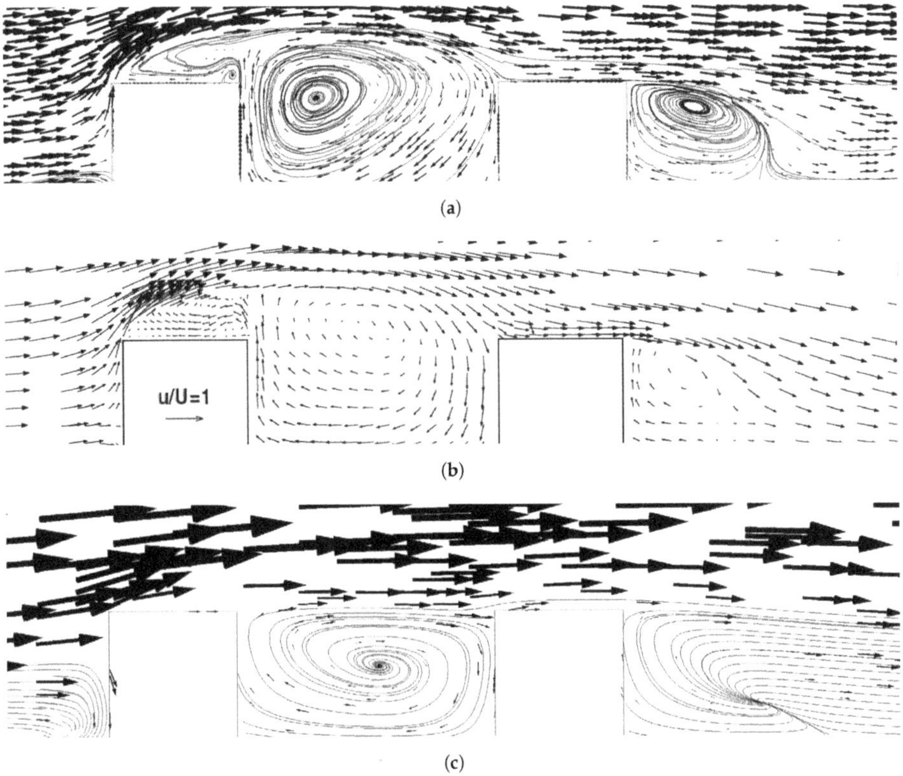

Figure 1. Comparison of the velocity fields of the simulations using the SST $k - \omega$ turbulence model (**a**) to experimental wind-tunnel data [15] (**b**). Simulations with the atmospheric inflow boundaries of the scaled-up geometry are shown in (**c**); these panels are not to scale. The size of the arrows has been adjusted to aid visualisation, and streamlines have been added to enhance the patterns in which the arrows are small.

3.3. Warsaw

Accurate 3D representations of buildings are needed to simulate the atmospheric properties in the vicinity of those structures. For the Parade Square in Warsaw, the geometry is characterised by a central landmark building in an otherwise open square, followed by several large and tall buildings that are downwind. The upwind regime consists of smaller rectangular buildings. This gives rise to the simulated wind field seen in Figure 2, with several recirculation and deflection zones, as well as separation lines. Clearly, the CFD simulation captures the physical flow behaviour around the central landmark building, as well as the smaller and larger circulations throughout the domain. This complex flow pattern influences the transport of a simulated release. The release point was chosen to be at the rooftop of a smaller building that was upwind (marked by a star in Figure 2) to capture the urban turbulence effects in the close vicinity of the source. The concentration also increases upwind from the source due to the wake patterns, which are similar to the those of the double-bluff experiment. The central building then deflects the transport of the pollutant from the general wind direction as the plume spreads and travels downwind, which is in agreement with the results of other simulations [3] and the MUST field trials [32].

Figure 2. Visualisation of the ground concentration of a pollutant on a logarithmic scale that was released on a roof top (white star), which is in the bottom part of the figure. The complex wind field is visualised using streamlines, and the inflow wind direction is given by the white arrow. The geometry is that of the Parade Square in Warsaw, Poland.

3.4. Parallel Scaling

The parallel performance of the RANS turbulence model was investigated on an Intel(R) Xeon(R) Gold 6230 CPU at 2.10 GHz with four sockets containing 20 cores each. The mesh was decomposed using *scotch* [33], which aims to minimise the number of processor boundaries. Two meshes with different coarseness were produced—one with 10,839,462 and the other with 2,292,898 cells. In both cases, 1000 iterations were performed with the *simpleFoam* solver in OpenFOAM. The computational domain had a footprint with a size of 958×758 m^2 and a height of 300 m.

As seen in Figure 3, for this particular problem, the solver showed good parallel behaviour. A further increase in the processor number would not significantly increase the total clock time. Up to about 35 cores, the solver exhibited strong, excellent scaling for these problem sizes; however, above this number, the communication between different cores became significant. The full capacity of the server gave approximately 135,000 or 28,000 cells for each core to handle, and the total clock time was 954 or 133 s, respectively. To further reduce the computation time, a smaller area or coarser mesh is needed, and the one-minute mark is viable.

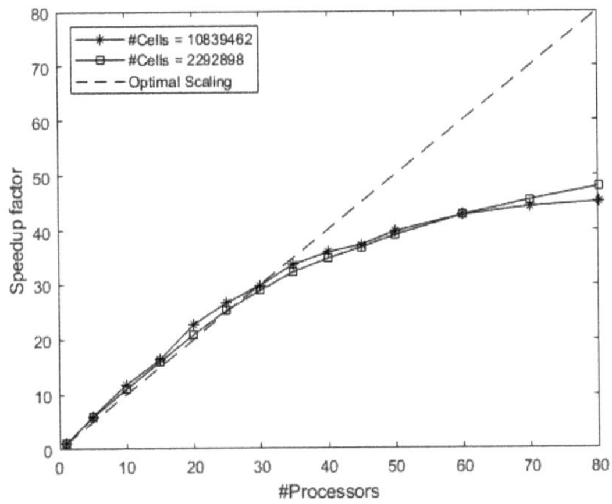

Figure 3. Parallel scaling using two different mesh sizes of the RANS solver on the real geometry of the Parade Square in Warsaw, Poland.

From the perspective of situational awareness, this means that with pre-generated meshes, in terms of time, RANS modelling is within the reach of the actions first responders in moderate-sized urban environments. Larger scales that approach the whole of even small cities will currently be too large, considering that the number of cells increases with the size of the footprint modelled at a constant mesh resolution.

4. Conclusions

Here, it was shown that parallel scaling in *OpenFOAM* increases the speed enough that the wind and dispersion over a local urban environment can be simulated in close to real time with the SST $k-\omega$ Reynolds-averaged Navier–Stokes (RANS) model. Nevertheless, length scales that cover whole cities would still be too time consuming. Building geometry and ground topography can, respectively, be incorporated by using the CityGML standard and laser-scanning data if available.

It is suggested for future work to include validation of the model with a large-scale field experiment. Furthermore, the inflow profiles used here are for neutral stratification, but stable and unstable stratification would also be of interest. The vertical mixing processes can then be further studied for all stratification types, and, since the RANS framework does not seem to be sufficiently fast for large scales, such as whole cities, a more accurate hybrid approach could be foreseen by combining the outcomes from the RANS model with those of either a street network model or a Gaussian puff model, as in [3]. Another possible research direction would be to compare the SST $k-\omega$ turbulence closure to the cheaper zero-equation approach used in [1] for urban geometries.

Author Contributions: Conceptualisation, D.E. and C.L.; methodology, D.E. and C.L.; software, D.E; validation, C.L.; formal analysis, D.E. and C.L.; investigation, D.E. and C.L.; data curation, D.E.; writing—original draft preparation, D.E. and C.L.; writing—review and editing, C.L.; visualisation, D.E. and C.L. All authors have read and agreed to the published version of the manuscript.

Funding: The research was funded by the EU-SENSE project within the European Union's Horizon 2020 research and innovation programme under grant agreement No. 787031.

Institutional Review Board Statement: Not applicable.

Informed Consent Statement: Not applicable.

Data Availability Statement: Not applicable.

Acknowledgments: The authors are thankful to Anna Makowska at the Main Office of Geodesy and Cartography of Poland (abbr. GUGiK), Joanna Kozioł at the Main School of Fire Service (Szkoła

Główna Służby Pożarniczej), and Mateusz Oleś at the iTTi company for their help with accessing map data. The authors are also thankful to their colleagues at their own institution for the fruitful technical discussions and project administration.

Conflicts of Interest: The funders had no role in the design of the study; in the collection, analyses, or interpretation of data; in the writing of the manuscript, or in the decision to publish the results, other than their stated need for fast urban dispersion modelling.

References

1. Gowardhan, A.; Pardyjak, E.; Senocak, I.; Brown, M. A CFD-based wind solver for an urban fast response transport and dispersion model. *Environ. Fluid Mech.* **2011**, *11*, 439–464. [CrossRef]
2. Patnaik, G.; Moses, A.; Boris, J. Fast, Accurate Defense for Homeland Security: Bringing High-Performance Computing to First Responders. *J. Aerosp. Comput. Inf. Commun.* **2010**, *7*, 210–222. [CrossRef]
3. Oldrini, O.; Armand, P.; Duchenne, C.; Perdriel, S.; Nibart, M. Accelerated Time and High-Resolution 3D Modeling of the Flow and Dispersion of Noxious Substances over a Gigantic Urban Area—The EMERGENCIES Project. *Atmosphere* **2021**, *12*, 640. [CrossRef]
4. Brown, M. Urban dispersion—Challenges for fast response modeling. In Proceedings of the 5th AMS Symposium on the Urban Environment, Vancouver, BC, Canada, 23–26 August 2004.
5. Röckle, R. Bestimmung der Strömungsverhältnisse im Bereich komplexer Bebauungsstrukturen. Ph.D. Thesis, Technical University Darmstadt, Darmstadt, Germany, December 1990
6. Kopka, P.; Potempski, S.; Kaszko, A.; Korycki, M. Urban Dispersion Modelling Capabilities Related to the UDINEE Intensive Operating Period 4. *Bound.-Layer Meteorol.* **2019**, *171*, 465–489. [CrossRef]
7. Hertwig, D.; Soulhac, L.; Fuka, V.; Auerswald, T.; Carpentieri, M.; Hayden, P.; Robins, A.; Xie, Z.T.; Coceal, O. Evaluation of fast atmospheric dispersion models in a regular street network. *Environ. Fluid Mech.* **2018**, *18*, 1007–1044. [CrossRef]
8. Soulhac, L.; Lamaison, G.; Cierco, F.X.; Salem, N.B.; Salizzoni, P.; Mejean, P.; Armand, P.; Patryl, L. SIRANERISK: Modelling dispersion of steady and unsteady pollutant releases in the urban canopy. *Atmos. Environ.* **2016**, *140*, 242–260. [CrossRef]
9. Weller, H.G.; Tabor, G.; Jasak, H.; Fureby, C. A tensorial approach to computational continuum mechanics using object-oriented techniques. *Comput. Phys.* **1998**, *12*, 620–631. [CrossRef]
10. Menter, F.R. Two-equation eddy-viscosity turbulence models for engineering applications. *AIAA J.* **1994**, *32*, 1598–1605. [CrossRef]
11. Menter, F.; Esch, T. Elements of industrial heat transfer predictions. In Proceedings of the 16th Brazilian Congress of Mechanical Engineering (COBEM 2001), Minas Gerais, Brazil, 26–30 November 2001; Volume 109, p. 650
12. Menter, F.R.; Kuntz, M.; Langtry, R.; Nagano, Y.; Tummers, M.J.; Hanjalic, K. Ten Years of Industrial Experience with the SST Turbulence Model. In *Turbulence, Heat and Mass Transfer*, 4th ed.; Internal Symposium, Turbulence, Heat and Mass Transfer; Begell House: New York, NY, USA, 2003; Volume 4, pp. 625–632
13. Yu, H.; Thé, J. Validation and optimization of SST $k-\omega$ turbulence model for pollutant dispersion within a building array. *Atmos. Environ.* **2016**, *145*, 225–238. [CrossRef]
14. Trindade da Silva, F.; Reis, N.C.; Santos, J.M.; Goulart, E.V.; Engel de Alvarez, C. The impact of urban block typology on pollutant dispersion. *J. Wind. Eng. Ind. Aerodyn.* **2021**, *210*, 104524. [CrossRef]
15. Martinuzzi, R.J.; Havel, B. Vortex shedding from two surface-mounted cubes in tandem. *Int. J. Heat Fluid Flow* **2004**, *25*, 364–372. [CrossRef]
16. Schmitt, F.G. About Boussinesq's turbulent viscosity hypothesis: Historical remarks and a direct evaluation of its validity. *C. R. MéCanique* **2007**, *335*, 617–627. [CrossRef]
17. Tominaga, Y.; Stathopoulos, T. Turbulent Schmidt numbers for CFD analysis with various types of flowfield. *Atmos. Environ.* **2007**, *41*, 8091–8099. [CrossRef]
18. Gualtieri, C.; Angeloudis, A.; Bombardelli, F.; Jha, S.; Stoesser, T. On the Values for the Turbulent Schmidt Number in Environmental Flows. *Fluids* **2017**, *2*, 17. [CrossRef]
19. OpenFOAM: User Guide v2012. Available online: https://www.openfoam.com/documentation/guides/latest/doc/ (accessed on 23 July 2021).
20. Doormaal, J.P.V.; Raithby, G.D. Enhancements of the simple method for predicting incompressible fluid flows. *Numer. Heat Transf.* **1984**, *7*, 147–163. [CrossRef]
21. Richards, P.; Hoxey, R. Appropriate boundary conditions for computational wind engineering models using the k-ϵ turbulence model. *J. Wind. Eng. Ind. Aerodyn.* **1993**, *46-47*, 145–153. [CrossRef]
22. Geoportal–Spatial Information for Everyone. Available online: https://www.gov.pl/web/archiwum-inwestycje-rozwoj/geoportal-informacja-przestrzenna-dla-kazdego (accessed on 20 November 2019).
23. Geoportal Infrastruktury Informacji Przestrzennej. Główny Geodeta Kraju (eng. "Principal Surveyor of The Country") Is the Holder of the Data. Available online: http://www.geoportal.gov.pl (accessed on 4 February 2020).
24. Gröger, G.; Plümer, L. CityGML—Interoperable semantic 3D city models. *ISPRS J. Photogramm. Remote Sens.* **2012**, *71*, 12–33. [CrossRef]
25. Schroeder, W.; Martin, K.; Lorensen, B. *The Visualization Toolkit*, 4th ed.; Kitware, Inc.: New York, NY, USA, 2006

26. Martinuzzi, R.J.; Havel, B. Turbulent Flow Around Two Interfering Surface-Mounted Cubic Obstacles in Tandem Arrangement . *J. Fluids Eng.* **1999**, *122*, 24–31. [CrossRef]
27. Cui, P.-Y.; Li, Z.; Tao, W.-Q. Numerical investigations on Re-independence for the turbulent flow and pollutant dispersion under the urban boundary layer with some experimental validations. *Int. J. Heat Mass Transf.* **2017**, *106*, 422–436. [CrossRef]
28. Hargreaves, D.; Wright, N. On the use of the k–ε model in commercial CFD software to model the neutral atmospheric boundary layer. *J. Wind. Eng. Ind. Aerodyn.* **2007**, *95*, 355–369. [CrossRef]
29. Milliez, M.; Carissimo, B. Numerical simulations of pollutant dispersion in an idealized urban area, for different meteorological conditions. *Bound.-Layer Meteorol.* **2007**, *122*, 321–342. [CrossRef]
30. Yang, Y.; Gu, M.; Chen, S.Q.; Jin, X.Y. New inflow boundary conditions for modelling the neutral equilibrium atmospheric boundary layer in computational wind engineering. *J. Wind. Eng. Ind. Aerodyn.* **2009**, *97*, 88–95. [CrossRef]
31. Lateb, M.; Meroney, R.; Yataghene, M.; Fellouah, H.; Saleh, F.; Boufadel, M. On the use of numerical modelling for near-field pollutant dispersion in urban environments—A review. *Environ. Pollut.* **2016**, *208*, 271–283. [CrossRef] [PubMed]
32. Yee, E.; Biltoft, C.A. Concentration Fluctuation Measurements in a Plume Dispersing Through a Regular Array of Obstacles. *Bound.-Layer Meteorol.* **2004**, *111*, 363–415. [CrossRef]
33. Pellegrini, F.; Roman, J. Scotch: A software package for static mapping by dual recursive bipartitioning of process and architecture graphs. In *High-Performance Computing and Networking*; Liddell, H., Colbrook, A., Hertzberger, B., Sloot, P., Eds.; Springer: Berlin/Heidelberg, Germany, 1996; pp. 493–498.

atmosphere

MDPI

Article

Atmospheric Wind Field Modelling with OpenFOAM for Near-Ground Gas Dispersion

Sebastian Schalau [1,*], Abdelkarim Habib [1] and Simon Michel [2]

1 Bundesanstalt für Materialforschung und—Prüfung (BAM), 12205 Berlin, Germany; karim.habib@bam.de
2 Fakultät für Mathematik, Informatik und Naturwissenschaften, Meteorologisches Institut, Universität Hamburg, 20146 Hamburg, Germany; simon.michel@uni-hamburg.de
* Correspondence: sebastian.schalau@bam.de

Abstract: CFD simulations of near-ground gas dispersion depend significantly on the accuracy of the wind field. When simulating wind fields with conventional RANS turbulence models, the velocity and turbulence profiles specified as inlet boundary conditions change rapidly in the approach flow region. As a result, when hazardous materials are released, the extent of hazardous areas is calculated based on an approach flow that differs significantly from the boundary conditions defined. To solve this problem, a turbulence model with consistent boundary conditions was developed to ensure a horizontally homogeneous approach flow. Instead of the logarithmic vertical velocity profile, a power law is used to overcome the problem that with the logarithmic profile, negative velocities would be calculated for heights within the roughness length. With this, the problem that the distance of the wall-adjacent cell midpoint has to be higher than the roughness length is solved, so that a high grid resolution can be ensured even in the near-ground region which is required to simulate gas dispersion. The evaluation of the developed CFD model using the German guideline VDI 3783/9 and wind tunnel experiments with realistic obstacle configurations showed a good agreement between the calculated and the measured values and the ability to achieve a horizontally homogenous approach flow.

Keywords: atmospheric boundary layer; OpenFOAM; gas dispersion; CFD; turbulence model; hazard assessment; horizontal homogeneity; wind field

Citation: Schalau, S.; Habib, A.; Michel, S. Atmospheric Wind Field Modelling with OpenFOAM for Near-Ground Gas Dispersion. *Atmosphere* **2021**, *12*, 933. https://doi.org/10.3390/atmos12080933

Academic Editor: Patrick Armand

Received: 2 July 2021
Accepted: 15 July 2021
Published: 21 July 2021

1. Introduction

When assessing hazards from industrial plant applications, calculating the gas dispersion is one of the main tasks for determining safety distances. Simple models for calculating the gas dispersion are generally not able to account for obstacles such as buildings or the topography of the dispersion area. To simulate dispersion scenarios in complex areas, computational fluid dynamics (CFD) codes can be used. By solving the full Navier–Stokes equations, CFD codes can simultaneously model the wind field and the dispersion of pollutants, taking into account obstacles, topography, and thermal stratification, as well as the type of release (with or without momentum). Simulating gas dispersion is mainly influenced by the wind field. The main problem when using CFD codes is the cost–benefit ratio, as the computational time is exceedingly high compared to simpler models. To reduce the computational time, relying on RANS (Reynolds-averaged Navier–Stokes) equations is a common practice. To close the RANS equations, the k-ε turbulence model is used in a wide number of publications for calculating the atmospheric wind field.

When simulating near-ground atmospheric boundary layer phenomena, fully developed wind and turbulence profiles are specified as inlet boundary conditions. Due to numerical reasons, every CFD simulation requires a computational domain of spatial extensions much above the dimensions of the area of interest. Therefore, an obstacle free approach flow region is required between the inlet and the first obstacle [1]. To ensure that the posed inlet boundary conditions reach the area of interest, where the release and dispersion occurs, a CFD model should be chosen that is able to conserve the inlet conditions in the approach flow region, resulting in a horizontally homogenous boundary layer flow. Otherwise, the turbulence model of the CFD code will modify the approach flow depending on the domain length and grid resolution, resulting in different extents of hazardous areas for the same release scenario [2].

Several publications [3–6] state that reaching a horizontally homogenous boundary layer flow with RANS CFD models is not evident. Richards and Hoxey [7] state that a horizontally homogenous boundary layer flow can only be reached if there is a consistency between the turbulence model and the boundary conditions, such as the atmospheric inlet condition and wall functions. They propose inlet conditions on the basis of a logarithmic wind profile, as well as necessary modifications to the k-ε turbulence model. Several other authors [8–10] present comparable solutions for the problem of horizontal homogeneity. Whilst all aforementioned publications try to achieve a solution by modifying the constants of the k-ε model, [11] they achieved the horizontal homogeneity by introducing an additional source term to the dissipation equation [12] and by introducing it in the turbulent kinetic energy equation.

To reduce the computational cost, wall functions are used for describing the near wall area using a roughness factor, to avoid modelling even the smallest details (e.g., vegetation, fences, cars). Different types of roughness parameters and their suitability for modelling horizontally homogenous boundary layer flows were discussed [13–15]. However, for all it was stated that the general formulations of wall functions for rough walls (based on the equivalent sand grain roughness k_s or the roughness length z_0) introduce a dependency of the wall-adjacent cell size from the defined roughness value. This results in wall-adjacent cell sizes that are much too big for simulating near-ground or near-obstacle effects with acceptable accuracy. If a near-wall cell size smaller than the requirements due to the defined roughness is chosen, the simulation will not result in a horizontally homogenous boundary layer flow [13].

For near-ground gas dispersion scenarios, a coarse wall-adjacent mesh will lead to an unrealistic prediction of the gas cloud sizes, as the concentration gradients near ground cannot be resolved with sufficient accuracy.

In this work, a CFD model based on the open source CFD code OpenFOAM is presented, which was developed with the aim to compute a horizontally homogenous boundary layer flow and to overcome the wall resolution restrictions due to wall roughness. In this model the wind profile is described by a power law instead of a log wind profile. The used flow solver is a transient, compressible flow solver, to also be able to simulate in the future gas dispersion, gas cloud explosions, or gas cloud combustion. In this work we will focus on the wind field modelling. The model will be evaluated systematically using the German Guideline VDI 3783/9 [16,17]. This guideline provides test cases in which generic obstacle configurations are investigated and objective evaluation criteria are provided. To evaluate the model performance for a more realistic obstacle configuration comparable to an industrial area, wind tunnel measurements are used. In opposite to free field trials, they show a high reproducibility, and the average boundary conditions can be kept constant over a long time period to produce a statistically sound database. The wind tunnel data used were generated at the University of Hamburg in the context of an actual research project. The quality of the presented model to simulate a horizontally homogenous boundary layer approach flow and the flow around obstacles will be demonstrated and evaluated with the VDI Guideline 3783/9 and the aforementioned wind tunnel experiments.

2. Turbulence Model and Boundary Conditions for Neutral Stratification

The presented CFD model is based on the transient and compressible solver rhoReactingBuoyantFoam [18,19], as well as on the k-ϵ-turbulence model of Launder and Spalding [20] and El Tahry [21], implemented in OpenFOAM 5.0. The differential equations describing the turbulent kinetic energy k (1) and the dissipation ε (2) are coupled to the Navier–Stokes equations through the turbulent viscosity ν_t (3):

$$\frac{\partial(\rho k)}{\partial t} = \nabla \cdot \left(\rho\left(\frac{\nu_t}{\sigma_k} + \nu\right)\nabla k\right) - \nabla \cdot \left(\rho k \vec{U}\right) + P_k - \frac{2}{3}\rho\left(\nabla \cdot \vec{U}\right)k - \rho\epsilon\,, \tag{1}$$

$$\frac{\partial(\rho\epsilon)}{\partial t} = \nabla \cdot \left(\rho\left(\frac{\nu_t}{\sigma_\epsilon} + \nu\right)\nabla \epsilon\right) - \nabla \cdot \left(\rho\epsilon \vec{U}\right) + \frac{C_1\epsilon}{k}P_k - \frac{2}{3}C_1\,\rho\left(\nabla \cdot \vec{U}\right)\epsilon - C_2\rho\frac{\epsilon^2}{k} + S_\epsilon\,, \tag{2}$$

$$\nu_t = C_\mu \frac{k^2}{\epsilon}\,. \tag{3}$$

The values of the model constants C_1, C_2, C_μ, σ_k and σ_ε in Equations (1)–(3) are based on [22]. The variable ρ is the density of the fluid and $\vec{U}$ is the velocity vector, P_k describes the production of the turbulent kinetic energy and S_ε is an additional source term, which will be derived in the following.

With the idealized assumption of a horizontally homogenous boundary layer flow and with the additional assumptions that:

- the flow is stationary,
- the flow fields are free of horizontal gradients,
- the vertical velocity is zero,

Equations (1) and (2) can be simplified to:

$$0 = \frac{\partial}{\partial z}\left(\rho\frac{\nu_t}{\sigma_k}\frac{\partial k}{\partial z}\right) + P_k - \rho\epsilon \tag{4}$$

$$0 = \frac{\partial}{\partial z}\left(\rho\frac{\nu_t}{\sigma_\epsilon}\frac{\partial \epsilon}{\partial z}\right) + \frac{C_1\epsilon}{k}P_k - C_2\rho\frac{\epsilon^2}{k} + S_\epsilon \tag{5}$$

The dynamic viscosity is assumed to be negligible ($\nu_t \gg \nu$).

Assuming a two-dimensional horizontally homogenous boundary layer flow, the turbulence production is defined as

$$P_k = \rho\nu_t\left(\frac{\partial U_x}{\partial z}\right)^2 \tag{6}$$

Based on Equations (4) to (6), models have been developed to achieve a horizontally homogenous boundary layer solution (e.g., [7,10,12]). All these models are based on a logarithmic wind velocity profile. This assumption leads to a dependency of the grid resolution near ground from the roughness length [13]. To overcome this restriction, in this work a power-law wind profile is used as inlet boundary condition

$$U_x = U_{ref}\left(\frac{z}{z_{ref}}\right)^m \tag{7}$$

The value of the exponent m of the wind profile is set in accordance to the roughness length z_0, which is now only indirectly linked to the governing model equations, allowing us to choose the grid resolution independently of the ground roughness.

The vertical profile of the turbulent kinetic energy of the incident flow, as described by, e.g., Richards and Hoxey [7], is dependent on the wall shear stress velocity which can be defined as nearly constant with height for a neutrally layered atmosphere

$$k = \frac{u_*^2}{\sqrt{C_\mu}} \tag{8}$$

$$u_* = \frac{U_{ref}\,\kappa}{ln\left(\frac{z_{ref}+z_0}{z_0}\right)} \tag{9}$$

To achieve a horizontally homogenous atmospheric boundary layer flow, it is mandatory that the flow profiles at the inlet boundary satisfy the equations of the turbulence model. This is achieved for the turbulent kinetic energy Equation (4) with the condition that $\partial k/\partial z = 0$ for

$$\rho\epsilon \stackrel{!}{=} P_k \tag{10}$$

From Equations (3), (6)–(8), and (10), the vertical profile of the turbulent dissipation ε at the inlet boundary is

$$\epsilon = \frac{u_*^2 \cdot m \cdot U_{ref}\left(\frac{z}{z_{ref}}\right)^m}{z} \tag{11}$$

The inlet boundary conditions (7), (8), and (11) only satisfy the equation for the turbulent dissipation (5) with

$$S_\epsilon = -\frac{\partial\rho}{\partial z}\frac{\nu_t}{\sigma_\epsilon}\frac{\partial\epsilon}{\partial z} - \frac{\rho}{\sigma_\epsilon}\frac{\partial\nu_t}{\partial z}\frac{\partial\epsilon}{\partial z} - \rho\frac{\nu_t}{\sigma_\epsilon}\frac{\partial^2\epsilon}{\partial z^2} - \frac{C_1\epsilon}{k}\rho\nu_t\left(\frac{\partial U_x}{\partial z}\right)^2 + C_2\rho\frac{\epsilon^2}{k} \tag{12}$$

With this modification, horizontally homogenous boundary layer flows can be simulated for near-ground grid resolutions that are not dependent on the roughness length. This source term is evaluated solely based on the vertical inlet profiles at the beginning of the simulation and is therefore time independent. As this source term is relevant and designed to achieve a horizontally homogenous flow without obstacles, this term is only active in the upstream region when simulating obstacles in the domain. In the vicinity of obstacles, the turbulence model (Equations (1)–(3)) is used with $S_\varepsilon = 0$. The term S_ε compensates the expected dissipation of turbulent kinetic energy with the standard k-ϵ-model in an obstacle-free flow domain.

Besides the described modifications to the turbulence model, the simulation of a horizontally homogenous boundary layer flow also requires a formulation for wall functions consistent to the modifications made. Wall functions available in OpenFOAM describe the turbulent energy dissipation and production for hydraulically smooth walls (epsilonWallFunction), as well as the turbulent viscosity based on the roughness length (nutkAtmRoughWallFunction). To avoid reintroducing here the dependency of the possible grid resolution on the roughness length, and at the same time to satisfy Equation (11), the turbulent dissipation at the wall-adjacent cell center is defined as

$$\epsilon_p = \frac{u_*^2 \cdot m \cdot u_{ref}\left(\frac{z_p}{z_{ref}}\right)^m}{z_p} \tag{13}$$

Coupling of the wall function with the flow field occurs in analogy to the OpenFOAM epsilonWallFunction [18], by replacing the friction velocity by the wall adjacent turbulent kinetic energy through Equation (8)

$$\epsilon_p = \frac{\sqrt{C_\mu} \cdot k_p \cdot m \cdot u_{ref}\left(\frac{z_p}{z_{ref}}\right)^m}{z_p} \tag{14}$$

In accordance with Equation (6), the turbulence production in the wall adjacent cell is

$$P_{k,p} = \mu_t\left(\frac{\partial u_x}{\partial z}\right)^2 = \mu_t\frac{\partial u_x}{\partial z}\frac{\partial u_x}{\partial z} \tag{15}$$

In analogy to the OpenFOAM nutkAtmRoughWallFunction [18], one of the two velocity gradients of the wall function is replaced by the analytical derivative of the power law profile and the other by its spatial discretization. Thus, the turbulence production can be rewritten as

$$P_{k,p} = \mu_t \left(\frac{\partial u_x}{\partial z} \right)^2 \approx \mu_t \frac{m \cdot u_{ref} \left(\frac{z_p}{z_{ref}} \right)^m}{z_p} \frac{\Delta u_x}{\Delta z} \tag{16}$$

The wall sheer stress τ_0, considered as constant within the Prandtl layer, is

$$\tau_o = \rho u_*^2 = \mu_t \frac{\partial u_x}{\partial z} \tag{17}$$

and by integration with respect to the height and assuming a no slip wall, Equation (17) becomes

$$\rho u_*^2 = \mu_t \frac{u_p}{z_p} \tag{18}$$

resulting in the wall function for the turbulent viscosity, by substituting u_p with Equation (7):

$$\nu_t = \frac{u_*^2 z_p}{u_{ref} \left(\frac{z_p}{z_{ref}} \right)^m} \tag{19}$$

$$u_* = C_\mu^{0.25} k^{0.5} \tag{20}$$

For the turbulent kinetic energy, a zero gradient is specified at the wall, in accordance with the inlet boundary condition.

Studies [13,23] observed that choosing fixed shear stress as a boundary condition on the top of the domain instead of a symmetry boundary condition leads to a less inhomogeneous horizontal boundary layer. This was also confirmed by [24] who noticed a better agreement with experimental values when setting a fixed shear stress as the top boundary condition. Therefore, in the presented model, constant values for the wind speed and the turbulent values based on Equations (7)–(9) are set as the top boundary conditions. All other values are set to zero gradient.

3. Results

Several investigations have been carried out to assess the accuracy of the presented model. Beginning with a study of the horizontal homogeneity of the boundary layer flow when using the original and the modified code over a benchmark with the VDI 3783/9 [16,17], to a comparison with wind tunnel experiments for complex terrain, the quality of the new model is shown.

3.1. Horizontally Homogeneous Boundary Layer

By using a power law to describe the vertical velocity profile and introducing the additional source term S_ε, it should be possible to simulate a horizontally homogenous boundary layer flow with a grid resolution not depending on the roughness length. Figure 1 shows the wind velocity in the main direction U_x calculated by the k-ϵ turbulence model without modification (standard model) and with the model presented here (modified model) at a height of 1 m above the ground from the inlet up to a distance of 200 m. A power law with a reference velocity of 3 m/s at a reference height of 10 m was specified as the inlet boundary condition and a roughness length of 1.2 m was assumed. For the standard model, a wall function based on the roughness length is chosen (OpenFOAM nutkAtmRoughWallFunction), whilst the roughness length is approximated by the exponent of the power law in the modified model. The simulations were carried out in a 400 m × 8 m × 100 m (l × w × h) pseudo 2D domain without obstacles. In Figure 1, it can be clearly seen that the standard model, as expected, does not conserve the wind profile over the distance, whereas the modified model shows insignificant deviations from the given inlet velocity.

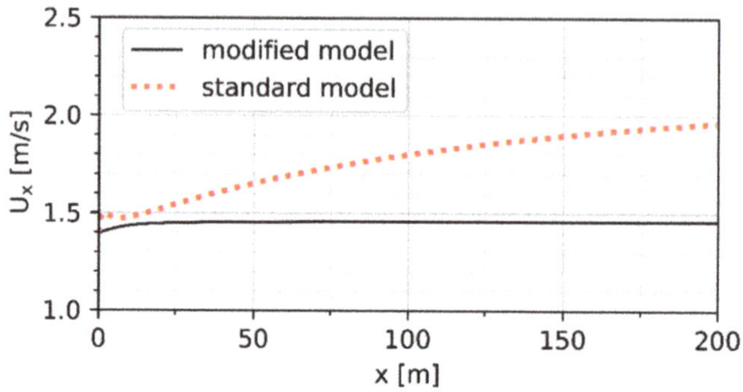

Figure 1. Development of the boundary layer for the standard model and the model presented.

To ensure the presented model being able to carry out simulations of a horizontally homogenous boundary layer flow where the grid resolution can be chosen independently of the surface roughness, which is not possible for standard turbulence models in CFD [13], further simulations were carried out in the same domain. In these simulations, the wind speed at 10 m in height over the ground was varied in a range of 1 m/s up to 5 m/s, and for wind profiles corresponding to roughness lengths between 0.1 m and 1.2 m, covering a typical range for hazard assessment scenarios. The wind profiles were investigated after a distance of 300 m, a typical distance from the inlet according to the best practice guidelines [1], where distances from the inlet of two to eight times the obstacle height are recommended as the approach flow distance. The grid resolution ranged from 0.5 m down to 0.125 m for the wall-adjacent cell, corresponding to cell midpoint heights z_p of 0.25 m down to 0.0625 m.

Figure 2 shows the relative deviation between the calculated vertical wind profile in the main direction (U_x) and the inlet boundary condition. As no significant changes were observed over the range of wind speeds, the figure shows exemplarily the results for a wind speed of 3 m/s at 10 m over the ground. The wind profiles after the investigated distance of 300 m do not differ from the inlet profile by more than +/−5% close to the ground. At a height of 1 m or more from the ground, the deviation between the inlet profile and the computed values is negligible. Similar observations can be made for the eddy viscosity profiles (Figure 3). The maximum deviation reached is around +/−10%, which seems small enough to be qualified as acceptable.

As these observations for the wind velocity and eddy viscosity profiles are valid for any of the investigated combinations of roughness length and grid resolution, it can be concluded that the modified turbulence model and wall functions presented here are able to simulate a horizontally homogenous boundary layer flow for near-wall grid resolutions that can be chosen independently to the roughness length.

3.2. VDI 3783/9 Benchmark

The German Guideline VDI 3783/9 [16,17] is used to evaluate the quality of wind field models for a built-up environment. In the guideline, a number of test cases are defined to evaluate different characteristics of the wind field model. In some test cases, comparisons of the model have to be performed with wind tunnel experiments that were carried out for generic obstacle configurations (scale 1:200) by the meteorological institute of the University of Hamburg [25]. As the aim of this work is to provide a wind field model for, e.g., dispersion scenarios in built up environment, the evaluation of the model presented here is carried out with these test cases.

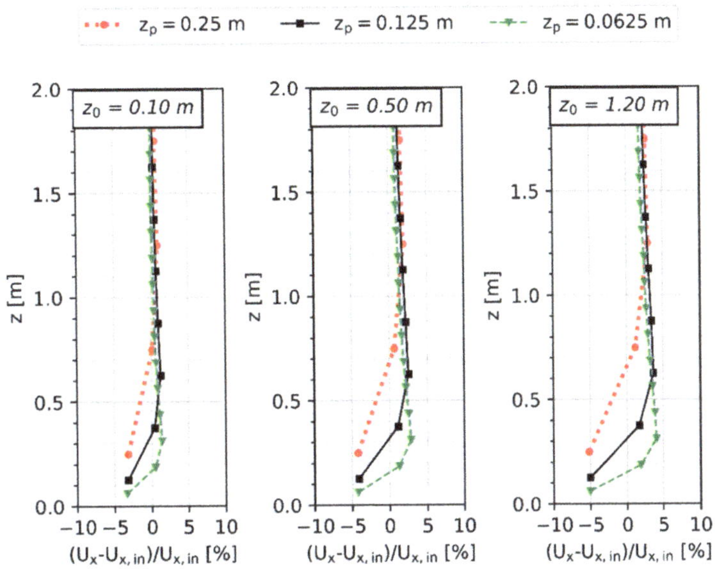

Figure 2. Wind profiles at 300 m distance from the inlet for varying surface roughness and grid resolution.

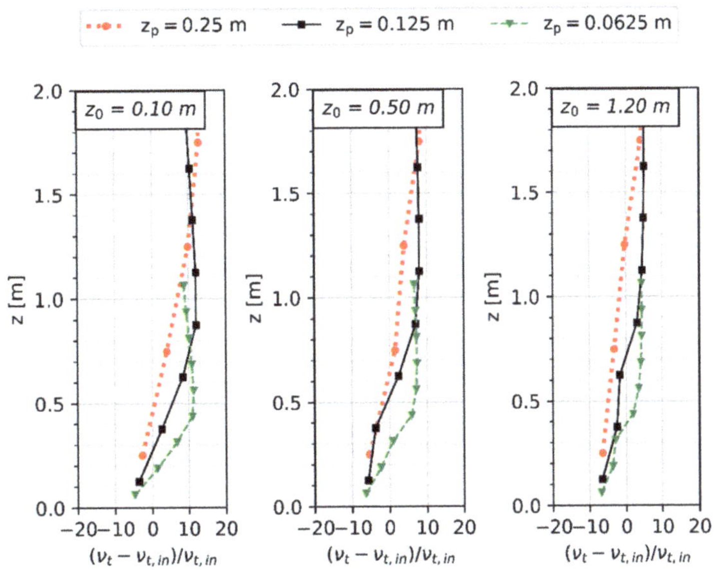

Figure 3. Eddy viscosity profiles at 300 m distance from the inlet for varying surface roughness and grid resolution.

The evaluation method of the guideline VDI 3783/9 gives a quantitative estimate of the model quality by determining a hit ratio q

$$q = \frac{n}{N} = \frac{\sum n_i}{N} \tag{21}$$

It indicates in percent the proportion of the total of correctly predicted values n in the total number of comparison values N. For successful validation in comparison with wind tunnel experiments, it is required that $q > 66\%$. The number of correctly predicted values n_i results from the comparison of the normalized results P_i and the normalized comparison values O_i [17]:

$$n_i = \left\{ \begin{array}{c} 1,\ if \left| \frac{P_i - O_i}{O_i} \right| \leq D\ or\ |P_i - O_i| \leq W \\ 0,\ otherwise \end{array} \right\} \tag{22}$$

In Equation (22), W is the absolute permitted difference and D the permitted difference, which are defined separately for each case in [16,17].

The focus is laid on the test cases c3, c4 and c6 of the 2005 version of the guideline [16], as a simple generic case for a multiple building scenario (test case c6) is available. Figure 4 shows the geometrical representation of the test cases, and the corresponding parameters are given in Table 1. Parameters of each test case. For the test cases c3 and c4 (single obstacle), the guideline requires evaluating the wind field in the near field (grey dotted region in Figure 4a,b), as well as in the whole computational domain. For test case c6, only the whole domain has to be considered.

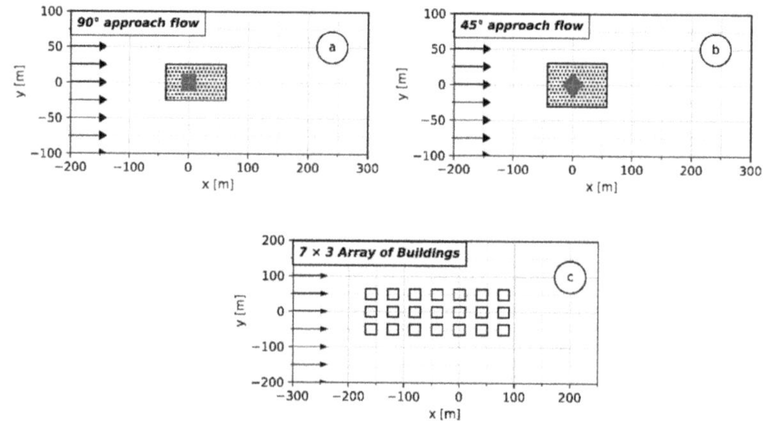

Figure 4. Geometrical representation of the test cases (**a**) test case c3 cube with 90° approach flow, (**b**) test case c4 cube with 270° approach flow, (**c**) test case c6 7 × 3 array of buildings.

Table 1. Parameters of each test case.

	Test Case C3	Test Case C4	Test Case C6
Domain in m (l × w × h)	500 × 200 × 200	500 × 200 × 200	520 × 400 × 200
Nr. of Buildings	1 Building	1 Building	3 × 7 Buildings
Building Size in m (l × w × h)	25 × 25 × 25	25 × 25 × 25	30 × 20 × 25
Mesh size	1.60 Mio cells	1.80 Mio cells	7.30 Mio cells
Atmospheric Inlet Conditions	$z_0 = 0.10$ m $m = 0.22$ $U_{ref} = 5.04$ m/s $z_{ref} = 75$ m $u_* = 0.31$ m/s	$z_0 = 0.10$ m $m = 0.21$ $U_{ref} = 4.43$ m/s $z_{ref} = 75$ m $u_* = 0.27$ m/s	$z_0 = 0.10$ m $m = 0.21$ $U_{ref} = 6.30$ m/s $z_{ref} = 132$ m $u_* = 0.36$ m/s
Definition of near fields	upwind of the building: 25 m downwind of the building: 50 m each side transverse to the direction of flow: 12.5 m above the building: 12.5 m		-
Permitted difference D	25%	25%	25%
Absolute permitted difference W	0.06	0.06	0.10

During a grid sensitivity study, the grid was refined from near-wall mesh sizes of 1.0 m down to 0.25 m, showing that the hit ratio is not very sensitive to the grid refinement, as only negligible differences for all refinements can be observed. In the following, only results for a mesh with a near-wall mesh size of 0.5 m will be discussed, although the observations made are also valid for all investigated meshes.

Table 2 Hit ratio in % for all three test cases. shows the hit ratio q for all three test cases, subdivided into hit ratios for q_{Ux}—corresponding to the main wind direction, q_{Uy}—the crosswind direction, and q_{Uz}—the vertical to the main wind direction. For all test cases and all three wind direction components, the required minimum hit ratio of 66% is reached or exceeded when evaluating the whole computational domain.

Table 2. Hit ratio in % for all three test cases.

	q_{Ux}		q_{Uy}		q_{Uz}	
	Whole Domain	**Near Field**	**Whole Domain**	**Near Field**	**Whole Domain**	**Near Field**
Test case c3	83	72	92	88	91	85
Test case c4	81	70	69	62	77	61
Test case c6	78	-	87	-	80	-

While the near-field values for test case c3 meet the requirements of the guideline, test case c4 does not reach the required minimum value for the hit ratio for some velocity components. Due to the high velocity gradients occurring at obstacles where a flow detachment or recirculation can be observed, the choice of the turbulence model has a significant influence on the quality of the result, whilst, in general, a higher deviation between computed values and experimental data is to be expected in these cases. Although the minimum criterion for the hit ratio is reached for test case c3, in test case c4 with a 45° approach flow, this can only be achieved for the hit ratio of the main wind direction component. Study [26] stated that the approach flow in the experiments for test case c4 was not exactly 45° but showed a slight deviation of 2°. It is still to be examined whether correcting the simulation values for the 2° deviation might lead to better hit ratios.

3.3. Wind Tunnel Experiments

After having compared the model results to such generic cases as the ones defined in [16,17], the model is still to be validated for a more realistic obstacle configuration. Due to the high reproducibility and to a lack of field experiments, wind tunnel data generated at the University of Hamburg in the context of an actual research project are used. The experiments were carried out at the EWTL large boundary layer wind tunnel of the University of Hamburg in a scale of 1:100. The 25 m long wind tunnel has an 18 m long, 4 m wide and approx. 3 m high test section. The investigated model area (Figure 5) was planned with aerial photos and building data from real industrial areas. In real scale, it covers about 300 m × 300 m. The model consists of 20 cuboid shaped buildings which have varying side lengths of 20 m to 40 m. Each building has a height of 15 m. The spaces between the buildings range between 10 m and 80 m. In the middle of the model area, a large building is located. This main building has a base area of 30 m × 60 m. The main building has three openings on the sides and one opening on the roof with a size of 4 m × 4 m. In the future, it is planned to carry out gas dispersion experiments with this setup where the mentioned openings on the buildings will be used as gas inlets.

Flow measurements were performed with Laser-Doppler Anemometry (LDA). The LDA measures two components of the turbulent wind vector simultaneously with a spatial resolution of approx. 110 mm in full scale. To receive statistical representative time series in a turbulent flow, measurements are performed for several minutes. In real scale, this represents a measurement taken for 30 h under stationary meteorological conditions.

Figure 5. Wind tunnel and obstacle configuration.

Two different boundary layer approach flows are modelled at a scale of 1:100 to investigate the influence of the flow on the measurements inside the model area. One boundary layer flow represents the conditions of a "moderately rough" boundary layer. This type of boundary layer flow develops over grasslands or farmlands. The second wind boundary layer represents a flow that develops over suburban area and is described by the definition of a "rough" boundary layer. The roughness classes are defined by VDI 3783/12 [27]. The similarity between wind tunnel experiments and real scale has been ensured by checking that the turbulence intensity, the wind profiles, the spectral distribution of turbulent energy and the lateral homogeneity are similar to nature-given profiles and distributions. In addition, the functional relationship between the roughness length z_0 and the wind profile exponent m has been checked, according to [28].

In the following, the simulated wind fields are compared to the profiles measured in the wind tunnel for the "moderately rough" case. For the considered case, the roughness length for the "moderately rough" case is $z_0 = 0.06$ m, corresponding to an exponent of the wind profile power law of $m = 0.17$. The same approach flow (Figure 6a) was used for two different wind directions, case M1 and case M2 (with respect to the obstacles, Figure 6b). The horizontal flow components (U_x, U_y) were measured around the obstacles at 327 positions for the direction M1 and at 383 positions for M2, at a height over the ground varying from 2 m up to 20 m. Figure 6a shows a good agreement of the measured and calculated normalized vertical wind velocity profile of the approach flow. For each horizontal, homogeneity was verified.

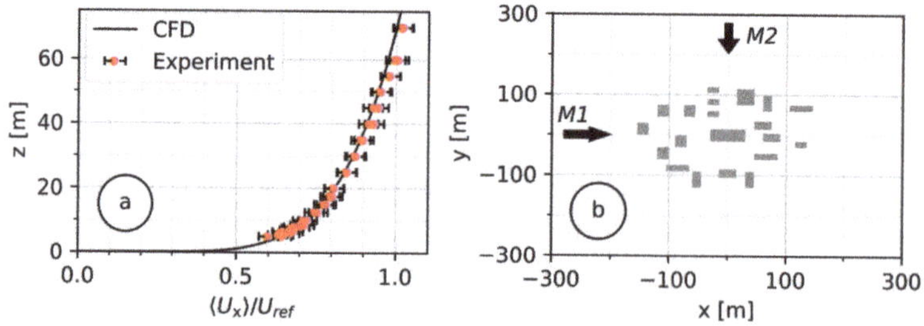

Figure 6. (**a**) Comparison of the boundary layer flow measured in the wind tunnel and calculated with the presented model; (**b**) schematic representation of the obstacle configuration and the approach flows M1 and M2.

For the following investigations of the model performance, an objective evaluation method comparable to the VDI Guideline 3783/9 is desirable. In the VDI Guideline 3783/9, in the 2017 version [17], a test case is defined for a real built-up area. The evaluation criteria for this case (D = 25% and W = 0.08, threshold for q = 66%) are applied to the cases investigated here, as they also represent realistic topographies. It should be noted that the evaluation criteria of the VDI Guideline are defined as case-specific and therefore might not be fully adapted to the cases M1 and M2. Nevertheless, as the permitted differences (D and W) and the hit ratio q do not change significantly for all cases of the guideline, this approach appears to be permissible.

For the cases M1 and M2, grid refinements were carried out to investigate if the hit ratio might be independent of the resolution, as this was the case in Section 3.2. For both cases, it is noticed that, with each refinement of the grid, the hit ratio increased up to a point where a grid resolution of 0.25 m adjacent to the obstacles finally led to a successful "validation" in terms of reaching the hit ratio threshold. These simulations already showed a total number of cells > 20 Mio. As further refinement would have led to domains with at least 80 Mio. cells, no further refinement was investigated due to the resulting extremely long simulation time.

Figures 7 and 8 show the repartition of the measuring points at heights of 2 m (Figures 7a and 8a) and 8 m over the ground (Figures 7b and 8b), respectively, for cases M1 and M2. Whilst in the VDI Guideline the experimental data used to define the thresholds are evenly spread over the whole domain, the measuring points in cases M1 and M2 are concentrated around buildings and close to the ground, making it harder to reach the threshold. Both figures represent the investigated area in such way that the approach flow is oriented parallel to the x-axis coming from the left. Full black bullets show a point where the calculated hit ratios q_{Ux} and q_{Uy} reach values higher than the threshold of 66%, half-filled bullets show points where the hit ratio of one velocity component did not reach the threshold, and for the blank bullets, no component of the horizontal velocity reached the threshold.

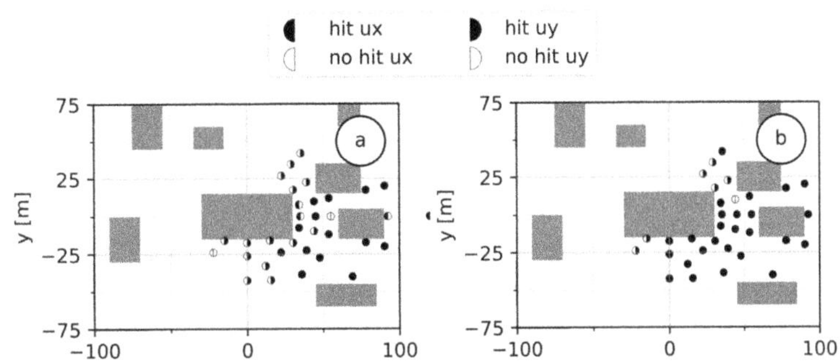

Figure 7. Illustration of the hit ratio for case M1 (**a**) at 2 m height over ground, (**b**) at 8 m height over ground.

With increasing height over the ground, the number of points which show a hit ratio higher than the threshold increases in both cases. This is due to the fact that the closer to the ground, the more complex the turbulence-influenced effects are on the flow. To resolve these effects in a satisfactory manner, at least a LES (Large-Eddy Simulation) would be required. Table 3 Hit ratio for the whole domain of case M1 and M2. shows that the hit ratio for the whole domain exceeds the threshold value of 66% for both cases. For the aim of providing an adequate wind field for hazard assessment purposes, the reached accuracy is therefore sufficient.

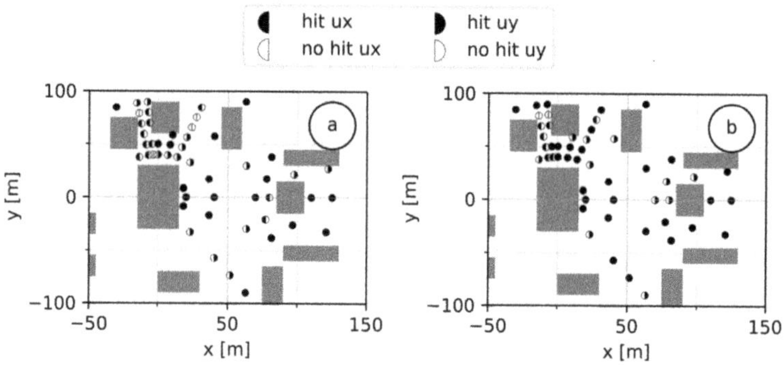

Figure 8. Illustration of the hit ratio for case M2 (**a**) at 2 m height over ground, (**b**) at 8 m height over ground.

Table 3. Hit ratio for the whole domain of case M1 and M2.

	q_{Ux}	q_{Uy}
M1	71	89
M2	67	82

4. Discussion

The transient, compressible flow solver rhoReactingBuoyantFoam of OpenFOAM 5.0 in combination with the standard k-ε turbulence model is promising for simulating the dispersion process of pollutants in the atmosphere. Due to inconsistencies between the required boundary conditions for modelling atmospheric boundary layer flow and the formulation of the turbulence model, simulating a horizontally homogenous boundary layer flow is not possible.

As the horizontally homogenous approach flow is mandatory for reliable estimations of the safety distances for hazard assessment purposes, when a pollutant is released near ground in the atmosphere, a modification to the turbulence model has been made and was presented in this work. Introducing an additional source term in the equation for the turbulent dissipation and replacing the log law profile by a power law profile for the wind field, in combination with a new formulation of the wall functions, showed the ability to simulate a horizontally homogenous boundary layer flow. Additionally, the grid resolution adjacent to the walls can now be chosen independently of the roughness length defined on the ground. For gas dispersion purposes in a built-up domain, the grid resolution in the vicinity of buildings is crucial to achieve a satisfactory resolution of the concentration gradients, leading to a reliable and reproducible prediction of the safety distances. The model's performance for this application is currently under investigation.

The evaluation of the presented model occurred in three steps. In the first step, it was checked against the original model formulation for its performance to produce horizontally homogenous wind and turbulence profiles, and showed a clear advantage compared to the original formulation by achieving the homogeneity. In the second step, the model was evaluated considering the rules defined by the VDI Guideline 3783/9 [16] where generic obstacle configurations are presented, and an objective statistical method for evaluation using a so-called hit ratio is defined. The model was evaluated against three test cases of the guideline with a single cube in a 90° approach flow and a 45° approach flow, as well as for an array of 7 × 3 obstacles. Considering the whole domain, the model performed well in all cases and reached hit ratios clearly fulfilling the defined threshold. For the two single cube cases, an evaluation of the "near-field" had to be carried out. The hit ratio showed lower values than for the whole domain, due to restrictions of the *k*-ε turbulence model in the vicinity of obstacles. The third evaluation step consisted of comparing the presented model to wind tunnel measurements and using the evaluation criteria of the VDI Guideline 3783/9 [17] to obtain an objective measure of the performance, even though the criteria might not fully match the considered case. Nevertheless, the presented model showed again a good performance by reaching the required thresholds.

Author Contributions: Conceptualization, S.S. and A.H.; model development and validation, S.S.; wind tunnel measurements, S.M.; All authors were actively involved in writing and revising the manuscript. All authors have read and agreed to the published version of the manuscript.

Funding: The IGF-Project No.: 20719 BG of the Research Association Society for Chemical Engineering and Biotechnology (DECHEMA), Theodor-Heuss-Allee 25, 60486 Frankfurt am Main, was funded by the German Federation of Industrial Research Associations (AiF) within the framework of the Industrial Collective Research (IGF) support program by the Federal Ministry for Economic Affairs and Energy due to a decision of the German Bundestag.

Institutional Review Board Statement: Not applicable.

Informed Consent Statement: Not applicable.

Conflicts of Interest: The authors declare no conflict of interest.

References

1. Franke, J.; Hellsten, A.; Schlünzen, H.; Carissimo, B. *Best Practice Guideline for the CFD Simulation of Flows in the Urban Environment: COST Action 732 Quality Assurance and Improvement of Microscale Meteorological Models*; Meteorological Institute University of Hamburg: Hamburg, Germany, 2007.
2. Batt, R.; Gant, S.E.; Lacome, J.-M.; Truchot, B. Modelling of stably-stratified atmospheric boundary layers with commercial CFD software for use in risk assessment. *Chem. Engineer. Trans.* **2016**, *48*, 61–66.
3. Richards, P.; Younis, B. Comments on "prediction of the wind-generated pressure distribution around buildings" by E.H. Mathews. *J. Wind. Eng. Ind. Aerodyn.* **1990**, *34*, 107–110. [CrossRef]
4. Quinn, A.D.; Wilson, M.; Reynolds, A.M.; Couling, S.B.; Hoxey, R.P. Modelling the dispersion of aerial pollutants from agricultural buildings—An evaluation of computational fluid dynamics (CFD). *Comput. Electron. Agric.* **2001**, *30*, 219–235. [CrossRef]
5. Zhang, C.X. Numerical predictions of turbulent recirculating flows with a k–ε model. *J. Wind. Eng. Ind. Aerod.* **1994**, *51*, 177–201. [CrossRef]
6. Riddle, A.; Carruthers, D.; Sharpe, A.; McHugh, C.; Stocker, J. Comparisons between FLUENT and ADMS for atmospheric dispersion modelling. *Atmos. Environ.* **2004**, *38*, 1029–1038. [CrossRef]
7. Richards, P.J.; Hoxey, R.P. Appropriate boundary conditions for computational wind engineering models using the k-ε mod-el. *J. Wind. Eng. Ind. Aerod.* **1993**, *46–47*, 145–153. [CrossRef]
8. Alinot, C.; Masson, C. k–ε model for the atmospheric boundary layer under various thermal stratifications. *J. Sol. Energy Eng.* **2005**, *127*, 438–443. [CrossRef]
9. Yang, W.; Quan, Y.; Jin, X.; Tamura, Y.; Gu, M. Influences of equilibrium atmosphere boundary layer and turbulence param-eter on wind loads of low-rise buildings. *J. Wind. Eng. Ind. Aerodyn.* **2008**, *96*, 2080–2092. [CrossRef]
10. Yang, Y.; Gu, M.; Chen, S.; Jin, X. New inflow boundary conditions for modelling the neutral equilibrium atmospheric boundary layer in computational wind engineering. *J. Wind. Eng. Ind. Aerodyn.* **2009**, *97*, 88–95. [CrossRef]
11. Pontiggia, M.; Derudi, M.; Busini, V.; Rota, R. Hazardous gas dispersion: A CFD model accounting for atmospheric stability classes. *J. Hazard. Mater.* **2009**, *171*, 739–747. [CrossRef] [PubMed]
12. Van Der Laan, M.P.; Kelly, M.C.; Sørensen, N.N. A new k-epsilon model consistent with Monin-Obukhov similarity theory. *Wind. Energy* **2016**, *20*, 479–489. [CrossRef]
13. Blocken, B.; Stathopoulos, T.; Carmeliet, J. CFD simulation of the atmospheric boundary layer: Wall function problems. *Atmos. Environ.* **2007**, *41*, 238–252. [CrossRef]
14. Zhang, X. *CFD Simulation of Neutral ABL Flows*; Danmarks Tekniske Universitet, Risø Nationallaboratoriet for Bæredygtig Energi: Kongens Lyngby, Denmark, 2009.
15. Parente, A.; Gorlé, C.; van Beeck, J.; Benocci, C. Improved k-ε model and wall function formulation for the RANS simula-tion of ABL flows. *J. Wind. Eng. Ind. Aerodyn.* **2011**, *34*, 107–110.
16. *Environmental Meteorology—Prognostic Microscale Windfield Models—Evaluation for Flow around Buildings and Obstacles*; VDI Guideline 3783, Part 9; Beuth Verlag: Berlin, Germany, 2005.
17. *Environmental Meteorology—Prognostic Microscale Windfield Models—Evaluation for Flow around Buildings and Obstacles*; VDI Guideline 3783, Part 9; Beuth Verlag: Berlin, Germany, 2017.
18. OpenFOAM. User Guide. Version 7. OpenFOAM—The OpenFOAM Foundation. 2019. Available online: https://openfoam.org (accessed on 24 July 2020).
19. Weller, H.G.; Tabor, G.; Jasak, H.; Fureby, C. A tensorial approach to computational continuum mechanics using ob-ject-oriented techniques. *Comput. Phys.* **1998**, *12*, 620–631. [CrossRef]

20. Launder, B.E.; Spalding, D.B. The numerical computation of turbulent flows. *Comput. Method Appl. Mech. Eng.* **1974**, *3*, 269–289. [CrossRef]
21. El Tahry, S.H. k-epsilon equation for compressible reciprocating engine flows. *J. Energy* **1983**, *7*, 345–353. [CrossRef]
22. Jones, W.; Launder, B. The prediction of laminarization with a two-equation model of turbulence. *Int. J. Heat Mass Transf.* **1972**, *15*, 301–314. [CrossRef]
23. Hargreaves, D.; Wright, N. On the use of the k–ε model in commercial CFD software to model the neutral atmospheric boundary layer. *J. Wind. Eng. Ind. Aerodyn.* **2007**, *95*, 355–369. [CrossRef]
24. Franke, J.; Sturm, M.; Kalmbach, C. Validation of OpenFOAM 1.6.x with the German VDI guideline for obstacle resolving micro-scale models. *J. Wind. Eng. Ind. Aerodyn.* **2012**, *104–106*, 350–359. [CrossRef]
25. Leitl, B. Compilation of Experimental Data for Validation of Micro Scale Dispersion Models (CEDVAL). Available online: https://mi-pub.cen.uni-hamburg.de/index.php?id=433 (accessed on 1 May 2021).
26. Eichhorn, J. MISKAM Version 5.02—Evaluierung nach VDI-RL 3783/9. Available online: http://www.lohmeyer.de/ (accessed on 1 May 2021).
27. *Environmental Meteorology—Physical Modelling of Flow and Dispersion Processes in the Atmospheric Boundary Layer—Application of Wind Tunnels*; VDI Guideline 3783, Part 12; Beuth Verlag: Düsseldorf, Germany, 2000.
28. Counihan, J. Adiabatic atmospheric boundary layers: A review and analysis of data from the period 1880–1972. *Atmos. Environ.* **1975**, *9*, 871–905. [CrossRef]

MDPI
St. Alban-Anlage 66
4052 Basel
Switzerland
Tel. +41 61 683 77 34
Fax +41 61 302 89 18
www.mdpi.com

Atmosphere Editorial Office
E-mail: atmosphere@mdpi.com
www.mdpi.com/journal/atmosphere

www.ingramcontent.com/pod-product-compliance
Lightning Source LLC
Chambersburg PA
CBHW061809210926
51850CB00006B/650
* 9 7 8 3 0 3 6 5 6 5 8 1 1 *